Climate Change Education for Sustainable Development

Jared Keengwe
University of North Dakota, USA

Benard O. Nyatuka
Kisii University, Kenya

A volume in the Advances in Educational Marketing, Administration, and Leadership (AEMAL) Book Series

IGI Global
PUBLISHER of TIMELY KNOWLEDGE

Published in the United States of America by
IGI Global
Information Science Reference (an imprint of IGI Global)
701 E. Chocolate Avenue
Hershey PA, USA 17033
Tel: 717-533-8845
Fax: 717-533-8661
E-mail: cust@igi-global.com
Web site: http://www.igi-global.com

Library of Congress Cataloging-in-Publication Data

Names: Keengwe, Jared, 1973- editor. | Nyatuka, Benard Omenge, editor.
Title: Climate change education for sustainable development / edited by Jared Keengwe and Benard Nyatuka.
Description: Hershey, PA : Information Science Reference, 2024. | Includes bibliographical references and index. | Summary: "Various international organizations and individuals have made passionate appeals to governments to empower and draw in various partners in policy formulation and implementation, and activities concerning the climate change crisis. The logic is that learning institutions serve as convenient avenues of promoting climate change education. To confront the impacts of climate change and related environmental issues, individuals must not only be provided with requisite education, but also enhanced awareness to drive such change. Although some effort has been made to offer such knowledge, skills and values, there are existing challenges to realize the anticipated objectives. These include limited understanding of the weather change phenomenon, less effective pedagogical approaches and weak partnerships among the critical stakeholders"-- Provided by publisher.
Identifiers: LCCN 2023025866 (print) | LCCN 2023025867 (ebook) | ISBN 9781668490990 (hardcover) | ISBN 9781668491003 (ebook)
Subjects: LCSH: Climatic changes--Study and teaching--Developing countries. | Environmental education--Developing countries. | Sustainable development.
Classification: LCC QC903.2.D44 H36 2024 (print) | LCC QC903.2.D44 (ebook) | DDC 551.6071--dc23/eng/20230816
LC record available at https://lccn.loc.gov/2023025866
LC ebook record available at https://lccn.loc.gov/2023025867
This book is published in the IGI Global book series Advances in Educational Marketing, Administration, and Leadership (AEMAL) (ISSN: 2326-9022; eISSN: 2326-9030)

British Cataloguing in Publication Data
A Cataloguing in Publication record for this book is available from the British Library.
All work contributed to this book is new, previously-unpublished material.
The views expressed in this book are those of the authors, but not necessarily of the publisher.
For electronic access to this publication, please contact: eresources@igi-global.com.

Advances in Educational Marketing, Administration, and Leadership (AEMAL) Book Series

ISSN:2326-9022
EISSN:2326-9030

Editor-in-Chief: Siran Mukerji, IGNOU, India, Purnendu Tripathi, IGNOU, India

MISSION

With more educational institutions entering into public, higher, and professional education, the educational environment has grown increasingly competitive. With this increase in competitiveness has come the need for a greater focus on leadership within the institutions, on administrative handling of educational matters, and on the marketing of the services offered.

The **Advances in Educational Marketing, Administration, & Leadership (AEMAL) Book Series** strives to provide publications that address all these areas and present trending, current research to assist professionals, administrators, and others involved in the education sector in making their decisions.

COVERAGE

- Consumer Behavior
- Advertising and Promotion of Academic Programs and Institutions
- Technologies and Educational Marketing
- Educational Finance
- Students as Consumers
- Educational Marketing Campaigns
- Enrollment Management
- Direct marketing of educational programs
- Governance in P-12 and Higher Education
- Academic Administration

IGI Global is currently accepting manuscripts for publication within this series. To submit a proposal for a volume in this series, please contact our Acquisition Editors at Acquisitions@igi-global.com or visit: http://www.igi-global.com/publish/.

The Advances in Educational Marketing, Administration, and Leadership (AEMAL) Book Series (ISSN 2326-9022) is published by IGI Global, 701 E. Chocolate Avenue, Hershey, PA 17033-1240, USA, www.igi-global.com. This series is composed of titles available for purchase individually; each title is edited to be contextually exclusive from any other title within the series. For pricing and ordering information please visit http://www.igi-global.com/book-series/advances-educational-marketing-administration-leadership/73677. Postmaster: Send all address changes to above address.

Titles in this Series

For a list of additional titles in this series, please visit:
http://www.igi-global.com/book-series/advances-educational-marketing-administration-leadership/73677

Improving Learning Through Assessment Rubrics Student Awareness of What and How They Learn
Chahna Gonsalves (King's Business School, King's College London, UK) and Jayne Pearson (King's Academy, King's College London, UK)
Information Science Reference • © 2023 • 380pp • H/C (ISBN: 9781668460863) • US $270.00

Culturally Responsive Leadership for Social Justice and Academic Equity for All
Bethel E. Cager (Xavier University of Louisiana, USA) Jill Tussey (Buena Vista University, USA) Leslie Haas (Xavier University of Louisiana, USA) and Monica Galloway Burke (Western Kentucky University, USA)
Information Science Reference • © 2023 • 330pp • H/C (ISBN: 9781668474822) • US $215.00

Canadian Education Law and Global Comparative Studies for Teachers and Administrators
Xiaobin Li (Brock University, Canada)
Information Science Reference • © 2023 • 225pp • H/C (ISBN: 9781668441633) • US $230.00

Promoting Inclusive Education Through the Integration of LGBTIQ+ Issues in the Classroom
Francisco Javier Palacios-Hidalgo (University of Córdoba, Spain) and Cristina A. Huertas-Abril (University of Córdoba, Spain)
Information Science Reference • © 2023 • 428pp • H/C (ISBN: 9781668482438) • US $215.00

Enrollment and Retention Strategies for 21st Century Higher Education
Rayshawn L. Eastman (Mount St. Joseph University, USA)
Information Science Reference • © 2023 • 300pp • H/C (ISBN: 9781668474778) • US $215.00

For an entire list of titles in this series, please visit:
http://www.igi-global.com/book-series/advances-educational-marketing-administration-leadership/73677

IGI Global
PUBLISHER of TIMELY KNOWLEDGE

701 East Chocolate Avenue, Hershey, PA 17033, USA
Tel: 717-533-8845 x100 • Fax: 717-533-8661
E-Mail: cust@igi-global.com • www.igi-global.com

Table of Contents

Preface xv

Chapter 1
A Review of Integration of Environment Education Into Teacher Preparation Programs 1
Jenny Bladow, University of North Dakota, USA

Chapter 2
Integrating Climate Change Education Into the Curriculum in Kenya 18
Maureen Syallow, Strathmore University, Kenya
Magdalene Dimba, Strathmore University, Kenya
Esther Muthoni King'ori, Strathmore University, Kenya

Chapter 3
Pre-Service Teachers' Pedagogical Knowledge and Perceptions of Climate Change Education 40
Evelyn W. Njurai, Kisii University, Kenya
Billiah Gisore, Kisii University, Kenya

Chapter 4
Importance of Gender Inclusiveness in Sustainable Climate Change Education 63
Falak Shad Memon, Institute of Business Management, Karachi, Pakistan

Chapter 5
Ethical Dimensions of Sustainable Development: The Philosophy, Logistics, and Climate Change Education 92
Ebua Jarvis Ebua, University of Bamenda, Cameroon

Chapter 6
Climate Change Reporting and the Role of Media in Shaping Public Discourse in Nepal: Evidence From a Literature Review 125
Sudeep Uprety, Prince of Songkla University, Thailand
Arkom Palamanit, Prince of Songkla University, Thailand
Kuaanan Techato, Prince of Songkla University, Thailand

Chapter 7
Climate Change and Its Implications on School Education in Kenya 144
George Mokua Ogendi, Kisii University, Kenya

Chapter 8
Building Bridges to Sustainability: Overcoming Barriers and Promoting Integration in Architectural Education 168
Seyda Emekci, Ankara Yildirim Beyazit University, Turkey

Chapter 9
Promoting Environmental Sustainability Among Students and Citizens for a Sustainable Future 187
Seema Yadav, Bhopal School of Social Sciences, India

Chapter 10
Pedagogical Strategies to Enhance Climate Change Education Outcomes 212
Billiah Nyamoita Gisore, Kisi University, Kenya
Evelyn Njurai, Kisii Univrsity

Chapter 11
Being a Rural Woman During Climate Change: An Analysis Based on the Concept of Ubuntu 235
Bilge Abukan, Ondokuz Mayıs University, Turkey
Kübranur Görmüş, Maltepe University, Turkey
Filiz Yildirim, Ankara University, Turkey

Chapter 12
Best Practice in the Use of the Water Sources of Portoviejo Canton, Ecuador, and Its Parishes 263
Arturo Luque González, Universidad Tecnica de Manabí, Ecuador & Observatorio Euromediterráneo de Espacio Público y Democracia, URJC, Spain
José Humberto Ponce Ponce, Universidad Tecnica de Manabí, Ecuador & Observatorio Euromediterráneo de Espacio Público y Democracia, URJC, Spain
Cristina Raluca Gh. Popescu, University of Bucharest, Romania & The Bucharest University of Economic Studies, Romania

Chapter 13
Dynamics of Teaching in the Era of Climate Change Education 288
John M. Onyango Yambo, Kisii University, Kenya

Compilation of References 307

About the Contributors 362

Index 369

Detailed Table of Contents

Preface xv

Chapter 1
A Review of Integration of Environment Education Into Teacher Preparation Programs 1
Jenny Bladow, University of North Dakota, USA

Compared to previous generations, children today spend less time outdoors interacting with nature and all it offers. The current leisure trends are encompassed in technology as it becomes more accessible to children. Perhaps there is a way to bring back the importance of nature and intertwine environmental education concepts into teacher preparation programs. In that case, there is hope for future generations of children to reap the benefits of nature and more community members who are environmentally literate. This chapter reviews the importance of incorporating environmental education into teacher preparation. The review of literature also discusses common approaches to addressing the current gap in environmental education implementation along with barriers that have been encountered.

Chapter 2
Integrating Climate Change Education Into the Curriculum in Kenya................ 18
Maureen Syallow, Strathmore University, Kenya
Magdalene Dimba, Strathmore University, Kenya
Esther Muthoni King'ori, Strathmore University, Kenya

The effects of climate change are a major global concern. In Kenya, the impact of climate change has been felt in perennial droughts which sometimes result into conflict as communities fight for pasture and water. In a bid to address climate change issues, Kenya has developed policies that focus on mitigation and adaptation measures borrowed from ideas outlined in global climate change strategies. One such strategy is the employment of education as a tool to create awareness on climate change issues. This chapter will examine the climate change policies developed by bodies in charge of higher education in Kenya. The chapter will further analyze attempts that have been made to integrate climate change education into curricula. In addition, the chapter will trace the history of discourse that saw the introduction of climate change education in institutions of higher learning. Finally, the chapter will analyze pedagogical strategies that can be employed in teaching climate change content.

Chapter 3
Pre-Service Teachers' Pedagogical Knowledge and Perceptions of Climate Change Education 40
Evelyn W. Njurai, Kisii University, Kenya
Billiah Gisore, Kisii University, Kenya

Climate change education (CCE) has been proposed to be a driver for the much-needed knowledge and skills on climate change (CC). Teachers, and consequently learners, are key change agents in any society; therefore, it is necessary that they have knowledge and skills on CC and how to mitigate and adapt to these changes. This chapter discusses pre-service teachers' pedagogical knowledge and perceptions of CCE. The discussion is contextualized through research conducted in Kenya. While it was evident that the pre-service teachers had basic knowledge and skills about CC, they also had misconceptions. Further, CCE was not documented in their curricula. They propose that CCE can be integrated with other subjects or implemented as a stand-alone subject or both. The study consequently identifies pre-service teachers' knowledge gaps and reports on their mixed perceptions of CCE. The need to include CCE in the curriculum is then underscored in view of targeted agencies.

Chapter 4
Importance of Gender Inclusiveness in Sustainable Climate Change Education 63
Falak Shad Memon, Institute of Business Management, Karachi, Pakistan

This study investigates the importance of gender-inclusive climate change education programs and the challenges involved with designing such curriculum in rural Sindh, Pakistan. The study addresses three research areas: (1) perspectives and consequences of gender inclusivity in climate change education, (2) gender inclusivity's effect the efficacy and success of Sindh's rural climate change education initiatives, and (3) the hurdles to attaining gender equality in sustainable climate change education and its effect on the overall effectiveness of such initiatives. The qualitative research method was applied, and interviews were conducted from eight leading educationist in Pakistan. The findings show that women face various hurdles in climate change education, including restricted access to education, societal norms, and cultural barriers. However, research identifies various tactics and recommendations to improve gender inclusion, such as the promotion of female role models from best practices, the incorporation of gender viewpoints in curricula, as well as programs to empower women.

Chapter 5
Ethical Dimensions of Sustainable Development: The Philosophy, Logistics, and Climate Change Education....................92
Ebua Jarvis Ebua, University of Bamenda, Cameroon

This chapter examines the ethical dimensions of sustainable development within the context of climate change education. It explores the intersection of philosophy, logistics, and climate change education. The chapter sheds light on ethical implications and considerations arising in pursuing sustainable development. It seeks to provide valuable insights for climate change analysts, policymakers, curriculum and instructional designers, faculty, administrators, teacher educators, and other stakeholders interested in averting the impact of climate change and fostering sustainable development.

Chapter 6
Climate Change Reporting and the Role of Media in Shaping Public Discourse in Nepal: Evidence From a Literature Review....................125
Sudeep Uprety, Prince of Songkla University, Thailand
Arkom Palamanit, Prince of Songkla University, Thailand
Kuaanan Techato, Prince of Songkla University, Thailand

Though the role of media in climate change messaging has been understood in academic and public spheres, limited studies have been conducted and published, especially in the context of Nepal. A review of 65 readings highlights increased focus on behavior change communication than media advocacy, lack of innovation scope in funded media projects, absence of empirical research. The study recommends comprehensive study climate change reporting through framing and factor analysis to identify reasons and incentives.

Chapter 7
Climate Change and Its Implications on School Education in Kenya....................144
George Mokua Ogendi, Kisii University, Kenya

Climate change is the greatest challenge of the 21st century. It is threatening all aspects of societal life. The right to education of children is negatively impacted by these climate-induced challenges. This chapter presents findings of a study on the effects of climate change on the education of the younger generation. The study used a mixed research design method and was done within Kisii Municipality, Kenya. The findings were that climate change and its effects have brought about increased waterborne diseases, reduced available income for families, and compromised living standards, thereby negatively impacting on children's education. To address the effects of climate change, however, provision of appropriate education can go a long way in mitigating these climatic challenges. Secondly, topics on climate change should be included in the school curriculum.

Chapter 8
Building Bridges to Sustainability: Overcoming Barriers and Promoting Integration in Architectural Education..168
Seyda Emekci, Ankara Yildirim Beyazit University, Turkey

The consideration of environmental factors plays a key part in the design of sustainable buildings within their respective contexts. The global community is currently grappling with the adverse impacts of climate change, which can be attributed to the release of greenhouse gases into the atmosphere. Architecture is a unique area of study that enables the development of spatial solutions to address human requirements, while also carrying the essential duty of ensuring the creation of a sustainable built environment. Although the inclusion of sustainability in architectural education is widely acknowledged, the majority of curricula lack a comprehensive integration of this topic. In order to conceptualize a sustainable built environment, it is imperative for the architectural curriculum to incorporate a foundational comprehension of the interconnection between sustainability, technology, and design. This chapter analyzes the barriers to sustainable environmental design in architectural education. It also identifies worldwide integration promotion efforts.

Chapter 9
Promoting Environmental Sustainability Among Students and Citizens for a Sustainable Future...187
Seema Yadav, Bhopal School of Social Sciences, India

Climate change is a dangerous problem that endangers the continuation of life. The sixth major extinction catastrophe in recorded history is being brought on by climate change. There is an existential threat to life on Earth from the climate changes brought on by the use of fossil fuels. Education increases young people's knowledge while simultaneously fostering awareness, enabling informed decision-making, and fostering changes in attitudes and behaviors. Education is essential to the global response to climate change. Climate change education (CCE) is one of the most significant approaches to increasing awareness of mitigating climate change. The majority of students must receive the knowledge, critical thinking skills, scientific understanding, and ethical frameworks necessary to prevent, adapt, and reverse climate change as a result of climate change education. Climate change education initiatives can play a significant role in developing communities of the kind of pragmatic visionaries needed to address new issues as they emerge and build on the successes of others.

Chapter 10
Pedagogical Strategies to Enhance Climate Change Education Outcomes.......212
Billiah Nyamoita Gisore, Kisi University, Kenya
Evelyn Njurai, Kisii Univrsity

Global temperatures have soared since the Industrial Revolution some 140 years ago, with the ecosystems having a severe impact globally. Environmental issues are changing not only the normative lifestyles, but also the ethics and understanding of the world. Educating people on ways to relate to the habitat calls for adopting sustainable practices that reduce the impact on climate. Climate change education (CCE) is a new field of pedagogy that addresses the issues of climate change, but has no established model for implementation. The need for transformative learning with pedagogical approaches that incorporate diverse climate change learning experiences is required. The chapter therefore addresses the need for various aspects of climate change to be addressed as part of educational pedagogy and the need for a curriculum with climate-friendly, age-appropriate content. The possible strategies and the barriers that seem to hinder the achievement of educational outcomes are also examined.

Chapter 11
Being a Rural Woman During Climate Change: An Analysis Based on the Concept of Ubuntu...235
Bilge Abukan, Ondokuz Mayıs University, Turkey
Kübranur Görmüş, Maltepe University, Turkey
Filiz Yildirim, Ankara University, Turkey

Climate change affects all living things, and especially women in rural areas. Focusing on human-environment interaction, ubuntu offers a holistic perspective to interventions to reduce the destructive effects of not only the social environment but also the physical environment. The main purpose of the present study is to evaluate the impact of climate change on the lives of rural women with a bio-psycho-social perspective, in the focus of ubuntu philosophy, and offer social work intervention suggestions.

Chapter 12
Best Practice in the Use of the Water Sources of Portoviejo Canton, Ecuador, and Its Parishes 263
Arturo Luque González, Universidad Tecnica de Manabí, Ecuador & Observatorio Euromediterráneo de Espacio Público y Democracia, URJC, Spain
José Humberto Ponce Ponce, Universidad Tecnica de Manabí, Ecuador & Observatorio Euromediterráneo de Espacio Público y Democracia, URJC, Spain
Cristina Raluca Gh. Popescu, University of Bucharest, Romania & The Bucharest University of Economic Studies, Romania

Water is a fundamental element for the sustenance of all living beings in nature, hence the importance of its conservation and the correct management of its sources. In the case of the Portoviejo River, water is used as a means of irrigation in plantations, a drinking water supply to the city, and a space for developing tourism. However, new settlements along the banks of the Portoviejo River have led to a rise in pollution levels and other misuses of the water. Owing to these circumstances, there was a clear need to evaluate the historical background and conservation mechanisms of water sources used by the canton authorities and its control bodies. The main objective of this study was to carry out an exhaustive analysis of good practices in the use of water sources within the Portoviejo Canton. The study used a hermeneutic research approach, which allows the researcher to perform a qualitative analysis of various interpretations of different texts through a dialectical process, which supports the investigation and substantiates the results obtained.

Chapter 13
Dynamics of Teaching in the Era of Climate Change Education 288
John M. Onyango Yambo, Kisii University, Kenya

This chapter addresses the dynamics of teaching in the era of climate change education. It equally looked at encounters of mobile schools and teaching and facilities in mobile schools in mitigating climate change. It necessitates that the education researchers engage in inquiry and involvement of other stakeholders. Two methods were used, partial relevant literature review and qualitative assessment of teaching methods. The study targeted policy makers, curriculum and instructional designers, faculty, administrators, teacher educators, and other stakeholders interested in averting the impact of climate change and promoting sustainable development in Kenya and beyond.

Compilation of References 307

About the Contributors 362

Index 369

Preface

In a world grappling with unprecedented challenges brought about by the complex web of climate change, our collective future hinges on our ability to impart knowledge, foster understanding, and cultivate a shared commitment to sustainable development. This volume emerges as a response to the urgent need for comprehensive guidance in the realm of climate change education – a cornerstone of our journey toward a more resilient and harmonious planet.

The modern global landscape is marked by an intricate interplay of forces, where the consequences of climate change resonate deeply in every facet of our existence. As stewards of knowledge, we recognize that education serves as a potent catalyst for change, empowering individuals to navigate the intricate nexus of environmental, social, and economic dynamics that define our times. To this end, the *Climate Change Education for Sustainable Development* is aimed at to equipping educators, policymakers, practitioners, and researchers with a robust toolkit to address the challenges and seize the opportunities presented by the climate change crisis.

CHAPTER OVERVIEW

Chapter 1: A Review of Integration of Environment Education Into Teacher Preparation Programs

In the ever-evolving landscape of education, a pressing concern emerges – the detachment of modern children from the natural world. As technology's allure captivates young minds and outdoor interactions dwindle, the imperative to reintroduce nature's significance gains prominence. The author's exploration resonates with Evans' compelling assertion that Education for Sustainability (EfS) is not merely an option but a necessity, considering the heightened uncertainties and complexities facing our world (Evans, 2020, p. 819). The chapter traverses the essential terrain of why and how environmental education should find its abode in teacher preparation curricula. The investigation extends beyond the 'why,' encompassing a detailed analysis of prevailing approaches that address the existing chasm in environmental education integration. Moreover, the chapter candidly confronts the barriers that have hindered the realization of this ambitious integration, offering readers a comprehensive insight into the challenges that await innovative solutions. This chapter sets the stage for a broader discourse on nurturing a new breed of educators – ones equipped to instill a deep reverence for the environment and inspire future generations to become stewards of our planet's well-being.

Chapter 2: Integrating Climate Change Education Into Curriculum in Kenya

This chapter delves into the intricate tapestry of Kenya's response to the global imperative of climate change education. The effects of climate change reverberate as profound global concerns, with Kenya bearing witness to the tangible repercussions of its erratic patterns, manifesting as recurrent droughts that ignite conflicts over dwindling resources. In this context, the chapter navigates the reader through Kenya's multifaceted journey, where policies of mitigation and adaptation have taken root, drawing inspiration from the bedrock of international climate change strategies.

Nestled within these initiatives is the pivotal role of education as a catalyst for fostering awareness and empowering change. With a keen eye on higher education, the chapter meticulously examines the policies that have germinated under the stewardship of educational bodies in Kenya, tracing their evolution and impact. Furthermore, the chapter unveils the diligent attempts made to weave climate change education into the fabric of curricula, epitomizing Kenya's commitment to nurturing a generation cognizant of their environmental responsibilities.

Embedded within the narrative is the historical trajectory that has shaped the discourse on climate change education within Kenya's institutions of higher learning. From its nascent origins to the present, this historical reflection illustrates the tenacity with which educators, policymakers, and visionaries have championed the integration of climate change education.

The chapter culminates in a thought-provoking exploration of pedagogical strategies poised to transform the teaching and learning of climate change content. Additionally, the chapter beckons readers to explore not only the tangible policies and practices, but also the intangible spirit of resilience and enlightenment that define Kenya's quest to address climate change through the powerful medium of education.

Chapter 3: Pre-Service Teachers' Pedagogical Knowledge and Perceptions of Climate Change Education

This chapter delves into the vital realm of Climate Change Education (CCE), exploring its potential to cultivate the essential knowledge and skills required to navigate the intricate landscape of Climate Change (CC). Central to this exploration is the recognition that teachers, bearing the mantle of change agents, are pivotal in guiding societies toward sustainable futures. This chapter offers a nuanced understanding of pre-service teachers' pedagogical knowledge and their perceptions of CCE, rooted in the unique context of Kenya. The backdrop of this discourse is illuminated by empirical research conducted within the Kenyan educational milieu.

The findings of the research reveal a complex tapestry. While pre-service teachers exhibit a foundational grasp of CC, the presence of certain misconceptions within their cognitive framework is unveiled. The chapter unearths a stark reality – the absence of CCE within their formal curricula. In response to this critical observation, the chapter ponders innovative pathways forward. It delves into the prospect of integrating CCE with existing subjects, weaving it seamlessly into the fabric of education, or even envisioning it as a distinct and stand-alone subject.

A pivotal highlight of this chapter is the impassioned plea for the integration of CCE within curricula. It underscores the imperative to empower future educators with the tools to champion informed environmental stewardship. In the broader context, this chapter serves as a compelling call to action as well as readers are invited to reflect upon the transformative role of education in shaping a future where climate change education resonates as an indomitable force for change.

Chapter 4: Importance of Gender Inclusiveness in Sustainable Climate Change Education

This chapter examines the significance of integrating gender perspectives within the realm of sustainable climate change education. The focus of this inquiry is rooted in the unique context of rural Sindh, Pakistan, where the nuances of gender inclusiveness intersect with the imperatives of climate change education.

Drawing upon the qualitative research paradigm, this study embarks on a comprehensive exploration of three pivotal research areas. The first-dimension casts light on the perspectives and consequences that emanate from weaving gender inclusivity into the fabric of climate change education. As the tapestry unfolds, we delve into the second realm, a captivating voyage into the realm of efficacy and success - where gender inclusivity becomes an indomitable catalyst for Sindh's rural climate change education initiatives.

However, the path toward gender inclusivity is not without its challenges, and this chapter unfurls the intricate hurdles that have beset the journey. These hurdles encompass a multitude of societal norms, cultural barriers, and restricted access to education, all of which resonate as impediments in the path of women's participation in climate change education. This chapter stands as a testament to the pressing need for holistic and inclusive educational landscapes, where the narrative of climate change education harmoniously intertwines with the narratives of gender equality and empowerment.

Chapter 5: Ethical Dimensions of Sustainable Development – The Philosophy, Logistics, and Climate Change Education

With a philosophical lens, this chapter delves into the intricate interplay between philosophy, logistics, and climate change education. It unravels the complex web of ethical implications and considerations that emerge when the pursuit of sustainable development converges with the urgent imperative of addressing climate change. As the narrative unfolds, readers are invited to navigate the dynamic landscape where philosophy becomes a compass, guiding the ethical trajectories of climate change education.

The chapter offers a compass to navigate the intricate ethical labyrinth that shapes the trajectory of climate change education and, by extension, the ethos of sustainable development. Additionally, it resonates as a call to action, an invitation to intertwine ethics with education, philosophy with practice, and sustainability with the very fabric of our collective future.

Chapter 6: Climate Change Reporting and the Role of Media in Shaping Public Discourses in Nepal – Evidence From a Literature Review

This chapter embarks on a scholarly journey into the dynamic interplay between climate change reporting and media's pivotal role in Nepal's context. While the role of media in climate change communication has captured both academic and public attention, the context of Nepal beckons with limited studies that delve into this profound nexus. Anchored in a review of 65 readings, the narrative unveils an intriguing landscape marked by nuanced findings.

Notably, the chapter highlights a shift towards behavior change communication, a trajectory that mirrors the dynamic nature of societal response to climate change imperatives. Yet, within the tapestry of media advocacy, the review unravels limitations and gaps, shedding light on the need for innovative scopes in funded media projects and the dearth of empirical research. This inquiry, informed by framing and factor analysis. beckon readers to contemplate the intricate pathways that media weaves, shaping perceptions, influencing behaviors, and catalyzing a broader societal response to the climate change narrative.

Chapter 7: Climate Change and Its Implications on School Education in Kenya

This chapter navigates through the intricate web where climate change and education intersect. Drawing upon a mixed research design method, the chapter unfolds findings culled from the crucible of Kisii Municipality, Kenya. These findings lay bare a tapestry of consequences - from increased prevalence of waterborne diseases to the erosion of family income and living standards, each thread underscoring the pervasive and far-reaching impacts of climate change on children's education.

There are two key issues explored in this chapter. First, the profound potential of education as a harbinger of change resonates. In the battle against the effects of climate change, education emerges as a potent weapon to mitigate its onslaught, offering a pathway toward resilience and adaptation. Second, the chapter underscores the urgent necessity of weaving climate change topics into the fabric of school curricula. This call for integration stands as a testament to the power of education to shape future generations that are not only cognizant of the perils of climate change but also equipped with the knowledge and tools to combat it.

Chapter 8: Building Bridges to Sustainability – Overcoming Barriers and Promoting Integration in Architectural Education

In a world ensnared by the far-reaching impacts of climate change, the chapter underscores architecture's unique role in combating its adverse effects, attributing them to the release of greenhouse gases into the atmosphere. The narrative unfurls with a discerning gaze, revealing a critical chasm - the often-tenuous link between sustainability and architectural education. While the importance of sustainability within architectural practice is widely acknowledged, the chapter candidly delves into the limitations of current curricula, where the comprehensive integration of this imperative topic remains elusive.

In response to this challenge, a clarion call emerges - the imperative for architectural education to embrace a foundational comprehension of the nexus between sustainability, technology, and design. Within this chapter, readers are beckoned to confront the barriers that beset the path toward sustainable environmental design within architectural education. A tapestry of insights emerges as these barriers are illuminated, paving the way for a comprehensive understanding of the transformative potential that lies within the integration of sustainability and architectural pedagogy.

Chapter 9: Promoting Environmental Sustainability Among Students and Citizens for Sustainable Future

This chapter embarks on an impassioned exploration into the transformative potential of education in the context of climate change and environmental preservation.

The narrative unfurls with a stark acknowledgment - climate change stands as a perilous precipice, threatening life's very continuity. Against the backdrop of the sixth major extinction catastrophe in recorded history, the chapter casts a sobering light on the existential threat posed by climate change, an outcome of the rampant use of fossil fuels. As the repository of knowledge, education equips young minds with the facts, enabling informed decision-making and fostering shifts in attitudes and behaviors. It has become an essential instrument within the global response to climate change, underscoring the profound potential of climate change education (CCE) as a transformative force.

Chapter 10: Pedagogical Strategies to Enhance Climate Change Education Outcomes

This chapter delves into the dynamic realm of pedagogy, casting a spotlight on pedagogical strategies that hold the potential to amplify the outcomes of climate change education. It uncovers the profound impacts of climate change, a force that has not only reshaped normative lifestyles but also prompted a reevaluation of ethics and our understanding of the world. As ecosystems bear the brunt of this transformation, education emerges as a potent tool, fostering an ethos of sustainable practices that mitigate the far-reaching impacts on our climate.

At the heart of this inquiry is Climate Change Education (CCE), a nascent field of pedagogy poised to tackle the complexities of climate change. Yet, within this burgeoning field, the chapter unravels a pivotal challenge - the absence of an established model for implementation. Through a comprehensive analysis, readers are invited to confront the obstacles that hinder the realization of enhanced educational outcomes. As the narrative unfolds, strategies crystallize - strategies poised to dismantle these barriers, fortifying the foundation of climate change education.

Chapter 11: Being a Rural Woman During Climate Change – An Analysis Based on the Concept of Ubuntu

This chapter embarks on a contemplative journey into the intersection of climate change, rural women, and the profound philosophy of ubuntu. At its core, this exploration is rooted in the recognition that climate change does not discriminate, yet its effects are often magnified for women in rural areas. Focusing on the intricate interplay between human and environment, the chapter casts a gaze towards ubuntu, a holistic philosophical concept that transcends the boundaries between the social and physical realms.

The narrative unfurls with a poignant acknowledgment - that climate change, with its far-reaching tendrils, impacts not only the environment but also the very essence of humanity. Within this understanding, the chapter delves into the lives of rural women through a bio-psycho-social perspective, where the nuances of their experiences, challenges, and resilience are illuminated.

Drawing upon ubuntu philosophy, the narrative navigates the delicate balance between human and environment, offering a tapestry of insights that transcend mere analysis. Within this framework, social work interventions emerge as a clarion call, beckoning for compassionate and informed strategies that empower rural women to navigate the challenges posed by climate change.

Chapter 12: Best Practice in the Use of the Water Sources of Portoviejo Canton, Ecuador, and Its Parishes

This chapter explores the delicate interplay between water resources, conservation, and sustainable management within the unique context of Portoviejo Canton, Ecuador. The narrative opens with a profound recognition - that water, as a fundamental element, serves multifarious roles within the ecosystem. From nurturing plantations through irrigation to quenching the thirst of urban populations, from supporting vibrant tourism to embodying the essence of life itself, the Portoviejo River stands as a quintessential example of water's diverse utility.

Yet, amidst these roles, a sobering reality emerges - the burgeoning settlements along the river's banks have heralded a rise in pollution levels and the erosion of water's integrity. Within this context, the chapter embarks on a vital journey - a quest to evaluate the historical backdrop, the mechanisms of conservation, and the governance of water resources within the purview of Portoviejo Canton.

Employing a hermeneutic research approach, the narrative navigates the intricate web of interpretations, extracting qualitative insights through a dialectical process. This process weaves a comprehensive narrative, substantiating the investigation and yielding compelling results. This chapter resonates as a call to action, an invitation to nurture practices that honor water's profound role in shaping the present and safeguarding the future.

Chapter 13: Dynamics of Teaching in the Era of Climate Change Education

This chapter examines the dynamic interplay between teaching methodologies and the urgent imperatives of climate change education. In a world marred by uncertainty, the chapter casts a discerning gaze towards mobile schools, exploring the nuanced dynamics of teaching and the facilities within these mobile institutions as they engage in the noble endeavor of climate change mitigation. Through a dual approach encompassing partial literature review and qualitative assessment of teaching methods, the chapter navigates the complex terrain, yielding insights that resonate across a spectrum of stakeholders.

In Summary

Principles of teaching sustainable and climate change education form the foundational bedrock of this volume, setting the stage for a comprehensive exploration of pedagogical perspectives, theoretical underpinnings, and effective instructional strategies. From educators' professional orientation to the intricate dance between learning resources, ICT frameworks, and cultural orientations, the chapters within this book delve into the intricate tapestry that shapes the landscape of climate change education.

With each chapter, a new facet of this intricate tapestry emerged - from the role of education in confronting the unpredictability of climate change to the integration of climate change principles into curricula, from the empowerment of educators as agents of change to the exploration of ethical dimensions within sustainable development. We delved into the impacts of climate change on communities, the invaluable perspectives of rural women, and the strategies to harness water resources amidst shifting paradigms.

As editors, we have assembled an esteemed group of contributors who are poised to guide readers through a myriad of topics. Through a blend of insightful analyses, innovative approaches, and reflective practice considerations, this handbook seeks to bridge the divide between knowledge and action, theory and practice. The journey traverses through the nuances of pedagogical strategies, customized learner support, and action research within the realm of sustainable and climate change education.

In the spirit of shared endeavor, we extend our gratitude to all those who have contributed to this undertaking. May the insights gleaned from these pages kindle a flame of inspiration, lighting the way toward a future where education and climate change intertwine, and where the vision of a sustainable world becomes a tangible reality.

Preface

Jared Keengwe
University of North Dakota, USA

Benard O. Nyatuka
Kisii University, Kenya

Chapter 1
A Review of Integration of Environment Education Into Teacher Preparation Programs

Jenny Bladow
University of North Dakota, USA

ABSTRACT

Compared to previous generations, children today spend less time outdoors interacting with nature and all it offers. The current leisure trends are encompassed in technology as it becomes more accessible to children. Perhaps there is a way to bring back the importance of nature and intertwine environmental education concepts into teacher preparation programs. In that case, there is hope for future generations of children to reap the benefits of nature and more community members who are environmentally literate. This chapter reviews the importance of incorporating environmental education into teacher preparation. The review of literature also discusses common approaches to addressing the current gap in environmental education implementation along with barriers that have been encountered.

INTRODUCTION

A critical issue facing teacher preparation programs in the United States is the gap in environmental education. In many education systems (both higher and P-12 education), there has been an increased shift toward the implementation of technological and online learning experiences. While these online collaborations are essential, how much is too much? How often have technology-based activities substituted environmental activities? How many hours a day do students spend indoors compared to outdoors?

DOI: 10.4018/978-1-6684-9099-0.ch001

Can P-12 teachers regularly expose students to nature once again? Ferguson et al. (2021) state, “Teachers, by virtue of the roles they perform, are critical change agents in schools and beyond” (p. 1355). Educators must reintroduce opportunities to get students outside for nature education. Otherwise, the current gap in environmental education will only continue to widen.

Now, more than ever, teachers need to provide opportunities for students to disconnect from technology and welcome the exposure to learning more about the world outside the four walls of the classroom. According to Franzen (2018), “Environmental education develops an environmentally literate citizenry that makes choices that are better for the health of the environment, leading to a more sustainable planet.” It is possible to accommodate for both technology and environmental enhancements within all levels of education. For teachers to have a dominant role in preparing P-12 grade students, they must first learn how to incorporate environmental education into their curricula. Teacher preparation programs are responsible for providing experiences for preservice teachers to learn different ways of implementing environmental education into everyday discussions and activities, regardless of the grade level or content area in which they will be teaching.

It is common knowledge that P-12 teachers and teacher preparation programs have considerable content to “fit” into the school year. Because of this, it is vital to find existing places for connections to environmental education instead of simply adding more units to the curriculum and expecting them to be covered each school year. Some teachers have discovered ways to implement environmental education, but most of them need further guidance in this critical undertaking.

REVIEW OF LITERATURE

When presenting the implementation of environmental education concepts to preservice teachers or within P-12 grade classrooms, educational leaders play an integral role. According to Mulvik et al. (2022), successful implementation of environmental education “requires the quality professional training of educators and school leaders, as well as opportunities for continuous professional development” (p. 9). In contrast, Gan (2021) discusses the idea of leadership existing in every person and promoting and fostering this thought “might help to integrate EE leadership in all school communities to include students, staff, and parents, both locally and globally” (p. 1462). Shriberg and MacDonald (2013) further explain:

While there is not currently a recognized model for teaching leadership for sustainability, nor a standard set of skills for these leaders, programs are rapidly emerging in response to the urgent call (and high student interest) to develop more sustainability leaders, typically using experiential learning as the primary methodology. (p. 3)

Educational leaders must demonstrate sustainable leadership and provide training that will aid educators in serving as change agents and masters of experiential learning (Shriberg & MacDonald, 2013).

Shriberg and MacDonald (2013) state, "Colleges and universities across the U.S. and world are rushing to respond to an increasingly urgent challenge: developing the next generation of sustainability leaders" (p. 2). Environmental education leaders need to be trained to emphasize pro-environmental behavior to combat the current environmental crisis (Gan, 2021). It starts with demonstrating the importance of environmental concepts before preservice teachers or leaders even begin their studies with teacher preparation or educational leadership programs. By conducting research and sharing findings on how to successfully facilitate environmental education, colleges and universities "can become crucial hubs for innovations that bring together different disciplines and stakeholders across the formal and non-formal education divide" (Mulvik et al., 2022, p. 11). Considering the current worldwide environmental issues, learning experiences relating to environmental education should carry the same weight as literacy, mathematics, or the subject-specific courses currently required for university admission (Smith & Stevenson, 2017).

One of the most pressing worldwide environmental issues is climate change. Favier et al. (2021) argue that "mitigating climate change is urgent and giant leaps forward are necessary to avoid catastrophic climate change" (p. 594). Students need to be guided in developing an awareness of the issues directly related to climate change and opportunities to create a sense of urgency and agency to make a change (Favier et al., 2021). White (2020) stresses that students must learn how to "take the necessary action to influence public opinion and pressure their governments and other authorities to take steps to cope with the emergency" (p. 871). Ferguson et al. (2021) state, "If the goal is to effectively develop citizens who are sustainability conscious, then this will only be achieved if teachers are themselves knowledgeable about sustainable development and have positive attitudes and perspectives about such" (p. 1348). While climate change cannot be resolved overnight, action must be taken immediately to begin moving in the right direction (White, 2020). Now is the time to support teachers in integrating important environmental education topics, such as climate change, with their students.

ENVIRONMENTAL EDUCATION IN HIGHER EDUCATION

If the goal is to expose P-12 students to knowledge, skills, and motivation to become more environmentally aware citizens, preparing preservice teachers to deliver such a curriculum is a logical place to start. Cheung et al. (2018) state, "Being an experienced educator does not necessarily mean that one is rich in environmental knowledge" (p. 26). Further explained by Murphy et al. (2021), "It is essential that teachers have the requisite subject and pedagogical knowledge and skills to implement sustainability practices in their classrooms" (p. 1012). Gordon (2009) states that preservice teacher education can "encourage teacher-candidates to become more critically aware of fundamental educational problems so that they will be better equipped to become change agents" (p. 53). Teacher education faculty possess the responsibility to be productive environmental education role models for preservice teachers. However, according to a study conducted by Crim et al. (2017), the implementation of environmental education into teacher preparation programs is limited and varies drastically across the United States and worldwide. McKeown-Ice (2000) further explains, "While many institutions of higher education are involved in environmental education, the extent of their involvement has not been documented" (p. 4).

Research efforts by McKeown-Ice (2000) discovered that "most colleges and universities have not institutionalized their commitment to environmental education in the ways that they have to reading, science, and special education, among others" (p. 9). Furthermore, Stanišić and Maksić (2014) argue, "It is difficult to accomplish the intended objectives of environmental education when the teachers are expected to teach new information in a novel, different way, without any appropriate education or training" (p. 124). The current gap is due to the expectation that P-12 educators should teach environmental education without the necessary training to do so successfully.

When considering environmental education implementation within any country's university systems, it is vital to brainstorm ways to engage preservice teachers in developing their curriculum to encourage their future students to be curious, ask questions, and seek out understandings about the environment. When preservice teachers are exposed to this type of learning structure, it models what their teaching style can look like to facilitate similar discussions and activities with their future students. Teachers' attitudes influence their classroom environment, and they are more likely to teach the curriculum that is consistent with their knowledge base and comfort level (Crim et al., 2017).

A study discussed by Crim et al. (2017) demonstrates that preservice teachers exposed to environmental education training were more likely to explore such topics within their classrooms. Similarly, Ferguson et al. (2021) state, "Teachers' views are often shaped by the beliefs they hold about curriculum, students, pedagogy and the learning process, which in turn influence learning outcomes" (p. 1348). Ferguson et al. (2021) explain that "sending clear and consistent signals to teachers about what to teach is an important component of teaching for sustainable development in schools" (p. 1348). As stated above, research shows that if preservice teachers are exposed to environmental education opportunities within their teacher preparation programs, they will be more comfortable and more likely to facilitate similar experiences with their future P-12 students.

In 1999, the United Nations Educational, Scientific, and Cultural Organization (UNESCO) and York University launched the University Twinning and Networking Program (UNITWIN/UNESCO) Chair on Reorienting Teacher Education to Address Sustainability (Sureda-Negre et al., 2014). The Chair established an International Network of 30 teacher education colleges from 28 countries to "reorient teacher education, decide which themes or sustainability goals to emphasize within their curricula, programs, practices, and policies to ensure that teacher-education programs fit the environmental, social, and economic conditions and goals of their communities, regions, and nations" (UNESCO, 2005). In response, a network of over 30 teacher preparation programs were established, along with additional resources such as guidelines and recommendations (Sureda-Negre et al., 2014). Below are examples of countries that were motivated to engage in environmental education due to the UNITWIN/UNESCO Chair's efforts.

In Canada, a network of educational professionals aims to enhance the development of high-quality environmental and sustainability education with preservice teachers across the country through ongoing research, new policies, and professional development (Inwood, 2020). In addition, the Ontario Association of Deans of Education has labored diligently to gain traction in integrating environmental and sustainability education with preservice teachers. As of 2020, five Ontario teacher education programs include core environmental and sustainability education components that all preservice teachers must experience prior to graduation (Inwood, 2020).

According to Sureda-Negre et al. (2014), Spanish universities require preservice teachers to attain three out of 79 competencies that are directly related to environmental sustainability. While environmental concepts have been woven into the curriculum for preservice teachers, it seems the process for consistently implementing those concepts has not occurred. Furthermore, Sureda-Negre et al. (2014) argue, "The curricula of future primary teachers in Spain does not seem to ensure good training with respect to environmental sustainability" (p. 281). It is encouraging that Spanish universities have inserted environmental education concepts into the required competencies for all preservice teachers. However, Sureda-Negre et al. (2014) imply that implementation methods need to be learned in order for teacher preparation programs to effectively incorporate environmental education concepts and experiences.

In 2018 in Africa, a Guidebook on Education for Sustainable Development for Educators was developed to operationalize effective teaching and learning strategies with preservice teachers. With the primary goal of Education for Sustainable Development (ESD), the guidebook aims to mainstream sustainability in all areas of preservice training, including teaching, learning, and assessment (UNESCO, 2018). This guidebook is a part of The Global Education 2030 Agenda, led by UNESCO, which includes a worldwide movement to eradicate poverty by 2030 and stresses the importance of preservice teachers promoting sustainable development within society (UNESCO, 2018). The guidebook serves as a valuable resource for teaching colleges in Spain as they move forward with their goals of preparing more environmentally aware citizens.

In 2011, Taiwan became the sixth country to enact a law for environmental education (Liu et al., 2015). The Environmental Protection Administration (EPA) established an Environmental Education Act, which states that all elementary and secondary students and teachers must participate in environmental education training, including at least four hours of related professional development annually (Liu et al., 2015). Even though the act has existed for over 12 years, the environmental education training sessions facilitated in each school district lack consistency. In addition, Liu et al. (2015) explain that some teachers do not regularly engage in environmental education types of activities with their students "due to their unfamiliarity of content knowledge and lack of teaching training related to environmental topics" (p. 115). This situation in Taiwan provides another example of how environmental education implementation needs to be at forefront of the streamlining efforts in teacher preparation programs.

A search on the North American Association for Environmental Education (NAAEE) website produces countless options for environmental education-related degree plans, ranging from endorsements to bachelor's and master's degrees. Some include teacher certification, and others do not. However, many degree plans are paired with science content area majors. For preservice teachers at the secondary level, only those specializing in science education are often introduced to environmental education (McKeown-Ice, 2000).

Colucci-Gray (2013) supports "the call for environmental literacy not as a specialist subject but as an integral part of science teacher preparation: environmental awareness is seen as a core competence of scientific education" (p. 246). After reading Colucci-Gray's thoughts on this concept, one might still ask why teacher preparation programs do not include environmental education in all majors. Exploring environmental education with preservice teachers across all concentration areas and grade levels of study would bring a holistic approach to their P-12 curricula. Similarly, Mulvik et al. (2022) explain that environmental education "should be a comprehensive education that closely follows learners through all phases and stages..."

(p. 9). Environmental education should not be taught at a single, isolated point in a student's educational career but instead in an ongoing and transformational manner. It is essential to consider how to meet the goals of environmental education through all school subjects (Matějček et al., 2020). Colucci-Gray (2013) stresses that the incorporation of environmental education with preservice teachers will ultimately help their future P-12 "pupils become critical learners of their own communities" (p. 247). Facilitating environmental education concepts with all preservice teachers can significantly impact future generations.

Environmental Education in P-12 Schools

Currently, the implementation of environmental education is vastly absent in P-12 schools throughout the country. Saylan and Blumstein (2011) posed a fundamental question: What can teachers in the United States do to improve students' understanding of the environment and its importance? The authors stressed the necessity for a full-scale implementation of environmental education. Saylan and Blumstein's (2011) discussion on shifting environmental awareness to a fundamental civic responsibility is essential in adopting long-lasting change. The authors explain the significance of ongoing, consistent, and widespread integration of environmental education into P-12 curricula.

Implementing environmental education has led to improved engagement, higher motivation, and better academic performance among students in P-12 classrooms. McCarty (2013) notes that "84% of students experiencing hands-on academic lessons in a [...] school garden report high levels of engagement, specifically in math and science" (p. 1). McCarty (2013) also discovered "on average, standardized science test score pass rates were 5.5% higher" (p. 1) with the implementation of environmental education. Furthermore, research conducted by Williams et al. (2018) concludes "garden-integrated science activities have the potential to rekindle student engagement and motivation for learning" (p. 11). After all, "wonder is more than just curiosity: It is about confusion and a willingness to learn more about the unknown and it comprises numerous emotions" (Sezen-Barrie et al., 2020, p. 571). Students need to be in the proper mindset to optimize the outcome of the conversations and experiences relating to environmental education. When educators facilitate appealing environmental opportunities in the curriculum, students effectively connect their learning to real life outside of school.

Additionally, Stanišić and Maksić (2014) state, "The question remains how to connect more school-based environmental education with real life, and to encourage young students who learn about ecological issues to be personally responsible and prepared to participate in solving ecological problems" (p. 124). There is a natural fit within science courses when it comes to finding organic connections between content area curriculum and environmental education. Many science courses involve insect collections, leaf identifications, and recycling activities, among many other associations with nature. However, many more subjects effectively pair with environmental education at both higher education and P-12 levels. For example, utilizing math is a logical practice when referring to all the calculations used to construct a learning garden space, run experiments on the plant growth rate, or calculate hours of appropriate sun exposure for flowers to efficiently grow.

Language arts, for instance, allows the opportunity for environmental concepts to be used as the focus for writing observations, conducting research, or creating poetry. Bringing students outside can enhance reading comprehension and engagement and can increase students' abilities "to express ideas and concepts through drawing" (Largo-Wight et al., 2018). Environmental education even applies to geography. When given a particular location, students learn what types of crops can flourish based on soil conditions, precipitation, and temperature. Social studies teachers could facilitate learning about past and present agriculture, climate, or natural resources in a specific area. Finally, environmental education can be investigated through physical education courses, where students could be actively engaged in outdoor exploratory activities.

There are opportunities to incorporate environmental education across multiple content areas simultaneously using pedagogies such as project-based, collaborative, and place-based learning (Ferguson et al., 2021). These types of teaching approaches allow for more inquiry-based experiences for students as they learn that "humans protect nature in order to protect themselves" (Stanišić & Maksić, 2014, p. 125). Simply getting students outside more often can increase on-task behavior and provide "motivation for reluctant learners, environmental ethics and knowledge, and reduce discipline and classroom management issues" (Largo-Wight et al., 2018, p. 655). An example of a preservice teacher project could involve tracing a commodity or product from nature through the process to consumption, while paying attention to both the sustainable and unsustainable ecological and economic processes along the way (Burns et al., 2015). The brainstorming list is endless for potential connections that could be made between environmental education and existing curriculum.

Learning gardens are increasingly popular in school districts worldwide, bridging the gap between a traditional school setting and a nature setting as students engage in environmental processes and systems. As stated by Eugenio-Gozalbo et al. (2020), "Gardens are appropriate real-life contexts for science teaching across educational stages" (p. 11). Multiple angles could be considered when developing and fostering a garden space, all of which are applicable to a wide range of grade levels and content areas. More specifically, Williams and Brown (2012) state, "School gardens are positioned as prime venues for nurturing children's tendency toward questions, which can be gently directed toward critical inquiry related to pressing environmental issues" (p. 81). Learning gardens allow ample opportunities for teachers to guide P-12 students through the process of questioning, analyzing, and interpreting, which eventually assists the P-12 students with understanding and addressing environmental issues.

Some P-12 schools and teacher preparation programs have successfully implemented environmental education within their communities. However, these practices are not as widespread as they could be. Several authors and researchers have validated the importance of environmental education for students of all ages. Williams and Brown (2012) discuss the loss of one of the seven areas of concern, curiosity, and wonder with current educational trends:

Education as sustainability, then, envisions education that is no longer conceived of as something that someone does to another person but is recognized as a lifelong process involving holistic experience including embodied participation and interaction with all members of the human/biotic community...teaching and learning become co-occurring processes. (p. 53)

The current youth, Generation Z, has suffered from a loss of experiences in nature without even knowing it. In his book, Louv (2006) discusses the research that indicates how exposure to a natural green environment can boost children's attention spans and alleviate Attention Deficit Disorder (ADD) and Attention Deficit Hyperactivity Disorder (ADHD) symptoms. For children with ADD or ADHD, "the move outdoors reduces stress and so enables better concentration as well as increasing motivation to learn" (Harris, 2018, p. 224). Similarly, Louv (2006) states, "Copious studies show a reduced amount of leisure time experienced by modern families, more time in front of the TV and the computer, and growing obesity among adults and children because of sedentary lifestyles" (p. 33). P-12 schools could infuse nature into the lives of the youth once again.

Children today play outside less often than previous generations, along with having fewer and less diverse playmates (Louv, 2006). In addition, Louv (2006) goes on to state, "Time in nature is not leisure time; it's an essential investment in our children's health" (p. 120). Many individuals see outdoor recess as a leisure time for students. While this is true and an essential component of elementary school days, recess should not be their only exposure to nature. Instead, schools should strive for additional opportunities for students of all ages to interact with the outdoors.

For P-12 schools to become more environmentally conscious, administrators and teachers must shift the focus from stand-alone field trips and prioritize inviting guest speakers, engaging students in hands-on learning experiences, facilitating ongoing environmental projects, and initiating learning gardens. Stanišić and Maksić (2014) state, "Teachers should not only transfer relevant information about environmental issues but also provoke students' personal interests and care for environmental issues, as well as their readiness for involvement in the process of solving environmental programs" (p. 125). These recommended actions, based on a curriculum infused with environmental education, would kickstart collaborative opportunities to improve community environmental issues.

BARRIERS IN THE IMPLEMENTATION OF ENVIRONMENTAL EDUCATION

Several roadblocks exist for attempting to implement environmental education into teacher preparation programs. First, international comparisons of student achievement, along with altered priorities and values, have de-emphasized most learning concepts outside the core tested areas of math, science, and literacy (Smith & Stevenson, 2017). As a result, P-12 schools primarily focus on limited subjects instead of including an expansive learning base for their students. In turn, pressure is placed on teaching colleges to emphasize the subject areas of math, science, and literacy to ensure preservice teachers are prepared for how P-12 students are taught currently. Consequently, teacher preparation programs are generally forced to omit the integration of environmental education concepts into all subject areas.

Second, the implementation of environmental education is not necessarily common knowledge for college professors. Because of this, preservice courses lack the transference of pedagogical knowledge to teach environmental education courses. Franzen (2018) validly states, "As we work to prepare preservice teachers to include environmental education in their future classrooms, we must also prepare teacher education faculty members to explicitly include environmental education." Faculty members in teacher preparation programs lack sufficient time and adequate support to collaborate on implementation (Crim et al., 2017). Teacher preparation program constituents may be eagerly invested in implementation if they would receive efficacious direction.

Recently, the North American Association for Environmental Education [NAAEE] (2017) created the Guidelines for Excellence: Professional Development of Environmental Educators, which address the barriers teacher preparation programs encounter when attempting to include environmental education. The guidelines describe what teachers should know and be able to do in relation to preparing lessons through the act of evaluating their students' understanding. These guidelines are integrated into several teacher preparation programs, including programs in states such as Illinois, Iowa, Minnesota, and Wisconsin. However, many states have not yet formally implemented these guidelines, along with the other sets of standards used by faculty in teacher preparation programs to design courses for preservice teachers.

Third, standards on environmental education are not commonly developed or maintained, and if they are, they can be found embedded within science standards. There are very few states that honor environmental education as a licensable area on teaching licenses. Only four states (Wisconsin, Pennsylvania, Kentucky, and Washington) have developed teacher preparation standards or licensure requirements related to environmental education (Crim et al., 2017). These barriers need considerable examination before expecting the widespread implementation of environmental education into teacher preparation programs.

Finally, lack of funding is a common barrier when considering the implementation of environmental education resources. Some P-12 school budgets allocate available funds to environmental education efforts due to federal, state, or local dollars. However, schools often utilize those dollars for other purposes. The same is often true for higher education. Local and national grant options are available for teacher preparation programs to implement environmental education topics and activities into their curricula. Several grant options are easily accessible online. Partnering with local businesses and state agencies is another option for gathering the necessary resources to engage in environmental education activities. For example, local greenhouses could donate plants for a learning garden in return for a sign that would be displayed at the garden site saying the business contributed to the garden. Communities that include universities could partner with teaching colleges, sustainability clubs, or STEM-related programs. In doing so, sufficient resources and college students could be identified to assist with developing activities like a learning garden space. Environmental education may be the bridge that creates new partnerships within a community.

SUGGESTIONS FOR FUTURE RESEARCH

In addition to the examination of the literature review, the researcher recognizes other areas for future research. How topics such as environmental education become "teachable still seems to be an open question worthy of study" (Sjögren, 2019, p. 1621). Potential research questions include the following: (a) What strategies have been successful in gaining buy-in from leadership teams and faculty at colleges of teacher education who have successfully implemented environmental education? (b) What are the key components of environmental education that need to be implemented in teacher preparation programs for those preservice teachers to effectively implement environmental education practices with their P-12 students? (c) To what extent do environmental education experiences provide positive outcomes with P-12 students ultimately becoming more environmentally literate? (d) What would encourage more countries worldwide and U.S. states to have teaching license options for environmental education? (d) Which countries and U.S. states have mandates for teachers to obtain some level of environmental education prior to obtaining a license? What did this process entail?

As stated by Evans (2020), "Issues such as climate change, land degradation, sea-level rise, poverty, food production and scarcity are as much about social justice as they are about ecology and, therefore, relevant to all educators' work" (p. 822). If all educators actively teach concepts to create more environmentally aware students, these efforts could "lessen environmental crises through pro-environmental behavior" (Gal & Gan, 2020, p. 288). How does the world of education achieve the goal of preparing all teachers to eagerly implement environmental education with learners of all ages?

First and foremost, national guidelines for integrating environmental education within preservice teacher education must be developed to institutionalize these efforts across all teaching colleges (McKeown-Ice, 2000). More recently, healthy green school certification has become a way to facilitate environmental education implementation worldwide (Gan, 2021). The Healthy Green Schools & Colleges program provides opportunities for P-12 grade school districts and higher education colleges with resources to transform the health and sustainability of their learning communities (Healthy Green Schools & Colleges, n.d.). The program assists schools and colleges with different phases of implementation, which are customized to their specific locations and needs. This is just one example of a support that is currently in place to ensure students of all ages can be further immersed in environmental education learning experiences. However, teaching colleges still need national guidelines in place before any institutionalized efforts can begin.

Second, teacher licensing agencies need to implement environmental education requirements for each state. McKeown-Ice (2000) suggests:

Because state guidelines for teacher certification have a major influence on preservice teacher education requirements and conflict with mandated course content is the major barrier to environmental education, environmental education groups should work to change teacher certification requirements in their state. (p. 10)

Finally, preservice teacher education programs should actively pursue faculty with interests and experiences in environmental education (McKeown-Ice, 2000). These faculty members will naturally approach environmental education implementation with a collaborative mindset. In turn, teacher preparation programs will flourish in implementing environment education strategies with preservice teachers, ultimately impacting future generations of P-12 learners.

CONCLUSION

Now, more than ever, education stakeholders need to facilitate teacher preparation programs that equip preservice teachers to engage their future P-12 students in impactful discussions and experiences relating to environmental education. As stated by Walshe and Tait (2019), "Children today face significant challenges in response to living in a globalized world and the predicted environmental threats to the planet" (p. 1731). At a time when technology is omnipresent and has consumed the youth's recreational time, the execution of a strategic plan is needed in order to equip preservice teachers with sufficient opportunities to facilitate environment education for P-12 students.

The achievement of quick and effective implementation of environmental education in teacher education programs and P-12 classrooms requires a massive cooperative effort among all participants in education. Inwood (2020) suggests, "In order to successfully shift education systems towards sustainability, its network of leaders, policy-makers, administrators, educators, researchers and other stakeholders need to be working in unison, talking change in each of their respective jurisdictions simultaneously" (p. 826). Instead of independent sectors attempting to accomplish separate tasks, a select group of stakeholders laboring collaboratively on these efforts would enact change more effectively and efficiently. Working on the same overarching goal, unified stakeholders from all areas of education would accomplish widespread integration in the near future. If not, successfully integrating environmental education with students of any age will remain a challenge. When considering the future of Earth, all people must collectively foster environmental education integration for students of all ages, ultimately leading to more environmentally aware citizens who care for others and their communities.

REFERENCES

Burns, H., Diamond-Vaught, H., & Bauman, C. (2015). Leadership for sustainability: Theoretical foundations and pedagogical practices that foster change. *International Journal of Leadership Studies*, *9*(1), 88–100.

Cheung, Y. T.-Y., Chow, C.-F., & So, W. W.-M. (2018). A train-the-trainer design for green ambassadors in an environmental education programme on plastic waste recycling. *International Research in Geographical and Environmental Education*, *27*(1), 24–42. doi:10.1080/10382046.2017.1285138

Colucci-Gray, L. (2013). The inclusion of environmental education in science teacher education. *Educational Review, 65*(2), 246-247. https://doi.org.ezproxy.library.und.edu/10.1080/00131911.2011.634173

Crim, M., Moseley, C., & Desjean-Perrotta, B. (2017). Strategies toward the inclusion of environmental education in educator reparation programs: Results from a national survey: Strategies toward the inclusion of EE. *School Science and Mathematics, 117*(3-4), 104–114. doi:10.1111sm.12211

Eugenio-Gozalbo, M., Aragón, L., & Ortega-Cubero, I. (2020). Gardens as science learning contexts across educational stages: Learning assessment based on students' graphic representations. *Frontiers in Psychology, 11*, 2226–2226. doi:10.3389/fpsyg.2020.02226 PMID:33013580

Evans, N. (2020). What ought to be done to promote "Education for Sustainability" in teacher education? *Journal of Philosophy of Education, 54*(4), 817–824. doi:10.1111/1467-9752.12482

Favier, T., Van Gorp, B., Cyvin, J. B., & Cyvin, J. (2021). Learning to teach climate change: Students in teacher training and their progression in pedagogical content knowledge. *Journal of Geography in Higher Education, 45*(4), 594–620. doi:10.1080/03098265.2021.1900080

Ferguson, T., Roofe, C., & Cook, L. D. (2021). Teachers' perspectives on sustainable development: The implications for education for sustainable development. *Environmental Education Research, 27*(9), 1343–1359. doi:10.1080/13504622.2021.1921113

Franzen, R. (2018). Environmental education in teacher education programs: Incorporation and use of professional guidelines. *The Journal of Sustainability Education.* http://www.susted.com/wordpress/content/environmental-education-in-teacher-education-programs-incorporation-and-use-of-professional-guidelines_2018_01/

Gal, A., & Gan, D. (2020). Transformative sustainability education in higher education: Activating environmental understanding and active citizenship among professional studies learners. *Journal of Transformative Education, 18*(4), 271–292. doi:10.1177/1541344620932310

Gan, D. (2021). Environmental education leadership: The perceptions of elementary school principals as expressed in their drawings and explanations. *Environmental Education Research, 27*(10), 1440–1466. doi:10.1080/13504622.2021.1959899

Gordon, M. (2009). Toward a pragmatic discourse of constructivism: Reflections on lessons from practice. *Educational Studies: A Journal of the American Education Studies Association, 45*(1), 39-58.

Harris, F. (2018). Outdoor learning spaces: The case of forest school. *Area, 50*(2), 222–231. doi:10.1111/area.12360

Healthy Green Schools & Colleges. (n.d.). https://www.healthygreenschools.org/

Inwood, H. (2020). Emerging praxis of environmental and sustainability education in teacher education in Canada. *Journal of Philosophy of Education, 54*(4), 825–831. doi:10.1111/1467-9752.12466

Largo-Wight, E., Guardino, C., Wludyka, P. S., Hall, K. W., Wight, J. T., & Merten, J. W. (2018). Nature contact at school: The impact of an outdoor classroom on children's well-being. *International Journal of Environmental Health Research, 28*(6), 653–666. doi:10.1080/09603123.2018.1502415 PMID:30047798

Liu, S.-Y., Yeh, S.-C., Liang, S.-W., Fang, W.-T., & Tsai, H.-M. (2015). A national investigation of teachers' environmental literacy as a reference for promoting environmental education in Taiwan. *The Journal of Environmental Education, 46*(2), 114–132. doi:10.1080/00958964.2014.999742

Louv, R. (2006). *Last child in the woods: Saving our children from nature-deficit disorder*. Algonquin Books of Chapel Hill.

Matéjček, T., Bartoš, J., & Kučerová, S. R. (2020). Which subjects contribute to the teaching of cross curricular topic environmental education at elementary schools in Czechia? *Acta Universitatis Carolinae. Geographica. Universita Karlova, 55*(2), 200–209. doi:10.14712/23361980.2020.14

McCarty, J. (2013). REAL school gardens program: Learning gardens and teacher training to improve student engagement and academic performance in low-performing elementary schools. *The Journal of Applied Research on Children, 4*(2), 1–8. doi:10.58464/2155-5834.1169

McKeown-Ice, R. (2000). Environmental education in the United States: A survey of preservice teacher education programs. *The Journal of Environmental Education, 32*(1), 4–11. doi:10.1080/00958960009598666

Mulvik, I., Pribuišis, K., Siarova, H., Vežikauskaite, J., Sabaliauskas, E., Tasiopoulou, E., Gras-Velazquez, A., Bajorinaite, M., Billon, N., Fronza, V., Disterhelf, A., & Finlayson, A. (2022). *Education for environmental sustainability: Policies and approaches in European union member states*. Final Report. European Commission.

Murphy, C., Mallon, B., Smith, G., Kelly, O., Pitsia, V., & Martinez Sainz, G. (2021). The influence of a teachers' professional development programme on primary school pupils' understanding of and attitudes towards sustainability. *Environmental Education Research, 27*(7), 1011–1036. doi:10.1080/13504622.2021.1889470

North American Association for Environmental Education (NAAEE). (2017). *Guidelines for excellence: Professional development of environmental educators*. North American Association for Environmental Education.

Saylan, C., & Blumstein, D. (2011). *The failure of environmental education (and how we can fix it)*. University of California Press. doi:10.1525/9780520948723

Sezen-Barrie, A., Miller-Rushing, A., & Hufnagel, E. (2020). "It's a gassy world": Starting with students' wondering questions to inform climate change education. *Environmental Education Research, 26*(4), 555–576. doi:10.1080/13504622.2019.1610158

Shriberg, M., & MacDonald, L. (2013). Sustainability leadership programs: Emerging goals, methods, & best practices. *Journal of Sustainability Education, 5*(1), 1–21.

Sjögren, H. (2019). More of the same: A critical analysis of the formations of teacher students through education for sustainable development. *Environmental Education Research, 25*(11), 1620–1634. doi:10.1080/13504622.2019.1675595

Smith, G. A., & Stevenson, R. B. (2017). Sustaining education for sustainability in turbulent times. *The Journal of Environmental Education, 48*(2), 79–95. doi:10.1080/00958964.2016.1264920

Stanišić, J., & Maksić, S. (2014). Environmental education in Serbian primary schools: Challenges and changes to curriculum, pedagogy, and teacher training. *The Journal of Environmental Education, 45*(2), 118–131. doi:10.1080/00958964.2013.829019

Sureda-Negre, J., Oliver-Trobat, M., Catalan-Fernández, A., & Comas-Forgas, R. (2014). Environmental education for sustainability in the curriculum of primary teacher training in Spain. *International Research in Geographical and Environmental Education, 23*(4), 281–293. doi:10.1080/10382046.2014.946322

United States Educational, Scientific and Cultural Organization (UNESCO). (2005). *Guidelines and recommendations for reorienting teacher education to address sustainability*. United Nations Educational, Scientific and Cultural Organization. https://unesdoc.unesco.org/ark:/48223/pf0000143370

United States Educational, Scientific and Cultural Organization (UNESCO). (2018). *Guidebook on education for sustainable development for educators: Effective teaching and learning in teacher education institutions in Africa.* United Nations Educational, Scientific and Cultural Organization. https://unesdoc.unesco.org/ark:/48223/pf0000367474

Walshe, N., & Tait, V. (2019). Making connections: A conference approach to developing transformative environmental and sustainability education within initial teacher education. *Environmental Education Research, 25*(12), 1731–1750. doi:10.1080/13504622.2019.1677858

White, J. (2020). The climate emergency and the transformed school. *Journal of Philosophy of Education, 54*(4), 867–873. doi:10.1111/1467-9752.12461

Williams, D. R., & Brown, J. D. (2012). *Learning gardens and sustainability education: Bringing life to schools and schools to life*. Routledge.

Williams, D. R., Brule, H., Kelley, S. S., & Skinner, E. A. (2018). Science in the learning gardens (SciLG): A study of students' motivation, achievement, and science identity in low-income middle schools. *International Journal of STEM Education, 5*(8), 1–14. doi:10.118640594-018-0104-9 PMID:30631698

KEY TERMS AND DEFINITIONS

Curriculum: A program of study that describes the standards, teaching methods, instructional resources, and ways to assess student knowledge.

Environmental Education: A process by which individuals gain awareness of their environment, explore environmental issues, and collaborate with others to solve problems with the shared goal of improving the environment.

P-12 Schools: Schools which facilitate learning for preschool through twelfth grade students.

Preservice Teachers: A higher education student enrolled in a teacher preparation program to learn the content and pedagogy required to obtain a P-12 teaching license.

Sustainability Education: The process of finding ways to ensure the needs of current generations are being met without compromising the needs of future generations.

Teacher Preparation Programs: A program of study for students in higher education which includes coursework and field experiences within P-12 schools to prepare preservice teachers to become licensed classroom teachers.

Chapter 2
Integrating Climate Change Education Into the Curriculum in Kenya

Maureen Syallow
https://orcid.org/0000-0002-9787-2467
Strathmore University, Kenya

Magdalene Dimba
Strathmore University, Kenya

Esther Muthoni King'ori
Strathmore University, Kenya

ABSTRACT

The effects of climate change are a major global concern. In Kenya, the impact of climate change has been felt in perennial droughts which sometimes result into conflict as communities fight for pasture and water. In a bid to address climate change issues, Kenya has developed policies that focus on mitigation and adaptation measures borrowed from ideas outlined in global climate change strategies. One such strategy is the employment of education as a tool to create awareness on climate change issues. This chapter will examine the climate change policies developed by bodies in charge of higher education in Kenya. The chapter will further analyze attempts that have been made to integrate climate change education into curricula. In addition, the chapter will trace the history of discourse that saw the introduction of climate change education in institutions of higher learning. Finally, the chapter will analyze pedagogical strategies that can be employed in teaching climate change content.

DOI: 10.4018/978-1-6684-9099-0.ch002

INTRODUCTION

According to the UN, climate change refers to long-term shifts in temperatures and weather patterns. These shifts may be natural, such as through variations in the solar cycle. But since the 1800s, human activities have been the main driver of climate change. Since the Earth is a system, where everything is connected, changes in one area can influence changes in all others. Therefore, the consequences of climate change include, among others, intense droughts, water scarcity, severe fires, rising sea levels, flooding, melting polar ice, catastrophic storms and declining biodiversity. These have been observed in different parts of the world. Unesco (2015) has described climate change as “one of the greatest threats to sustainable development” (Unesco, 2015, p. 9).

Africa has been projected as a region that is expected to face the greatest effect of climate change (Fahey, Verstraten & Berry, 2016). The Intergovernmental Panel on Climate Change (IPCC) has stated that the huge effect of climate change on the continent will result from inability to adapt (Fahey, Verstraten & Berry, 2016). Challenges of climate change have been seen in various ways such as food insecurity and drought. For example, Kenya has recently experienced drought due to “failure of four consecutive rainy seasons” (IFRC, 2022). The drought exposed millions to the risk of hunger and led to deaths of hundreds of livestock. If not tackled decisively drought will contine to cause a health crisis and impoverish Kenyans who engage in farming and those who do livestock business. In addition, drought in Kenya has on many occasions led to “disputes over pasture and water” and resulted into “students from nomadic communities” missing out on school (Njino, 2022).

Many climate change solutions are being explored. Some of the notable solutions are the global frameworks and agreements like the Sustainable Development Goals, the UN Framework Convention on Climate Change and the Paris Agreement. Education can be one of the most effective means of ensuring future mitigation against climate change. Climate change education can contribute to the development of the necessary knowledge, skills and attitudes so that students can become agents of change, as future consumers, policymakers, and social influencers. Climate change adaptation is an example of a wicked problem. Teachers need to have extensive knowledge to design high-quality education that addresses the wickedness and contributes to its problem-solving. Unesco argues that climate change education is critical as “it helps learners understand the causes and consequences of climate change. It prepares them to live with impacts of climate change and empowers learners to take appropriate actions to adopt more sustainable lifestlyes” (Unesco, 2015, p. 9). This chapter examines policies that guide the integration of climate change education into Kenyan curriculum. The focus will be on curricula of institutions of higher learning.

BACKGROUND

The impact of climate change has been a matter of public discourse in Kenya for a long time. In the 1980s, Kenya's then president, the late Daniel Arap Moi led citizens in building gabions to fight soil erosion. Subjects that comprised topics on conservation and soil erosion were taught in primary schools and songs creating awareness about environmental concerns were composed and popularized during national events. Besides the government, non-governmental organisations (NGOs) have also been major players in the promotion of environmental conservation. One such NGO is the Green Belt Movement founded by Professor Wangari Maathai in 1977 (www.greenbeltmovement.org). The organisation mobilized Kenyan women to plant trees as a way of mitigating climate change challenges such as the drying up of local rivers which are critical in reducing food scarcity (www.greenbeltmovement.org). Other strategies that have been used to promote environment conservation in Kenya have included the removal of communities that encroach on forest land.

However, environmental conservation efforts in Kenya have not produced a climate change consciousness in the larger populace. This has led experts in recommending that learning institutions develop curricula to help inculcate in the citizenry the importance of environmental conservation and the need to navigate climate change issues. Unesco has echoed these sentiments and pointed out that "climate change education is a powerful tool" that can be used to enlighten citizens on "the causes and consequences of climate change" (UNESCO, 2015, p. 9). This chapter focuses on the path that Kenya has travelled in the attempt to infuse climate change in higher education.

One legislation that has helped to shape Climate Change conversations in Kenya is the Environmental Management and Coordination Act of 1999. The Act was enacted with the aim of providing "for the establishment of an appropriate legal and institutional framework for the management of the environment" (National Council for Law Reporting, 2012, p. 9). The Act further established the National Environment Management Authority (NEMA). One of the functions of the authority is to "coordinate the various environmental management activities" so as to ensure "the proper management and rational utilization of environmental resources on a sustainable yield basis for the importance of the quality of human life in Kenya" (National Council for Law Reporting, 2012, p. 18). NEMA is further mandated to "undertake and co-ordinate research, investigation and surveys in the field of environment and collect, collate and disseminate information about the findings of such research, investigation or survey" (National Council for Law Reporting, 2012, p. 19).

Vision 2030 document was launched by Kenya's third President, Mwai Kibaki. The document outlines development steps that Kenya needs to take in order to become a middle income economy by 2030. The vision is laid out according to three pillars: economic, social and political (Government of Kenya, 2007). Climate change issues are outlined under the social pillar under a section titled Environment. According to the Government, "Kenya aims to be a nation that has a clean, secure and sustainable environment by 2030" (Government of Kenya, 2007, p. 19). The document states that Kenya will "improve capacity for adaptation to global climate change" (Government of Kenya, 2007, p. 19). The section further states that there will be harmonisation of laws that deal with environmental issues. An important point to note is that although the document is key as it incorporates environmental issues into Vision 2030, it does not elaborate the place of education in climate change issues.

In 2010 the Kenyan government made another big attempt to tackle climate change by coming up with the National Climate Change Response Strategy (NCCRS). The launch of the strategy was the first nationwide attempt to formulate policy that would decisively tackle climate change issues. According to the then Minister for Environmental and Mineral Resources, John Michuki, the strategy was developed through "a participatory process" that involved the government, parliament, "private sector, civil society organisations, development partners, youth groups, women's groups, faith-based organisations, and the media, among others" (Government of Kenya, 2010, p .3). The NCCRS was to address some omissions that had been identified in the Vision 2030 document. The drafters of the NCCRS argued that Vision 2030 did not adequately tackle the climate change issue. They proposed that climate change concerns be integrated into strategies that the country would adopt in its efforts to propel Kenya into middle income economy status by 2030. The mission of the NCCRS was to "strengthen nationwide focused actions towards adapting to and mitigating against a changing climate by ensuring commitment and engagement of all stakeholders while taking into account the vulnerable nature of our natural resources and society as a whole" (Government of Kenya, 2010, p .45). The strategy recommended the enactment of laws that would specifically deal with climate change matters.

In page eight of the NCCRS document, the goverment emphasizes the importance of improving climate change awareness in the country and acknowledges that "the level of understanding of climate change and its impacts is very low countrywide" (Government of Kenya, 2010, p .8). The strategy elaborated the need to simplify environmental science for citizens to comprehend climate change, the risks involved and the mitigating efforts that can be put into place. The document also expressly outlines the role of education in climate change awareness and mitigation efforts. In addition to using the media to create awareness, the government would ensure that "climate change information" would be integrated into "schools, and colleges' curricula and syllabus"(Government of Kenya, 2010, p .8). The NCCRS proposed a number of measures to help in incorporate climate change into curricula. These included: updating school curricula with climate change content; updating textbooks and learning materials with climate change content; and updating/developing courses that contain climate change issues (Government of Kenya, 2010, p .69).

One of the major laws enacted to tackle Climate Change in Kenya is the Climate Change Act, 2016. The law was enacted so as "to provide a regulatory framework for enhanced reponse to climate change, to provide for mechanism and measures to achieve low carbon climate development, and for connected purposes" (The Climate Change Act, 2016, p. 81). The Act established the National Climate Change Council which is chaired by the President, who is deputised by the Deputy President. One key function of the Council is to "provide policy direction on research and training on climate change..."(The Climate Change Act, 2016, p. 185). Section 21(2) of the Act mandates the Council to "advise the public agencies responsible for regulating universities and tertiary institutions on integration of climate change into curricula" (The Climate Change Act, 2016, p. 81). This section of the law describes the role that government and bodies such as the Commission of University Education should play in ensuring that insitutions of higher learning develop curricula that focuses on climate change. Prior to the enactment of the Climate Change Act there was no legal framework to help mainstreaam "climate change across different sectors" (Ministry of Environment and Forestry, 2018, p. 6).

Another key initiative created to tackle issues of sustainability and climate change in the country is the Kenya Green University Network (KGUN). The network which brings together local univerisities was created so as to provide "a knowledge and innovation hub and support the sharing of best practices to support achievement of Sustainable Development Goals (SDGs) and commitments under the Paris Agreement on Climate change"(Unesco, 2016). The network was launched by UNEP in collaboration with NEMA and the CUE. According to its website, KGUN members aim at integrating sustainability and climate change issues into education and training strategies (https://www.kgun.org/about-us/). KGUN is focused on enhancing "student engagement" in tackling climate change issues (https://www.kgun.org/about-us/).

From the above discussion, it is evident that environmental concerns have been part of Kenya's discourse for a long time. In addition, there have been various initiatives by various arms of government to integrate climate change in Kenya's education system and specifically in institutions of higher learning which are the focus of this chapter. Despite these efforts, climate change education is yet to be fully embraced at instutions of higher learning. This state of affairs has been illustrated by scholars whose studies have established that institutions of higher learning in Kenya have been slow in integrating climate change in curriculum. One such study conducted in two public universities in Kenya in 2013 found that while a majority of the students, 96.3 percent, "were aware of changes in climatic patterns" in the country, only a minority, 7.4 percent wanted climate change education to be included as a subject in university education (Huho, 2015, p. 227). Another study found that the existence of climate change policies in Kenya are yet to be translated into learning and teaching strategies in Kenyan univeristies (Nyerere, 2022). The CUE has in the recent past pointed out that there has been a decrease in the uptake of climate change programmes in public universities (UNCC:Learn, 2020).

A 2021 study looked at how government policies have impacted on the manner in which Kenyan institutions of high learning have responded to climate change. Nyerere, Gatwiri and Okinyi (2021) examined how national policies on climate change issues are linked to those adapted by the higher education sector. The scholars further analysed how Kenyan universities have been tackling issues of "climate change mitigation, adaptation, impact reduction and early warning" (Nyerere, Gatwiri & Okinyi, 2021, p. 4). The scholars hail efforts that "public and private universities" have made towards adoption of "some forms of greening approaches in their operations" (Nyerere, Gatwiri & Okinyi, 2021, p. 11). However, the scholars point out that there needs to be training on Greening Human Resource Management. The scholars made a key finding: the Kenyan government has established policies to tackle climate change but universities are slow in embracing these policies (Nyerere, Gatwiri & Okinyi, 2021, p. 11). The concern is that while existing policies highlight the role that universities can play in creating awareness and seeking solutions to climate change problems, they are silent on how institutions of higher learning can participate in the "governance of climate change affairs"(Nyerere, Gatwiri & Okinyi, 2021, p. 11). The researchers propose sensitisation campaigns be carried out to enlighten universities on ways of adopting the national policies on climate change.

INTEGRATING CLIMATE CHANGE INTO THE KENYAN CURRICULUM

Globally, the danger of climate change has continued to be felt, with the rising global temperatures causing various environmental disorders. Incidences of extreme weather conditions such as floods, droughts, cyclones and wildfires continue to rise and this has led to unfavorable living conditions all over. Climate change is therefore a weighty matter to be discussed in various forums including government institutions that direct important subjects of the nation, including education.

In Kenya, the ministry of education in conjunction with various associated institutions have been given the mandate for coordinating and providing quality education from basic to tertiary level. Consequently, knowing the impact that climate change has on daily lives, the government of Kenya got involved in signing various agreements and one of them culminated from the 2012, UN Conference on Sustainable Development-Rio+20 (Ministry of Education, 2017). From this agreement, *The Future We Want*, the Government of Kenya made a promise to support Education for Sustainable Development (ESD) and ensure that this concept is integrated actively into all levels of education. To this end, climate change education has been progressively motivated into various curricula of primary, secondary, universities and Technical & Vocational Education Training (TVETs) of Kenya.

This chapter examines how selected higher education governing bodies have attempted to direct universities and TVETs to integrate climate change into the Kenyan curriculum. This will be done by looking at two policy documents that have given directions on how climate change education needs to be integrated into higher institutions of learning and whether this has been implemented or not. It will also examine the guidelines that the Commission for University Education (CUE) has given in relation to the policy documents formulated by the Government of Kenya. The two policy documents include: The National Education Sector Strategic Plan (NESSP) 2018-2022; and The Education for Sustainable Development Policy for the Education Sector by the Ministry of Education 2017.

1. Education for Sustainable Development Policy for the Education Sector-Ministry of education (2017)

The document on education for sustainable development policy for the education sector in Kenya formulated by the ministry of education originated from a ministerial conference held in Arusha Tanzania in 2012 dubbed “The Africa Ministerial Conference on Environment (AMCEN)” which led to the formulation of the African Environmental Education and Training Action Plan (AEETAP 2015–2024) (AMCEN/SS/VI/6, 2012). The contents of these policy document are in sync with the agenda 2063 of the African union of which Kenya is an active member. The various policies formulated under Education for sustainable development and how this has been implemented include the following:

(a) **Policy statement:** The Ministry of Education in collaboration with other stakeholders shall provide an enabling environment for developing and implementing policies on ESD based on cross-sector and multi-stakeholder approaches (Ministry of Education, 2017). Some of the strategies that the policy has put forth to support this initiative include: Integrating education for sustainable development (ESD) into the various programmes in the education sector. For the higher education sector, the Commission for University Education (CUE) supported this policy and strategy specifically for SDG 13; Climate action, by emphasising that all universities in Kenya need to find ways of including SDG 13 into their programmes. This is a directive that the CUE is keen on following and in June 2020, the CUE through a decree on its website gave instructions which required all universities to integrate climate change education into their various programmes (CUE report, 2020).

All TVETs and other tertiary colleges were also required to incorporate Climate Change education into existing Courses. A statement from CUE website reads that: "The CUE encourages universities to integrate climate change into existing courses across different disciplines, including science, engineering, social sciences, and humanities' (CUE report, 2020).

A few examples to demonstrate how this policy statement and directive has been implemented in various universities in Kenya include: Kenyatta University which has The Department of Environmental Studies that offers a Bachelor of Environmental Studies degree program which includes a course on climate change. This allows students to study in detail matters surrounding climate change and their consequences. Another example is the University of Nairobi which offers a Master of Science degree programme in Climate Change and Adaptation, Jomo Kenyatta University of Agriculture and Technology which has a Department of Environmental Sciences offering a Bachelor of Environmental Science degree program that includes a course on climate change, Strathmore University which has an Institute for Public Policy and Governance that offers a course on Climate Change and Development, Egerton University which has a Department of Environmental Science that offers a Bachelor of Science degree programme in Environmental Science with a course on climate change and Maseno University which offers a Bachelor of Science degree programme in Environmental Science with a course on climate change. These are some examples of major universities in Kenya that have picked the directive from the Ministry of Education and the Commission of University Education and acted on it swiftly. Other universities in Kenya may be in the process of introducing the climate change courses or are integrating climate change courses into their existing programs.

The second strategy from the ESD policy is to promote participation of all formal and non-formal education organizations both from private sector and civil society, in the development and implementation of education plans. This is a strategy that is geared towards ensuring inclusivity in tackling the climate change disruptions. When education programmes are formulated, it is key to work with the communities that surround the particular institution as these are the people who experience the extreme weather conditions. They are quite instrumental in advising on how the education materials can be tailored to suit their conditions. This is where theory and practice work hand in hand. Therefore, the CUE has also supported this strategy by giving a directive on partnerships and collaboration by indicating that: There is need for partnerships and collaborations between universities, industry, government, and civil society in promoting climate change education and research. Looking at how this policy strategy has been implemented, various institutions of higher education in Kenya have actualized the strategy from ESD and the directive from CUE. The partnerships are in terms of research centers and institute partnerships between industry and institutions of higher learning including TVETs and universities. Some of these initiatives include: First, The African Climate and Development Initiative (ACDI) at the University of Cape Town which has a research partnership with Kenyatta University in Kenya, Second, The Institute for Climate Change and Adaptation (ICCA) at the University of Nairobi which is a multidisciplinary research center that focuses on climate change adaptation and mitigation, disaster risk reduction, and sustainable development, the third is The Centre for Climate Change Studies (CCCS) at the University of Nairobi which conducts research on climate change that focuses on adaptation and mitigation measures for sustainable development., fourth is The Pan African Climate Justice Alliance (PACJA) which is an alliance of over 1000 organizations and individuals from across Africa, including several universities in Kenya, the fifth, is The PACJA which conducts research on climate change and advocates for climate justice in Africa and sixth, is the Institute for Resource Assessment (IRA) at the University of Dar es Salaam in Tanzania which has a research partnership with Egerton University in Kenya, focusing on climate change and natural resource management in East Africa. However, it is worth noting that many other universities in Kenya may also have initiated conversation on climate change research centers or are already being involved in climate change research through other partnerships and collaborations.

Another policy statement from the ESD reads as follows:

(b) **Policy Statement:** The Ministry of Education, research institutions and other stakeholders shall mainstream sustainability into higher education curricula, research and community engagement activities. This policy establishes the regularity with which institutions of higher learning need to regularize conversations regarding climate change into their daily teaching and learning. Several strategies have been put forth to ensure this is maintained and one of them indicate that; Institutions of higher learning will promote green economy development opportunities, practices and technology (Ministry of Education, 2017) and establish higher education networks to foster student innovations and acquisition of relevant skills in ESD. Several universities in Kenya as mentioned earlier have organized awareness campaigns on climate change to sensitize the public on the effects of climate change, mitigation and adaptation measures, including the importance of climate change education. This allows both the education sector and the citizens to mainstream these ideas into their daily activities.

2. The National Education Sector Strategic Plan (NESSP) 2018-2022)

The National Education Sector Strategic Plan (NESSP) 2018-2022 is a comprehensive plan that describes policy strategies, priorities and programmes for the education sector in Kenya over a period of five years that started in 2018 and ended in 2022.

This plan recognized the significance of climate change education and emphasized the necessity to integrate climate change courses into the education system. Some of the key policies included in the NESSP 2018-2022 concerning climate change education in Kenya, in institutions of higher learning will be discussed in this chapter. Worth noting is that, while the ESD policy discussed above concerned strategies to integrate climate change into university education, the NESSP policy document contains strategies that apply mostly to Technical & Vocational Education and Training(TVET) institutions. The most crucial policies include;

a) Programme 3.6: Greening Technology in TVET; This policy states that:

Greening TVET (GTVET) is a programme aimed at creating awareness on conservation and sustainability of the environment. This has been necessitated by concerns about climate change, environmental degradation and scarcity of resources. Environmental degradation and effects of global warming require that deliberate efforts be put in place to enhance the total vegetation cover. Greening involves protection of environment through exploitation of opportunities available in green economy by employing environment friendly aspects (Kenya Ministry of Education, 2018).

To support this policy, various strategies were put forth to encourage TVETs to have climate change courses integrated into their programmes. One of the strategies indicates that: Implement waste management technology into the curricular which will ensure that students are equipped with the knowledge and skills needed to manage waste effectively. Waste management technology is such a crucial area of concern in Kenya, as there is a significant amount of waste being generated from households and industries that need to be properly managed to avoid environmental pollution that accelerates the effects of climate change. Some units offered under this course encompass recycling techniques and composting. Students learn about the various types of waste, their direct and indirect sources and how to dispose of them in the best manner. Waste reduction and how to recycle them are also part of what they learn. Some of the TVETs and universities that have implemented this strategy include: The Technical University of Mombasa (TUM), The Kenya

Water Institute (KEWI) in Nairobi, The Kenya Institute of Highways and Building Technology (KIHBT) and Jomo Kenyatta University of Agriculture and Technology (JKUAT) among many others.

The second strategy from the policy of Greening Technology in TVET is to have courses that train students on the use of renewable energy technologies. From this directive, the following are some of the courses that were introduced and are being offered in the country. They include: Diploma in Renewable Energy technology, Certificate in wind energy technology, Certificate in solar PV installation and certificate in Biogas technology among others. The TVETs that offer these courses include: The Nairobi Institute of Technology, The Kenya institute of Energy and environment, The Eldoret polytechnic and the Kabete National polytechnic among many others.

This is a clear indication that the greening Technology policy for TVETs has been actualized quite well for the sake of Climate change education. Apart from the above policy, TVETs have also been tasked to conduct Research on greening technology which will eventually inform policy on climate change education in higher institutions of learning in Kenya.

NESSP 2018-2022 has had several other approaches incorporated into various TVET curricula. They include:

i. *The Interdisciplinary Approach:* This approach encourages a multidisciplinary method of teaching climate change aspects whereby educators have to combine environmental, social, and economic perspectives into finding solutions to mitigate or adapt to climate change disruptions.
ii. *Advancement of Teaching and Learning Resources*: This approach emphasizes the need to design and publish teaching and learning materials that incorporate climate change education, such as textbooks, journals, and digital resources. These must be distributed widely for all to access the information easily and freely.
iii. Teacher Professional Training: This plan emphasizes that, for teachers to be effective in passing on climate change education, they need to be equipped with proper knowledge and skills in regards to this topic. Therefore, professional development and training opportunities must be passed to them to enrich their knowledge and skills. This is done through workshops and conferences.
iv. Research and Innovation: research and innovation that brings about new knowledge about climate change education is emphasized for all TVETs and universities. Therefore, while teaching climate change courses, facilitators are encouraged to advise students to pick topics that research on climate change matters.

b) Programme 1.5: Promote Education for Sustainable Development (ESD); This policy reads in part as below;

This programme aims at enhancing sustainable development through education and training to ensure that all learners acquire knowledge and skills on human rights, gender equality, promotion of a culture of peace and non-violence, global citizenship and appreciation of cultural diversity and cultures. Challenges of climatic change continue to exist such as droughts, floods, frost and heat waves. There is a need for increased awareness on environmental issues for sustainable development. The overall objective of this programme is to operationalize the ESD policy and equip learners with knowledge and skills needed to promote sustainable development. The following activities will be implemented:

- Develop ESD Action plan;
- Build capacity of education managers and stakeholders on integration of ESD and climate change in all learning institutions;
- Conduct awareness campaigns on ESD for learners and school communities;
- Monitor and evaluate ESD policy implementation in learning institutions. (Kenya Ministry of education,2018)

The above policy among many other SDGs recognizes that SDG 13 on climate action is important and that institutions of higher learning need to integrate environmental courses into the curricular so as to teach and create awareness on climate change issues.

It is evident that the Ministry of Education in conjunction with several other institutions of higher learning Kenya have done a commendable job in coming up with policies which incorporate strategies of furthering climate change education in Kenya. This is quite important because through climate change education, knowledge and skills of how to mitigate and adapt to climate change disruptions is being passed on. This is vital because individuals and communities must take action to reduce their carbon footprint and build resilience towards climate pressures. Secondly, climate change education helps to create awareness on the main causes and effects of climate change disruptions among students and staff. As it is, many people in our communities are still ignorant about these impacts and hence do very little to either adapt or mitigate. Therefore, education might enlighten the students who will in turn pass on the knowledge to those around them. Thirdly, climate change education is key as it helps students across various disciplines such as science, technology and policy to collaborate in coming up with interdisciplinary measures of tackling climate change impacts. Lastly, climate change education will aid in preparing students for green jobs such as renewable energy, sustainable agriculture, green infrastructure among many other opportunities in this area. Consequently, these policies on climate change education in higher institutions of learning in Kenya must be supported by all as it fosters best practices in not only climate change education but Education for sustainable development for all. Interdisciplinary solutions should also be encouraged as it brings both industry and learners to the same table for a better world.

PEDAGOGICAL SKILLS RELEVANT TO CLIMATE CHANGE EDUCATION

Climate change education largely refers to the learning activities that involve climate change matters. These activities range from climate change causes, effects and strategies to either mitigate or adapt. It also comprises of a variety of educational strategies from informal learning practices to class-room based trainings. The main goal for climate change education is to equip learners with skills, knowledge and values required to comprehend and counter the climate change crises. It involves teaching and learning about the science of climate change, its economic, social and political impacts as well as policies on solutions.

Karl et al. (2011) argue that the basis for tackling climate change is through collaborative, flexible and integrated decision-making processes at multiple scales that range from the local to the international. These processes require knowledge and skills among citizens and policymakers, as well as political and social will. To make change happen, awareness, a sense of urgency and of agency is needed. Climate change education can contribute to the development of the necessary knowledge, skills and attitudes so that students can become agents of change, as future consumers, policymakers, and social influencers. Given that young people are future citizens and policymakers, education should give them insight into the wicked problem of climate change, and develop the competencies to contribute to solving the issue. Climate change competencies will give graduates "workforce preparedness" (Maxwell & Blashki, 2016, p. 16).

As the local context matters, students should develop observation skills and curiosity (Arnold & Wade, 2015), and skills to enable them apply theoretical knowledge to real-world local problems (Lopatto, 2003). Scholars haved argued that wicked problem-solving requires collaboration skills (Cantor et al., 2015 ;Weber & Khademian, 2008) This could be stimulated by letting students work together, and to connect education with local stakeholders. It is important to bring climate change close to students: to engage them and to make it personally relevant for them. Climate change education should therefore focus on the visible effects of climate change in the daily environment of students, and focus on effects that can be noticed, such as water issues.

Fieldwork is essential to teaching. It is often emphasized that fieldwork can facilitate learning, for example, by connecting theory with the real geographical world outside, and by connecting cognitive with affective processes (Wilson et al., 2017). When well planned and executed, fieldwork can lead to deep learning. In addition, teaching climate change requires an interdisciplinary approach. As climate change issues contain many dimensions and are connected to other issues, it requires interdisciplinary thinking skills to solve them.

An academic database, EBSCOhost, has identified 959 unique citation records addressing climate change education. Two themes were identified that are common to most environmental education: (1) focusing on personally relevant and meaningful information and (2) using active and engaging teaching methods. For climate change the following were generated: (1) engaging in deliberative discussions, (2) interacting with scientists, (3) addressing misconceptions, and (4) implementing school or community projects.

Educators are faced with many challenges when teaching about climate change and there is no broad agreement about what strategies are most effective. One of the ways of dealing with the issues in climate education is to infuse the teachings and learnings into the already existing curriculum.

Alberta Council for Environmental Education (2017) identifies six key principles of excellent climate change education:

a. Frame climate change education in ways that focus on solutions, rather than on problems, build a positive narrative around shared identity. Focus on energy, conservation and outdoors education. Rely on pedagogies which engage in deliberative discussions, promote exchanges with scientists, address misconceptions, and implement school and community projects
b. Keep the audience in mind. Develop curriculum that is appropriate to the age of the child, support teachers.
c. Design programmes which are action oriented. Build the agency of students.
d. Develop activities that extend beyond climate science, including imagining a positive desired future, focus on local content, teach students how to think, not what to think, do not scare students.
e. Establish connections to the curriculum and identify competencies. Emphasize cross-curricular approaches, cultivate systems thinking, and help students understand the interdependencies between climate change mitigation, adaptation and resilience.
f. Evaluate for program improvement.

Specific Pedagogical skills

To develop specific climate change education strategies and to build the institutional capacity to implement them, the pedagogical skill used should encourage intentional engagement of universities and other institutions in partnership. This would serve the dual role of providing support for each other in advancing climate change education, while also educating higher education students on climate change through problem based, participatory and contextually situated approaches.

Developing the moral imagination of students through Human Rights education or education for social justice, cultivating their capacity to recognize how the impact of climate change varies for different people (women, minorities, the poor) is a necessary step to animating them to engage with the subject at greater levels of complexity and inventiveness.

Climate change education should go beyond equipping people with the skills to understand climate change, important as that is. It must equip them to understand tradeoffs, to make choices and to invent solutions that can help us integrate choices that are environmentally sustainable within a larger framework of how we live. A simple way to represent the choices involved in mitigation would be to ask how much people are willing to give up, or to pay, or to change their lifestyle, to reduce their own carbon footprint, for example demanding that their governments adopt cap and trade that limit the total emissions or a carbon tax. Educating people for adaptation involves knowledge and skills to change their lifestyles in ways that are responsive to the impact of climate, for example influencing how they build or rebuild their cities so they are more resilient to coastal flooding. Finally, educating them for climate change reversal requires developing the talent and ingenuity that can lead to technological innovations. This means education not just to understand, but to act effectively, in influencing not just personal patterns of consumption, but with the agency and efficacy to collaborate with others to influence the complex systems which undergird climate change.

Environmental themes can be included within scientific literacy. UNESCO's guide for whole school climate change education emphasizes a six-step process, and is relatively thin on curriculum and content. The process comprises creating a school climate action team, infusing sustainable development across all subjects, teaching creative and futures thinking, empowering students to take action, addressing facilities and operations and building community partnerships. In terms of curriculum, the guide offers two examples of activities which could be integrated into each of eleven different subjects, such as designing and maintaining a school garden and compost, creating maps showing areas of the world most at risk due to climate change and examining how societies throughout history have resolved conflicts and responded to environmental challenges (Gibb, 2016, p. 12).

Deliberative discussion: Discussions and conversations among learners were used to help them think more deeply about concepts, compare perceptions, understand different opinions, and reflect on what they know. There is need to emphasise the development of knowledge based on models about the factors which influence engagement with climate action. Emotions have been found to be important correlates of active engagement with climate. A study of the relationship of hope concerning climate change with pro-environmental behavior among Swedish youth and young adults found that hope plus worry about climate change was positively related to pro-environmental behavior, whereas hope plus lack of worry was not (Ojala, 2012). Given the important role of emotions such as hope and worry in pro-environmental action, Ojala has advocated that they should be cultivated to sustain actions that challenge existing norms and institutions that contribute to climate change. She explains that a transformative and transgressive climate change education that recognizes the worry and anxiety that disrupting and transgressing generates, can create hope through critical emotional awareness and through activities that develop visions of preferable futures (Ojala, 2016, p. 52).

School and community projects: educational intervention that reached beyond the classroom or training to provide learners with an opportunity to conduct a climate change project in the larger context of their school or community. students played a key role in reducing their school's energy use by learning about energy resources, collecting data and monitoring energy use in classrooms, and planning and implementing appropriate actions projects within the school.

Integrating the skills of educators and scientists provided multiple benefits. Building online information and databases into educational materials would enabled learners to access data across space and time, creating important opportunities to build understanding.

Various scholars have examined the challenges encountered in climate change education. Irwin (2020) has argued that many teachers do not have the skills to teach climate change. Anderson (2012) is of the view that climate change concepts must be ingrained in the value-system of educational institutions that teach the same. The scholar describes such learning institutions as those that are "climate-proofed and resilient as well as sustainable and green" (Anderson, 2012). Other scholars feel that there is still a lot that needs to be done to utilize education as a tool to deal with climate change (Mochizuki & Bryan, 2015). Regarding Kenya, Unesco (2011) argues that the main challenge of integrating aspects of sustainability in Kenyan education system is due to the focus on exams. At the same time, the Kenyan curriculum "is overloaded with content and learning outcomes," thus complicating the inclusion of sustainability aspects" (Unesco, 2011, p.71). Other scholars, Apollo and Mbah (2021) argue for well-designed strategies.

CONCLUSION

Climate change education is slowly being incorporated into Kenya's education system. The efforts to integrate climate change into the education system have been made through the enactment of laws and development of policies that highlight the role of education in creating awareness about climate change, its impact and the possible mitigation efforts. Governing bodies such as the Ministry of Education and the Commission for University Education have directed institutions of higher learning to incorporate climate change into their programmes. In addition, collaborative efforts such as that of Kenya Green University Network have been instrumental in promoting education and training in climate change issues. Over the years, a number of Kenyan universities have launched climate change courses. Some of these universities have institutes that focus on climate change and other environmental concerns.

However, some are of the view that Kenyan universities have not been very enthusiastic in entrenching climate change into curricula (Nyerere, 2022). This reluctance in embracing climate change education could be attributed to previous student apathy towards environmental courses. Climate change courses were previously associated with subject areas such as agriculture, a course seen in the past as unpopular with Kenyan students. This view has been illustrated by Huho (2015) who conducted a study that established some university students did not consider climate change as a subject that should be taught in institutions of higher learning. Another hindrance to the growth of climate change courses could be the lack of experts who can develop the curricula and teach the material at university level. Furthermore, the lack of awareness on the impact of climate change on society could be a factor that curtails the popularity of the courses. This could be addressed by demonstrating to students the importance of sustainability for current and future generations.

Adopting specific strategies could help universities integrate climate change in curricula. One strategy is for bodies such as CUE and the Ministry of Education to provide guidelines and training on development of climate change courses. In fact, some scholars such as Nyerere (2022) have suggested stricter rules of integration of climate change curricula by proposing to have "climate action" included as a criterion for "evaluating university programmes". Training on development of climate change curricula would help in translating into action the various government policies that have been outlined in various laws and strategic plans. A second strategy would be for government to provide funding for climate change research (Nyerere, 2022). The research could entail collaborative efforts between government agencies and various institutions of higher learning. In conclusion, introducing climate change as a common unit taught across disciplines could be one way of promoting the subject among the student body.

REFERENCES

Alberta Council for Environmental Education. (2017). *What is excellent climate change education?* Alberta Council for Environmental Education.

AMCEN. (2012). *The Africa Ministerial Conference on Environment (AMCEN).* Retrieved from https://www.unep.org/roa/amcen/

AMCEN. (2017). *The Africa Ministerial Conference on Environment* (AMCEN). Retrieved from https://www.unep.org/roa/amcen/

Anderson, A. (2012). Climate Change Education for Mitigation and Adaptation. *Journal of Education for Sustainable Development*, *6*(2), 191–206. doi:10.1177/0973408212475199

Apollo, A., & Mbah, M. F. (2021). Challenges and Opportunities for Climate Change Education (CCE) in East Africa: A Critical Review. *Climate (Basel)*, *9*(6), 93. doi:10.3390/cli9060093

Arnold, R. D., & Wade, J. P. (2015). A definition of systems thinking: A systems approach. *Procedia Computer Science*, *44*, 669–678. doi:10.1016/j.procs.2015.03.050

Cantor, A., Delauer, V., Martina, D., & Rogan, J. (2015). Training interdisciplinary "wickedproblem" solvers: Applying lessons from HERO in community-based research experiences from undergraduates. *Journal of Geography in Higher Education*, *39*(3), 407–419. doi:10.1080/03098265.2015.1048508

Climate Change 2014 Synthesis Report Commission for University Education. (2020, June 2). *Incorporation of Climate Change Education in University Curricula* [Press release]. Retrieved from https://www.cue.or.ke/

Fahey, S., Verstraten, L., & Berry, A. J. (2016). Education for Sustainable Development: Enhancing Climate Change Adaptation Expertise in Developing Countries. *Journal of Education for Sustainable*, *10*(1), 54–67. doi:10.1177/0973408215625534

Favier, T., Van Gorp, B., Cyvin, J. B., & Cyvin, J. (2021). Learning to teach climate change: Students in teacher training and their progression in pedagogical content knowledge. *Journal of Geography in Higher Education*, *45*(4), 594–620. doi:10.1080/03098265.2021.1900080

Gibb, N. (2016). *Getting climate-ready: A guide for schools on climate action.* UNESCO. doi:10.54675/MOYX1103

Government of Kenya. (2007). *Kenya Vision 2030: The popular version.* https://vision2030.go.ke/wp-content/uploads/2018/05/Vision-2030-Popular-Version.pdf

Government of Kenya. (2010). *National climate change response strategy*. https://www.greengrowthknowledge.org/sites/default/files/downloads/policydatabase//KENYA%29%20National%20Climate%20Change%20Response%20Strategy.pdf

Government of Kenya. (2015). *Addressing Climate Change: Success Stories from Kenya*. Author.

Government of Kenya. (2018). *National Climate Change Action Plan*. Author.

Government of Kenya. (2018). *National Education Sector Strategy Plan 2018 – 2022*. Author.

Government of Kenya. (2018). National Education Policy 2018–2022. *Vision (Basel)*.

Government Printer. (2016). *The Climate Change Act, 2016*. Nairobi: Government Printer. http://kenyalaw.org/kl/fileadmin/pdfdownloads/Acts/ClimateChangeActNo11of2016.pdf https://www.un.org/en/climatechange/what-is-climate-change http://www.greenbeltmovement.org/who-we-are/our-history https://www.kgun.org/about-us/

Huho, J. M. (2015). Climate change knowledge gap in education system in Kenya. *International Journal of Innovation and Research in Educational Sciences*, *2*(3), 227–232. https://www.researchgate.net/publication/279179079_Climate_Change_Knowledge_Gapin_Education_System_in_Kenya

IFRC. (2022). *Kenya hunger crisis: Voices from the drought*. https://www.ifrc.org/article/kenya-hunger-crisis-voices-drought

Irwin, R. (2020). Climate change and education. *Educational Philosophy and Theory*, *52*(5), 492–507. doi:10.1080/00131857.2019.1642196

Karl, H., Curtin, C., Scarlett, L., & Hopkins, W. (2011). Adapting to climate change, a wicked problem: A road map for action. In I. Linkov & T. S. Bridges (Eds.), Climate: Global change and local adaptation (pp. 237–254). Springer Science + Business Media. doi:10.1007/978-94-007-1770-1_13

Kuthe, A., Keller, L., Körfgen, A., Stötter, H., Oberrauch, A., & Höferli, K.-M. (2019). How many young generations are there? – A typology of teenagers' climate change awareness in Germany and Austria. *The Journal of Environmental Education*, *50*(3), 172–182. doi:10.1080/00958964.2019.1598927

Lopatto, D. (2003). The essential features of undergraduate research. *Council on Undergraduate Research Quarterly*, *24*, 139–142.

Maxwell, J., & Blashki, G. (2016). Teaching about climate change in medical education: An opportunity. *Journal of Public Health Research*, *5*(673), 14–20. doi:10.4081/jphr.2016.673 PMID:27190980

Ministry of Education. (2017). *Education for Sustainable Development Policy for the Education Sector*. Government Printer.

Ministry of Education. (2018). *National Education Sector Strategic Plan 2018-2022*. Government Printer.

Ministry of Environment and Forestry. (2018). *Thematic plan for climate change*. Ministry of Environment and Forestry.

Ministry of Environment and Natural Resources (MENR). (2017). *Climate Change Policy, Planning and Budgeting at National and County Level: Facilitators Manual*. Government of Kenya.

Mochizuki, Y., & Bryan, A. (2015). Climate Change Education in the Context of Education for Sustainable Development: Rationale and Principles. *Journal of Education for Sustainable Development*, *9*(1), 4–26. doi:10.1177/0973408215569109

Monroe, M. C., Plate, R. R., Oxarart, A., Bowers, A., & Chaves, W. A. (2019). Identifyingeffective climate change education strategies: A systematic review of the research. *Environmental Education Research*, *25*(6), 791–812. doi:10.1080/13504622.2017.1360842

National Council for Law Reporting. (2012). *Environmental Management and Co-ordination Act, 1999*. Nairobi: National Council for Law Reporting. http://kenyalaw.org/kl/fileadmin/pdfdownloads/Acts/EnvironmentalManagementandCo-ordinationAct_No8of1999.pdf

National Council for Science, Technology and Innovation. (2020). *Background Report on National Climate Change Priorities and Relevant Capacity Development Goals and Initiatives in Kenya 2020*. Retrieved from https://www.nccs.go.ke/

Njino, J. (2022, June 6). Why climate change is essential for Kenya's future. *Daily Nation*. https://nation.africa/kenya/blogs-opinion/blogs/why-climate-education-is-essential-for-kenya-s-future-3839492

Nyerere, J. (2022, May 26). Climate change isn't a priority for Kenyan universities: It should be. *The Conversation*. https://theconversation.com/climate-change-isnt-a-priority-for-kenyan-universities-it-should-be-181530

Nyerere, J., Gatwiri, W., & Okinyi, R. (2021, July). *Kenya's climate change policy actions and the response of higher education: Transforming universities for a changing climate, working paper series, No. 4*. London: Climate-U. https://www.climate-uni.com/_files/ugd/f81108_1fe481d96de94247b5b3a5baab268f33.pdf

Ojala, M. (2012). Hope and climate change: The importance of hope for environmental engagement among young people. *Environmental Education Research, 18*(5), 625–642. doi:10.1080/13504622.2011.637157

Ojala, M. (2016). Facing anxiety in climate change education: From therapeutic practice to hopeful transgressive learning. *Canadian Journal of Environmental Education, 21*, 41–56.

The Climate Change Act. (2016). Nairobi: Government Printer. http://kenyalaw.org/kl/fileadmin/pdfdownloads/Acts/ClimateChangeActNo11of2016.pdf

The Daily Nation. (2020, June 2). *Universities to teach climate change*. Retrieved from https://www.nation.co.ke/

UNCC Learn. (2020, July 22). *The journey begins towards developing Kenya's climate change learning strategy*. https://www.uncclearn.org/news/the-journey-begins-towards-developing-kenyas-national-climate-change-learning-strategy/

UNESCO. (2011). *National journeys towards education for sustainable development: Reviewing experiences from Chile, Indonesia, Kenya, Netherlands, Oman*. Paris: United Nations Educational, Scientific and Cultural Organisation. https://unesdoc.unesco.org/ark:/48223/pf0000192183

UNESCO. (2015). *Not just hot air: Putting climate change education into practice*. Paris: UNESCO. https://unesdoc.unesco.org/ark:/48223/pf0000233083

UNESCO. (2016, February 9). *UNEP launches Kenya Green University Network*. https://www.unesco.org/en/articles/unep-launches-kenya-green-university-network

UNESCO. (2022). *Kenya: Climate change communication and education*. https://education-profiles.org/sub-saharan-africa/kenya/~climate-change-communication-and-education

Weber, E. P., & Khademian, A. M. (2008). Wicked problems, knowledge challenges, and collaborative capacity builders in network settings. *Public Administration Review, 68*(2), 334–349. doi:10.1111/j.1540-6210.2007.00866.x

Wilson, H., Leydon, J., & Wincentak, J. (2017). Fieldwork in geography education: Defining or declining? The state of fieldwork in Canadian undergraduate geography programs. *Journal of Geography in Higher Education*, *41*(1), 94–105. doi:10.1080/03098265.2016.1260098

KEY TERMS AND DEFINITIONS

Climate Change: Long-term environmental changes that are to a large extent caused by human activities.

Climate Change Adaptation: Efforts directed towards mitigating consequences of climate change.

Climate Change Education: Education geared towards creating awareness about the causes of climate change, exposing the consequences of climate change and adopting relevant adaptation strategies.

Environmental Conservation: Strategies directed towards protection of environmental resources.

Higher Education: Education that is given by universities and other tertiary institutions.

Kenyan Curriculum: Syllabus used to teach in Kenyan institutions of higher learning.

Pedagogy: Teaching methods used to give instruction to students.

Sustainability: Efforts geared towards ensuring that everyday activities promote the preservation and conservation of the environment.

Chapter 3
Pre-Service Teachers' Pedagogical Knowledge and Perceptions of Climate Change Education

Evelyn W. Njurai
https://orcid.org/0000-0001-5261-6577
Kisii University, Kenya

Billiah Gisore
https://orcid.org/0000-0002-5116-7615
Kisii University, Kenya

ABSTRACT

Climate change education (CCE) has been proposed to be a driver for the much-needed knowledge and skills on climate change (CC). Teachers, and consequently learners, are key change agents in any society; therefore, it is necessary that they have knowledge and skills on CC and how to mitigate and adapt to these changes. This chapter discusses pre-service teachers' pedagogical knowledge and perceptions of CCE. The discussion is contextualized through research conducted in Kenya. While it was evident that the pre-service teachers had basic knowledge and skills about CC, they also had misconceptions. Further, CCE was not documented in their curricula. They propose that CCE can be integrated with other subjects or implemented as a stand-alone subject or both. The study consequently identifies pre-service teachers' knowledge gaps and reports on their mixed perceptions of CCE. The need to include CCE in the curriculum is then underscored in view of targeted agencies.

DOI: 10.4018/978-1-6684-9099-0.ch003

INTRODUCTION

Scientists have identified the actual and potential causes of Climate Change (CC), including its impact, mitigation, and adaptation measures in different global contexts. While these are concerns with lifelong effects, information on CC has not been embraced or understood and there are sometimes misconceptions about it (Bord, O'Connor & Fisher, 2000; Ikonomidis, Papanastasious, Mela & Avgoloupis, 2012) by many ordinary citizens in the world. This is partly because matters of CC have commonly been addressed in the public domain through awareness fora, media, or general readership where in-depth information may not be provided. Since education systems are seen to be effective means of developing capacities for addressing CC (Stevenson, Jennifer & Whitehouse, 2017), it is necessary that CC matters should be taught through formal education which provides a platform for longer learning periods and in-depth coverage of content and concepts. Further, there is a need to mainstream CC in school curricula in order to prepare the world for uncertain climate changes (United Nations Educational, Scientific and Cultural Organization (UNESCO), 2021b). One of the priorities of the UNESCO-led Global Action Program (GAP) on Education for Sustainable Development (ESD) is to build the capacity of educators (UNESCO, 2014) to influence how people think and act (UNESCO, 2017) on matters of global concern like climate change. Teachers, therefore, need to be equipped with the requisite knowledge and skills of and about CC and consequently Climate Change Education (CCE). The knowledge and skills are then passed on to learners who are change agents in current and future times.

This chapter discusses pre-service teachers' knowledge and perceptions on and about CC and CCE from past research. To contextualize the discussion, the chapter draws on research conducted in Kenya to assess knowledge of pre-service teachers on CC, preparation of pre-service teachers to teach CCE, suggested and appropriate pedagogies for CCE, and perceptions of pre-service teachers on CCE. The chapter finally proposes incorporating CCE into the pre-service teacher training curriculum to improve their knowledge and skills for impacting their learners.

KNOWLEDGE ON AND ABOUT CLIMATE CHANGE EDUCATION

Climate change is referred to as the variations in climate which are attributed directly or indirectly to human activities observed over a comparable period (UN, 1992). The variations alter the composition of the global atmosphere, hence affecting the natural climate. Climate change science draws from multiple subjects in schools and disciplines in tertiary education (Boon, 2016). Climate change education is consequently about teaching and learning matters of climate change. It aims to equip learners with knowledge, skills, attitudes, and values about CC so as to understand and address the causes and impacts of global warming, and consequently mitigation and adaptation measures. While it may introduce new content, CCE commonly uses existing content and concepts to explain matters of CC. (Intergovernmental Panel on Climate Change, 2014).

Climate Change Education is currently conceptualized and handled under the broad umbrella of two prevailing paradigms: Environmental Education (EE) and Education for Sustainable Development (ESD) (Hung, 2012). The UN Conference on Environment and Development of 1992 restated the importance of education in equipping the global youth and training teachers with the knowledge and skills necessary to manage the environment of the planet earth (UN, 1993). In more recent times, The Berlin Declaration of 2021 on ESD called for the introduction of EE as a core area of study in education systems (UNESCO, 2021a). In 2002, the UN adopted the ESD framework for CCE (UNESCO, 2011) for impacting CC. The framework is based on the principles and values that underlie sustainable development such as the use of a variety of pedagogical techniques that promote participatory learning and higher-order thinking skills, promote lifelong learning, and is of local relevance and is culturally appropriate. Further, it is based on local needs, perceptions, and conditions, and acknowledges that local activities may have global effects. It accommodates the evolving nature of the concept of sustainability, and addresses content, considering the context, global issues, and local priorities. UNESCO and UNEP guidebook of 2011 (UNESCO, 2011) maintains that CCE must emphasize causes and impacts as well as effective mitigation and adaptation measures using appropriate action-oriented and participatory pedagogies. Given the need to address CC, this diversified framework has been gaining global currency at all levels of education. The year 2020 marked the inception point of the Decade of Action to achieve the Sustainable Development Goals (SDG) by 2030 in which the entire global community is expected to catalyze actions in addressing urgent global challenges such as care for the environment and climate change in which ESD is a focal reference. A necessary question is, how much has been done in implementing CCE in learning institutions?

The subject matter knowledge on CC is organized around the macro-concepts of causality, consequence, and management. The causes of CC involve both natural factors and anthropogenic activities. Human activities have caused an increase in the concentration of Greenhouse Gases (GHG) and hence increasing the effect of natural GHG. Increased concentration and levels of GHG in the atmosphere cause depletion of the ozone layer and an increase in global temperatures. Consequently, the impacts of global warming are evident; they include increased frequency of, and sometimes prolonged, extreme weather events such as droughts; cyclones or heavy rainfalls, a rise in sea level all of which result in flooding in the low-lying lands; and changes in plant growth affecting agriculture and threat on food production (Hung, 2012; UNESCO, 2012) and increased intensity of the solar radiation on earth. There are also social, justice, and ethical (Walsh, 2022), psychological, health, and economic relations impacts. These and other observed and unobserved changes are expected to intensify and have a significant negative impact on human societies and the environment around the world in the present and the future. There is then a need for mitigation measures and adaptation strategies. Mitigation measures should involve either reducing sources of GHG such as transportation or enhancing the sinks that accumulate and store the gases such as the oceans, forests, and soil (US Environmental Protection Agency (EPA), 2023). Adaptation to CC involves adjusting to actual or expected future climate. The goal of adaptation is to reduce human risks from the harmful effects of CC and make the most of any potential beneficial opportunities associated with CC (Friedrich-Alexander-Universitat (FAU), 2022); Sohn, 2014).

There are gradual efforts to explore possibilities of human beings managing and taking control of their actions to mitigate and adapt to climate changes in the long term. School education is such a gradual effort that can be applied to inform and build capacities on CC. This is necessary because teachers and learners are a majority of present and future change agents (Stevenson, Nicholls & Whitehouse, 2017); they communicate about what they teach/learn in and out of school, and learners are the future environmentalist, policymakers, and community organizers among others. For this change to happen, it is necessary that pre-service teachers be exposed to knowledge and skills on CCE and that misconceptions about CC be addressed. They should explore and understand both anticipated and actual natural and anthropogenic causes of climate change, impacts on human lives and their environment, and measures for mitigation and adaptation strategies in the local, regional, and global contexts. The ESD framework becomes handy to achieve the training on CCE. In this regard, it is necessary to explore the knowledge of pre-service teachers on CC and CCE as they get prepared to actualize teaching.

PRE-SERVICE TEACHERS AND CCE

Universities' core mandate includes teaching, research, and community service. With such a mandate, efforts are expected and needed from universities the world over to develop programs, curricula, build capacities and engage in interdisciplinary collaborations to implement and support CCE (UNESCO, 2020). Further, universities could educate communities on matters of climate change through research and community service. A measure of whether universities perform research on CC, how they use low-carbon energy, and if they have education programs aimed at the achievement of carbon neutrality, is done using Times Higher Education Impact

Rankings. The tool describes the metric used by universities in implementing UN SDG with a focus on SDG 13 on climate action. The Impact Rankings of 2022 show universities in Europe, the US, and Australia as the majority highly rated implementers with no African university among the top 200 universities (The Impact Ranking, 2022). Research shows that some universities in the West have developed programs and curricula on climate action (Leal et al., 2021). This is through education courses being core or elective or integrated with climate change education content. For example, Climate Change Pedagogy is a course for pre-service teachers in one university in Canada (Berger, Gerum, & Moon, 2015). In other contexts, CC has been integrated with subjects (Hung, 2012: Berger, Gerum, & Moon, 2015) while in others it remains at a policy level. Further universities engage with unions and associations of staff and students, and leaders of both academic and non-academic departments and research at postgraduate levels on CCE so that all are involved (Leal et al, 2021). It is notable that the context of universities even in one country may differ, hence the need for each university or those in similar contexts to tailor make their programs and curricula and employ the appropriate pedagogies for CCE.

Teachers are motivated to teach matters of CC when topics on climate change are included in the curriculum (Wise, 2010). Past research shows that CCE has been implemented or is proposed to be included in teacher training curricula (Okoli, 2014) through an interdisciplinary approach (Belgrade Charter, 1975; Summers, Childs, & Corney, 2005). This approach requires a teacher to understand the philosophy of CCE, and the complexities of the science of climate change and to integrate these into the pedagogies of other disciplines. Pedagogical approaches may include transformative pedagogies such as learner-centered and innovative pedagogies, which may involve action research as a necessary component. To identify, develop and use a robust pedagogical approach for CCE, a teacher must first have a grasp of the "what" and "how" of it. This is necessary since what teachers impact on learners has a bearing on what and how they learn, apply and use it. Further, teacher education plays a necessary part in helping pre-service teachers shape their professional knowledge, skills, attitude, values, and practice for the purposes of teaching and their own use.

Where CCE is not included in the curriculum, teachers have been required to include and integrate topical areas on CC into existing curriculum and community service programs (Summers, Childs, & Corney, 2005). However, many teachers are pedagogically challenged to integrate CCE in their lessons; in fact, the aspect of integration is complex since it requires the teacher to identify the content in the subject area in which to integrate, the level of learners and the scope and sequence associated with such integration. Teachers may also have limited knowledge of the underlying climate concepts. Molthan-Hill, Jackson, Mifsud, and Odell (2020) propose how to integrate CCE across a wide range of subjects in university curricula based on content, pedagogy and contextual institutional and sector-wide constraints.

Hence, it is necessary that CCE, whether integrated or as a standalone subject be part and parcel of the curriculum so as to guide implementation.

Research shows lack of substance in understanding CC, and prevalence of misconceptions and misinformation about CC among teachers (Hung, 2012; University of Missouri-Columbia, 2017). These are probably due to, among other things, the complexity of climate science, the uncertainties and the controversies surrounding CC. The misconceptions may further be attributed to general public talks, and traditional, cultural and religious beliefs some of which have no basis in climate science. For example, the misconception that pollution causes CC (Bord, O'Connor & Fisher, 2000; Papadimitriou, 2004). These are subsequently imparted on learners and hence a spiral pattern of misconception, misinformation, and inadequate knowledge. Further, there has been confusion of the terms weather and climate among prospective primary school teachers in Greece (Papadimitriou, 2004). In investigating pre-service teachers' understandings of climate change, Boon (2016) found that even in a university with a specialist focus on sustainable education, pre-service teachers struggled with the scientific complexity of climate change. Boon pointed to the conceptual difficulty of climate change science being of interdisciplinary nature as a cause of the general low knowledge levels. Yet teachers are trained based on subject specialties, hence the challenge of engaging with other subjects. These issues need to be addressed in detail in teacher training so that misconceptions and misinformation are addressed, and in-depth knowledge is explored. Further, the traditional, cultural and religious beliefs, and traditional practices need to be addressed with regard to the science of climate change. There is also a need to redesign curricula to incorporate information on and about CC. Furthermore, with the continued evolution of CC, a CCE curriculum could be dynamic and evolving probably more than any other area of study.

PERSPECTIVES OF TEACHERS ON CCE

There is a general perception that climate change education should be taught in geography, agriculture, and environmental or science subjects (Hung, 2012; Wise, 2010; Siegner & Stapert, 2020). The authors note that CCE has generally been integrated with science. This is a traditional way of thinking about CCE. Although there is a fact that agriculture, geography, environmental studies, and the sciences are more conceptually connected to CCE than other subjects, looking at the causes and effects of climate change, all subjects should incorporate CCE. In fact, since CC is a human issue requiring social, technological, and scientific solutions, there is a need to integrate it with humanity subjects (Siegner & Stapert, 2020). Researchers have also identified how to integrate CC into mathematics lessons (Barwell, 2013;

Abtahi, Gøtze, Steffensen, Hauge & Barwell, 2017). There is also a need to introduce a compulsory subject on CCE. The effects of CC are all over the world, therefore, learning about CC from local to global contexts has the potential of promoting global citizenship among learners. Global citizenship is a competency that is key in the Competency-Based Education (CBE) that is being rolled out in many countries including Kenya (Kenya Institute of Curriculum Development (KICD), 2017).

CHALLENGES IN IMPLEMENTING CCE

Some of the challenges encountered in the implementation of CCE include a reluctance by teachers to expand their knowledge, skills, attitudes, practices, and values into new fields of study. Teachers also lack open-mindedness for innovative skills and competencies to engage learners on CC. Further, there is a conflict of time for extra courses in the already overburdened programs and a lack of goodwill from teachers (Siegner & Stapert, 2020). To address the challenges, both in-service and pre-service teachers should be motivated to acquire more education for everyday use as CC will take a long time to stabilize and bring forth more benefits than negative impacts. After all, being change agents, teachers should be at the forefront in delivering the much-needed education on CC. The curricula that are overburdened should be relieved of non-pressing content so as to allow space and time for CC implementation. A showcase of success stories to inspire teachers on instructing CC matters is also necessary. Further, teachers should be encouraged to participate in policy-related activities and curriculum development for CCE.

KENYA'S CONTEXT

Kenya is a signatory to various convections on CC and EE, for example, Berlin Conference (UNESCO, 2021a). Locally she is guided on matters of environmental management and coordination by the Environmental Management and Coordination Act of 1999. The National Environment Management Authority (NEMA) established under the Act, provides guidelines, and regulations on the management of the environment. Kenya Forest Service (KFS) established under the Forest Conservation and Management Act has a core mandate of conserving, protecting, and managing all public forests in the country. Some indicators of climate change in Kenya include uncertain weather patterns, more frequent flooding, and prolonged drought as is in North Eastern parts of the country. In practice, government and non-governmental agencies have been involved in addressing matters of the environment and climate change. For example, there has been the preservation of forests through an agroforestry approach, the "shamba system" also called "plantations establishment for livelihood improvement scheme". Under the system, farmers grow food and fodder crops on government forest land while tending to young tree seedlings till they are established. A review of past studies on the "shamba system" shows that success or failure depends on how well government guidelines on the same are implemented and enforced (Kagombe & Gitonga, 2005). Other initiatives are through school clubs such as environmental and wildlife clubs that deal with creating awareness of conservation among learners and communities. This is far from being sufficient to engage with matters on CC.

There has been research on CC in Kenya from various concerned sectors such as agriculture and education. Further, there has been a call by the Berlin Conference (UNESCO, 2021a) to introduce environmental education in school systems. With the start of CBC, Environmental Studies was introduced in 2017 as a core subject area from the early years of education through the primary level, and to the secondary level where it is an optional subject in the career pathways (KICD, 2017). Research at the tertiary level has identified CC knowledge gaps in Kenya as negative attitudes toward agriculture and bias in the integration of CCE into the education system (Huho, 2015). With these developments, there have been minimal attempts to include CCE wholesomely in the teacher training formal education curricula.

METHODOLOGY

To contextualize the pre-service teachers' pedagogical knowledge and perceptions of CCE, the chapter draws on research conducted in one public university in Kenya. The research engaged pre-service teachers at the university level guided by four questions: What is the knowledge of pre-service teachers on CC? How are the pre-service teachers prepared to teach CCE? What are the appropriate pedagogies for CCE? and What is the perception of pre-service teachers on CCE? A qualitative approach was adopted using Focus Group Discussions (FGD), interviews, and document analysis.

The target group was fourth-year pre-service teachers. Pre-service teachers training for secondary schools at university take two teaching subjects, and university, education, and departmental common courses at different levels of their 4-year program. The targeted group had done their teaching practice after their third year of study. FGDs were organized for the sampled students, a maximum of 8 participants per group, with balanced numbers of males and females. Participants in each group were of diverse subject combinations in both sciences and humanity subjects.

To triangulate the data collected, lecturers who taught university and education common courses, subject methods, and subject content areas were targeted and sampled. Interviews with lecturers were necessary to fill in gaps and/or confirm and bring in a different view from the pre-service teachers. The lecturers sampled were two subject matter lecturers, in mathematics and geography, and an education course lecturer who taught common courses in education and English subject method courses. Each of them had more than 8 years of teaching at the university. Further document analysis was done on the curriculum documents.

FINDINGS

The results reported and discussed are from two discussion groups and the three lecturers. They are presented thematically per the research questions. To start the FGDs, the participants gave a brief introduction of themselves including their home counties. They indicated that they were taking an education program for diverse reasons; the desire to engage with learners, family profession, and the fact that was the only option in career choice based on the criteria set by the joint admission board for admission to university.

Knowledge of Pre-Service Teachers on CCE

A variety of sub-questions were asked to unpack the knowledge of pre-service teachers on CC. An entry question on the meaning and difference between weather and climate saw some participants confuse or mix the two but through group discussion, there was consensus. A sample of responses are;

S18: *Weather is the climatic change, or the climatic condition of a given place for a period of 30 to 35 years.*

S26: *Weather refers to the atmospheric condition of a given place within a given period of time of 24hrs.*

One participant defined weather in terms of climate while another considered the aspect of atmospheric conditions within a day. While the confusion was unprecedented, it was notable that some students had a only studied geography up to the second year of their secondary education, hence a long time without formally engaging with the two terms. Others were taking geography as one of their teaching subjects at university. The confusion between the terms weather and climate is similar to earlier findings in Greece among prospective primary school teachers (Papadimitriou, 2004). This shows that the meanings and differences between the two terms should not be assumed.

The participants agreed that climate change is the unprecedented change in climatic patterns which are seen through; extreme rise in temperature; change in rain patterns sometimes causing floods and sometimes being too low for the growth of food crops; change in patterns of some activities that rely on climate like reduced rainy seasons per year. Another indicator of climate change was lowered water table which led to bole holes being sunk deeper than it was previously done. Frequent and prolonged drought periods made nomadic and pastoralist communities in Kenya migrate more frequently and increased the death of their livestock. The issue of nomadic and pastoralist lifestyles is common in Africa. The participants noted that climate changes are not the same across the country and globally. While the participants explained the meaning of climate change, their definition did not attribute it to human activities and duration of observance as that by the UN (1992). However, through subsequent explanations of the causes of CC, it was evident that human activities contribute to CC. The participants seemed to be more at ease with indicators of climate change and its effects than its definition.

On their knowledge of CCE, there was a minimal response that it is knowledge and awareness about CC while also agreeing that EE is awareness and knowledge of one's environment. This was in line with the lecturers' explanations and earlier findings by Boon (2016). All student participants had no idea about ESD while two lecturers indicated that it was education for future prosperity and was linked to SDGs. However, both pre-service teachers and lecturers were not aware of the ESD framework. This finding shows the need to bring to the fore more information about CCE and ESD to teacher training since the ESD framework is the necessary tool for implementing CCE (UNEP, 2008).

The participants identified some causes of climate change as natural factors and human activities. The natural factors are, for example, solar intensity, earthquakes, and volcanic eruptions. It was noted that the latter two release dust into the atmosphere causing pollution consequently affecting plant growth, clogging the pores of plants, and affecting the water cycle. The participants noted that natural causes are rare but have longer-term impacts. Human activities include deforestation and agricultural activities, for example, the use of synthetic fertilizers, pesticides, and cattle rearing increase nitrous oxide (NO_2), persistent pollutant pesticide molecules, and biogas (methane) respectively which are GHG. Industrial activities which lead to air, land, and water pollution cause climate change. Examples of intensive agricultural activities were in rural areas with high population densities such as Kisii and Kiambu counties. Another cause of CC was given as overgrazing. It was noted that pastoralist communities overgraze available vegetation leaving bare soils with the possibility of causing soil erosion and with minimal chances of growth of vegetation in the long term. Increased use of motorized means of transport common in urban areas has led to an increase in fumes and gases, mostly carbon monoxide (CO) and carbon dioxide (CO_2) released by engines as they burn petroleum products. Political influence was seen as an indirect cause of climate change where some politicians in Kenya have influenced citizens especially those who do not own land to invade forested areas like Mau and Mt Kenya Forests hence reducing vegetative land cover that absorbs some GHG. The increase in population was seen to lead to multiple causes of CC; competition for limited resources including land and water, and exploration and expansion of territories. It is notable that these are actually effects resulting from the impacts of CC but not causes. In analysis, it was notable that the issue of pollution as an eventual cause of CC is a misconception (Bord, O'Connor & Fisher, 2000; Papadimitriou, 2004; University of Missouri-Columbia, 2017), the dust in the atmosphere actually causes an increase in GHG hence contributing to global warming. The issue of overgrazing is also a misconception as a cause of CC, the fact being that overgrazing removes plants that absorb CO_2 which is a GHG.

In identifying which causes were more common between the natural and the human, the participants noted that human activities are likely to cause CC more than natural factors. This is because some of the human activities are necessary and probably unavoidable, such as transportation, industrialization, and farming which increase GHG. Further, to set up a settlement or industry, one may require clearing vegetation or landscape, therefore, reducing vegetative ground cover while also increasing the release of GHG. It was noted that human beings are responsible for taking care of the climate and can control their activities to some extent, but the majority do not out of ignorance.

In the context where the effects of climate change can be found, the participants indicated them to range from local through regional to global contexts. A further question asked to both pre-service teachers and lecturers was how climate change affects humans and their environment. This question was aimed at finding out the awareness of the participants on the effects of CC on the human environment. They identified some of the effects that were being experienced to include; prolonged drought which led to the migration of pastoralists with their livestock and sometimes loss of life of both fauna and flora. An example given was of wild rabbits that provide white meat, they were noted to be rare, and some food and fodder crops and indigenous trees almost going to extinction. There were also changes in agricultural patterns and general food shortages in the country. Increased rainfalls cause flooding in some of the low-lying areas which had led to the spread of waterborne diseases and irregular displacement of people. The pre-service teachers noted there was an increase in water levels in lakes, for example, Lake Naivasha where land along the shores of the lake has been submerged. The rise of water levels in lakes is, however, a case of continuing research, not yet confirmed. They said the depletion of the ozone layer may have led to skin diseases due to the direct heating of the human skin. Climate change was seen to have also affected the education calendar in some areas where learning was disrupted during the migration of pastoralists and when floods commonly displaced residents. These are common occurrences in the North Eastern, parts of Nyanza, and Western regions of Kenya. The lecturers added that climate change causes hurricanes, and high temperatures negatively affect plant growth and reduce dairy production. Further, there is a noted increase in crime and human conflicts due to reduced and inadequate resources. The lecturers noted that at school, there are increased student dropouts and poor retention due to lack of school fees which was attributed to hard economic times with the reduced resources resulting from CC. Rationing of water and hydroelectric power was more common, while an increase in harvesting and using solar power was noted. Solar power is one benefit of solar intensity (Sohn, 2014). The negative effects of CC were seen to continually cause psycho-socio-economic and ethical strain on people (Walsh, 2022).

The research also sought a response on mitigation and adaptation measures for CC. To mitigate, participants suggested a change from practices that cause negative climate change to those that would improve the climate such as afforestation and reforestation. The effects of CC such as flooding could see flood water redirected through dikes and canals which have been done in Bundalangi in Western part of Kenya or stored in dams for irrigation purposes. There is a need for change from fuels generating dangerous fumes and gases to green, solar, and biogas energies for both industrial and domestic use. The general population should be encouraged to use common transport and/or generally walk to do business or work and solar-powered vehicles should be introduced. There should be a reduction of over-reliance on firewood for domestic use and in industries such as the tea industry so that vegetative cover is maintained noting that Kenya was yet to attain the expected 10% forest cover. Industries and individuals should be encouraged to sort and treat their waste before disposing, so as to reduce the release of GHG, noting further that some waste may be recycled and reused. It was noted that there should be enforcement of policy directives and regulations on how to reduce the human-related causes of CC by agencies such as NEMA and KSF. There should be no settlement on wetlands or attempts to drain them. There is a need to curb corruption so that land in forested areas is not illegally allocated and the environment is destroyed. Instead, communities could use the "shamba system" to cultivate and grow trees at the same time. It is, however, notable that the "shamba system" is surrounded by controversies concerning climate change (Kagombe & Gitonga, 2005). To mitigate the many negative effects already in place, a multi-agency approach would be necessary, for example, the transport industry and forest departments would work together. It is notable that responses for mitigation were intertwined with those of adaptation; while mitigation deals with reducing emissions of GHG and increasing carbon sinks (US EPA, 2023), the participants indicated that dealing with flood water, an effect, is a mitigation measure. This shows that it was not out rightly clear what mitigation measures dealt with, however, strategies for adaptation show they understood what they should be.

To adapt to the changes, the participants suggested that there should be public participation in the drafting of policies on CC and matters of the environment. They noted that key education implementers who are the teachers hardly know or understand the policies in place. The participants suggested that people can adopt the "ujamaa system" where communities live in one place and have more land for common farming instead of fragmented pieces of farms against the capitalist system that is embraced in Kenya. They suggested that there should be a curriculum for both the public and learning institutions on CCE. The lecturers emphasized that there is a need to accelerate the implementation of CCE by including it in the formal curriculum as a common and core course in Early Years Education (EYE). The pre-service teachers noted that pastoralists can be encouraged to keep a few manageable livestock to be able to feed them locally and avoid overgrazing, to reduce the release of methane gas to the atmosphere and, ground vegetative cover respectively. The use, reuse, reduce, and recycle strategy of materials was also suggested. Further to adaptation, there should be reduced dependence on rain-fed crops for both humans and animals (US EAP, 2023). There is a further need to escalate the planting of trees to regulate gases and temperatures, and adherence to policies on environmental conservation. There should be educational policies on how to use and conserve the environment and how to manage water, especially during rainy seasons.

The findings from the question of knowledge of pre-service teachers on CC show that they had knowledge of causes, impacts, mitigations, and adaptation measures. However, there were misconceptions about pollution and overgrazing being causes of CC and that dealing with the effects of CC as a mitigation strategy. The participants were not aware of ESD and the associated framework for delivering CCE as well as carbon sinks as a mitigation measure.

Preparation of Pre-Service Teachers to Teach CCE

The preparation of pre-service teachers was explored through the content of courses and curriculum documents used in their training. Each participant explored at least one of their subject combinations and any other common course such as university, school, and department courses. In Religious Studies, matters of the environment were cited in the distribution of resources and traditional sacred places of worship which were under certain indigenous trees that were never cut down. The environments surrounding such places were maintained clean. The participants noted that one unit in History in the fourth year, "Themes in African Environmental History", focused on matters of environmental conservation and its impact, as well as the uses of the conserved forests for both fauna and flora, matters of soil erosion and challenges in post-independent Africa. Similar explanations were given for chemistry, biology, and geography. These findings are inclined to address matters of the environment which are in line with EE, one paradigm of addressing CCE (Hung, 2012). Further, the fact that the humanity subjects integrated CC in their content, is in line with earlier proposals (Siegner & Stapert, 2020). It was notable that no student linked mathematics to climate change. One education common course under education foundations; "Contemporary Issues in Education", highlighted climate change as one of the contemporary issues, briefly discussing causes, impacts, and mitigation. University courses such as philosophy and communication were not cited; neither did the curriculum document give an indication that they addressed or integrated CCE. It was notable that the participants mainly cited courses in their fourth year of study. The lecturers confirmed that aspects of CCE are not intentionally covered in the curriculum, noting that with a review of the curriculum, it should be incorporated since it is a serious emerging issue with lifelong effects on human beings and their environment. The inclusion of CCE issues during teaching by the lecturers was uncertain with two of them providing it as examples while one did not even mention it.

The participants indicated that there were a variety of extracurricular activities at the university that promoted and exposed them to the knowledge and skills necessary for CCE. There were clubs involved in tree planting and cleaning the environment within the university and nearby community areas. These were Environmental, Red Cross, and Wildlife clubs, and the Scouts movement. The Wildlife Club was also engaged in the conservation of both fauna and flora. The debate club often debated on emerging issues such as EE and CC. Home county associations, for example, Kakamega Students Association mentored their community members on environmental matters and planted trees in their environments. This resonated with the involvement of students' unions on matters of CCE in Canada (Berger, Gerum, & Moon, 2015) though only on environmental matters. The participants also noted that during their teaching practice, there were similar clubs in schools and that they were concerned with matters of the environment but not explicitly on CC.

The participants were asked if they were adequately equipped with the knowledge and skills required for CC, EE, CCE, and ESD. A majority said they were not equipped for the causes and interventions of climate change. While there were areas covered in the courses they took, they noted that the knowledge and skills they had were on environmental matters acquired from public discussions and extracurricular activities and were not detailed enough for them to instruct learners. On how to integrate CCE in everyday teaching in subjects, it was not clear for those in mathematics and humanities with the exception of geography and biology. They were of the view one impactful way of integrating CCE was through clubs. The participants expressed surprise saying that they did not know that they were supposed to have knowledge and skills on CCE for instruction purposes. The mathematics and English lecturers felt inadequately prepared to instruct on matters of CCE and therefore needed training and empowerment. The mathematics lecturer saw the need for pre-service teachers to be equipped with knowledge and skills on sanitation and waste disposal so that they can teach their students and also advise school administrators on the same. The English lecturer argued that pre-service teachers need practical skills in the conservation of the environment, and in teaching, they need demonstration and simulation skills. Further that educating pre-service teachers on CCE has a multiplier effect; first on immediate students and later on the students they would teach in the future and as well as their communities (Stevenson, Nicholls & Whitehouse, 2017). He, therefore, noted that if the government invested in CCE, pre-service teachers, and their future learners could be the greater voice for CC. The geography lecturer felt equipped to handle CCE. Their equipment on ESD did not arise since they earlier indicated that they did not know what it entailed.

The findings show that the pre-service teachers were not adequately prepared in their training to engage with matters of CCE. They had knowledge and skills on environmental matters more from extracurricular activities than from in-class instruction. This was evidenced by the fact that the lecturers were also not equipped with knowledge of CCE. Further, the pre-service teachers felt that the greatest impact on CCE would be through clubs, probably because that is how they got the knowledge and skills.

Appropriate Pedagogies for CCE

In response to pedagogical approaches used during teaching practice, the majority of learners responded with methods and approaches of teaching and research such as questions and answers, discussions, group work or teamwork, lecture methods, experiments, field trips and interviews, and questionnaires. On prompting them further on pedagogies, the participants responded with ways of being innovative, but they indicated they needed time to think through. The question of appropriate pedagogies was responded to with similar methods in teaching and research. This shows pre-service teachers' knowledge gap on pedagogies in general and hence appropriate pedagogies for CCE.

The lecturers informed on innovative pedagogies involving digital skills to facilitate simulations, an online search for relevant programs and documentaries, and hands-on skills to facilitate practical approaches to CCE. Further learner-centered pedagogies for self-discovery, and peer teaching and discussions among learners since the focus is on the learner to attain and utilize the CCE.

The finding on the question of appropriate pedagogies for CCE shows that the pre-service teachers could not differentiate pedagogies and methods of teaching and research. Hence, they could not respond to the appropriate pedagogies for CCE. This finding was unexpected and brought to the fore a pedagogical knowledge gap among the pre-service teachers.

Perceptions of Pre-Service Teachers on CCE

The pre-service teachers noted that CC was an urgent issue directly and indirectly affecting all life on earth and, therefore, there is a need to address it in both the short and long term. The participants noted that there were solutions to CC that could be detrimental, for example, growing of eucalyptus trees; while there were advantages of planting the tree, it was noted that it drains wetlands, lowers the water table, and renders soils to become very weak exposing them to none use and erosion.

The responses to which teachers should teach CCE in secondary school were mixed. The pre-service teachers observed that all teachers were responsible because CC is an emerging issue affecting humanity. Others indicated that teachers of History and Geography in the first and second years of secondary education should teach it as a compulsory subject noting that in the latter, matters of climate are detailed as a topic at different levels of schooling. It was further suggested that teachers of life skills should create awareness among learners. The proposal that CCE be taught by teachers of humanity subjects is in line with earlier proposals elsewhere (Wise, 2010; Hung, 2012; Siegner & Stapert, 2020). It was further suggested that extracurricular activities such as clubs be used to sensitize students on CC and that schools could invite experts in the area to educate learners on CC during off-lesson times.

On ways of integrating CCE with the school curriculum, the participants observed that it can and should be integrated with various subjects while others argued that it should be taught as a standalone subject. Integration was suggested as follows, in Physics, in the topic of radiation specifically the subtopic of sun and UV rays, and renewable and non-renewable energy. In Religion, in particular Christian Religious Education (CRE), in the subtopic of the role of man in taking care of the environment and God's punishment of the first parents. In Mathematics, data on climate can be used in the topic of Statistics and Probability. In History, integration can be in areas of migration and settlement of the people of Kenya caused by negative impacts of climate change, and further CC can be profiled to depict how the changes happened in the past and current times, and how communities were and are adapting to it. In Chemistry, CCE can be integrated with the topic of compounds and how these compounds affect the environment while in Kiswahili and English, there can be discussions and writing compositions, poems, and essays on 'Jinsi ya kutunza Mazingira', that is the preservation of the environment. In Geography, the participants noted that some matters of CC were covered but they need to be mentioned in the land-forming processes which are associated with natural causes of CC and their effects on the environment. In Biology under the topic of Ecology, CCE can be integrated with the subtopic of the environment and how living organisms interact with the environment. Here, matters of how to preserve the environment, and the effects of climate change on living organisms, both positive and negative, can be discussed. This finding on the need to integrate CCE with other subjects negates the bias towards integration reported as a challenge in implementing CCE (Huho, 2015) and strengthens the build-up of the call for all teachers to be involved in delivering CCE. It gives impetus to EE being implemented in the secondary section of the CBC in Kenya. To integrate, proposals such as those based on content, pedagogy and context of institutions (Molthan-Hill, Jackson, Mifsud, and Odell, 2020) could be used.

Considering CCE as a standalone subject, the arguments were that CC is an issue that needs urgency, therefore, a need to create a lot of awareness. It was also further noted that CCE is a broad area of study that needs a lot of time to cover, hence integration may not provide sufficient time. One lecturer felt that at the university level, it can be made into a program: to help explore multiple areas of CC and in fact conduct research. There was also the view that CCE could be both integrated and taught as a standalone subject.

The pre-service teachers indicated that they did not integrate CCE during their teaching practice since they did not know that it was necessary while some thought it was not in their domain to do so. If they did it was not out of intention and pedagogical knowledge of CCE. As noted earlier, they did not think they needed knowledge about it. They said, given the opportunity and with regard to the engagement in the current research discussion (FGD), they could look into their subjects at the secondary school level to see how and where they would integrate matters of CCE.

DISCUSSION

The findings and analysis show that the participants' responses were based mainly on their home localities and university environment, and on what was observable. The findings show that the pre-service teachers had considerable knowledge and skills about CC gathered from the public and in some instances from subjects taught or extracurricular activities done at the university. The geography students and lecturer seemed most advantaged in knowledge since there was the study of climate and its variations in the subject. This is in line with past research which suggests that geography teachers are better placed to teach CCE (Hung, 2012; Siegner & Stapert, 2020) because they had the knowledge. The pre-service teachers in the sciences, especially biology, saw a lot of connection between the subject and EE. Those in the humanity subjects also found substance in their subjects as in earlier findings (Siegner

& Stapert, 2020). The mathematics students initially did not link the subject to CC and their lecturer had scanty information about CC and CCE both in class and out of class which may be aligned to earlier finding that some subjects are not clearly related to CCE (Hung, 2012). It was then not surprising that the lecturer proposed that CCE should be taught as a standalone subject and not integrated. However, the students later identified that CCE could be integrated with mathematics through topics of Statistics and Probability which resonates with the works of Barwell (2013) and Abtahi, Gøtze, Steffensen, Hauge, and Barwell (2017).

The knowledge, and in fact pedagogical knowledge, and skills the pre-service teachers had were insufficient for teaching in class. They were not detailed. Further, there were knowledge gaps on CCE in the curriculum used to prepare the pre-service teachers in particular ESD and associated framework, and gaps in identifying pedagogies and hence appropriate pedagogies for CCE. It was notable that no student talked about establishing sinks for GHG, it could be a lack of knowledge on the same. There were also misconceptions such as pollution and overgrazing being a causes of CC, and confusion on and between weather and climate. The need for CCE to be included in the curriculum both as an integrated subject/course and as a standalone was presented. Thus, the need to train the pre-service teachers on CCE resonates with the goal of building the capacity of educators (UNESCO, 2014) so as to influence how learners and others out of school think about CC (UNESCO, 2017) using the already developed ESD framework (UNESCO, 2011) with an emphasis on identified gaps.

On perceptions on CCE, there was a general consensus that it was an agent issue that needed both short- and long-term solutions. It was clear that teachers of all subjects should engage with CCE either as an integrated or standalone course or both. Past research shows similar arguments (Huho, 2015; Hung, 2012; Siegner & Stapert, 2020; Wise, 2010) and it is a necessary discussion that should continue.

The findings in Kenya's context help to identify its unique context in regard to CC, these include pastoralism and the "shamba system". Hence mitigation and adaptation are also unique. Further, the pre-service teachers are equipped with knowledge and skills on matters of environment and not explicitly on climate change. Hence, they cannot comprehensively address CCE. Other contexts may also have their unique features in terms of causes, impacts, mitigation, and adaptation. Thus, the earlier argument that since CC is evolving and has geographical contexts, the CCE curriculum could be dynamic and based on context.

CONCLUSION

This chapter has analyzed pre-service teachers' pedagogical knowledge and perception on and about CCE through a literature review and empirical research. The empirical research focused on pre-service teachers' knowledge on CC, their preparation to teach CCE, their suggested and appropriate pedagogies for CCE, and their perceptions of CCE. While the participants displayed knowledge on and about CC, it mainly focused on environmental issues though not directly connected to education. Most of it had been acquired from out-of-class situations and they felt it was not adequate for instruction. They lacked knowledge on ESD and the associated framework and more particularly of appropriate pedagogies for CCE and in particular pedagogical knowledge on CCE. They had misconceptions about some causes of CC. They were of the view that CC being a global and lifelong concern, all teachers should be encouraged to acquire knowledge and skills for imparting on learners and for themselves. The question of whether to integrate CCE with other subjects or be a standalone subject yielded mixed responses and therefore should be a continuing debate with best practices adopted. An unexpected outcome was the dilemma of some mitigation measures and their eventual negative effects, which call do further research. The findings have opened up a pedagogical knowledge gap as well as perspective issue on CCE for the pre-service teachers and hence the need to include CCE in the teacher training curriculum is underscored.

REFERENCES

Abtahi, Y., Gøtze, P., Steffensen, L., Hauge, H. K., & Barwell, R. (2017). *Teaching climate change in mathematics classrooms: An ethical responsibility.* https://www.researchgate.net/publication/343385757

Barwell, R. (2013). The mathematical formatting of climate change: Critical mathematics education and post-normal science. *Research in Mathematics Education, 15*(1), 1–16. doi:10.1080/14794802.2012.756633

Berger, P., Gerum, N., & Moon, M. (2015). "Role up your sleeves and get at it": Climate Change Education in Teacher Education. *Canadian Journal of Environmental Education, 20*, 152–174.

Boon, H. J. (2016). Pre-Service teachers and climate change: A Stalemate? *The Australian Journal of Teacher Education, 41*(4), 39–63. Advance online publication. doi:10.14221/ajte.2016v41n4.3

Bord, R. J., O'Connor, R. E., & Fisher, A. (2000). In what sense does the public need to understand global climate change? *Public Understanding of Science (Bristol, England)*, *9*(3), 205–218. doi:10.1088/0963-6625/9/3/301

Friedrich-Alexander-Universitat (FAU). (2022). *What are the goals of climate change adaptation?* https://www.fau.eu/2022/03/15/news/

Huho, J. M. (2015). Climate change knowledge gap in education system in Kenya. *International Journal of Innovation and Research in Educational Science*, *2*(3), 227–232.

Hung, C. C. (2012). *Climate Change Education: knowing, doing and being*. Routledge.

Ikonomidis, S., Papanastasious, D., Mela, D., & Avgoloupis, S. (2012). Teacher preparation and climate change curriculum at university level in Nigeria. *International Journal of Multidisciplinary Academic Research*, *21*, 768–779.

Intergovernmental Panel on Climate Change (IPCC). (2014). *AR5 Synthesis Report: Climate Change 2014.* Contribution of working groups I, II and III to the Fifth Assessment Report of the Intergovernmental Panel on Climate Change. Geneva: IPCC. https://www.ipcc.ch/report/ar5/syr/

Kagombe, J., & Gitonga, J. (2005). *Plantation establishment in Kenya: The Shamba system case study*. Kenya Forest.

Kenya Institute of Curriculum Development (KICD). (2017). *Basic Education Curriculum Framework.* Nairobi, Kenya: Republic of Kenya.

Leal, F. W., Sima, M., Sharif, A., Luetz, J. M., Salvia, A. L., Mifsud, M., Olooto, O. F., Djekic, I., Anholon, R., Rampasso, I., Donkor, F. K., & Dinis, M. (2021). Handling Climate change education at university; an overview. *Environmental Sciences Europe*, *33*(1), 109–128. doi:10.118612302-021-00552-5 PMID:34603904

Molthan-Hill, P., Jackson, D., Mifsud, A., & Odell, V. (2020). *NTU carbon literacy Toolkit for all staff and students at universities- a systems thinking approach*. NTU Green Academy.

Okoli, N. J. (2014). Teacher preparation and climate change curriculum at university in Nigeria. *International Journal of Multidisciplinary Academic Research, 2*(3).

Papadimitriou, V. (2004). Prospective primary teachers' understanding of climate change, greenhouse effect and ozone layer depletion. *Journal of Science Education and Technology*, *13*(2), 299–307. doi:10.1023/B:JOST.0000031268.72848.6d

Siegner, A., & Stapert, N. (2020). Climate change education in humanities classroom: A case of Lowell School curriculum. *Environmental Education Research, 26*(4), 511–531. doi:10.1080/13504622.2019.1607258

Sohn, E. (2014). *2014 Climate change and the rise and fall of civilizations*. Climate NASA website. https://climate.nasa.gov/news/1010/climate-change-and-the-rise-and-fall-of-civilizations/

Stevenson, R. B., Nicholls, J., & Whitehouse, H. (2017). What Is Climate Change Education? *Curriculum Perspectives, 37*(1), 67–71. doi:10.100741297-017-0015-9

UNESCO. (2012). *Climate change education for sustainable development in small islands in developing states*. UNESCO.

UNESCO. (2014). *UNESCO Roadmap for implementing the Global Action Program on Education for Sustainable Development*. United Nation Educational Scientific and Cultural Organisation.

UNESCO. (2017). *Education for Sustainable Development Goal: Learning Objectives*. United Nations Educational Scientific and Cultural Organisation.

UNESCO. (2020). *Education for Sustainable Development: A Roadmap*. United Nations Educational, Scientific and Cultural Organization.

UNESCO. (2021a). *UNESCO World Conference on Education for Sustainable Education.* Berlin: UNESCO.

UNESCO. (2021b). *UNESCO declares environmental education must be a core curriculum component by 2025*. https://www.unesco.org/en/articles/unesco-declares-environmental-education-must-be-core-curriculum-component-2025

United Nations Educational, Scientific and Cultural Organization (UNESCO). (2011). *Climate Change Starters' Guidebook*. Retrieved from https://www.perlego.com/book/1669895/climate-change-starters-guidebook-an-issues-guide-for-education-planners-and-practitioners-pdf

United Nations Environment Programme (UNEP). (2008). *Nairobi City Development Strategy Top Priority for 21st Century Future of the Kenyan Capital.* https://www.unep.org/documents.multilingual/default.asp?documentid=504&articleid=5564&l=en

United Nations (UN). (1992). *United Nations Framework Convention On Climate Change (UNFCCC)*. FCCC/INFORMAL/84. GE.05-62220 (E) 200705. https://unfccc.int/resource/docs/convkp/conveng.pdf

United Nations (UN). (1993). *United Nations Conference on Environment and Development.* Rio de Janeiro: A/cnf.151/26/Rev.1.

University of Missouri-Columbia. (2017). *Climate change misconceptions common among teachers, study finds.* ScienceDaily. www.sciencedaily.com/releases/2017/06/170607123835.htm

US Environmental Protection Agency (EPA). (2023). *Inflation Reduction Act: Evaluating Greenhouse Gas (GHG) Reduction Measures.* https://www.epa.gov/inflation-reduction-act/evaluating-greenhouse-gas-ghg-reduction-measures

Wise, S. B. (2010). Climate Change in the Classroom: Patterns, Motivations and Barriers to instruction among Colorado science teachers. *Journal of Geoscience Education, 58*(5), 297–309. doi:10.5408/1.3559695

ADDITIONAL READING

Belgrade Charter. (1975). *A framework for environmental education.* UNESCO-UNEP.

Chitman-Booker, L., & Kopp, K. (2013). *The 5Es of inquiry-based science.* Shell Education.

Combes, B. P. Y. (2005). The United Nations Decade of Education for Sustainable Development (2005–2014): Learning to Live Together Sustainably. *Applied Environmental Education and Communication, 4*(3), 215–219. doi:10.1080/15330150591004571

Summers, M., Childs, A., & Corney, G. (2005). Education for sustainable development in initial teacher training: Issues for interdisciplinary collaboration. *Environmental Education Research, 11*(5), 623–647. doi:10.1080/13504620500169841

KEY TERMS AND DEFINITIONS

Climate Change: Is the long-term unprecedented changes in weather patterns of a place that lead to an increase in atmospheric temperatures directly or indirectly affecting human activities.

Climate Change Education: Is the knowledge, skills, values, and attitudes of changes in climate over a considerable time in a place. It aims to explore and explain the causes, impacts of climate change, and the necessary mitigation and adaptation measures.

Education for Sustainable Education: Is an education that enables individuals to acquire and develop knowledge, skills, attitudes, and values necessary to take part in decisions to improve the quality of life locally, regionally, and globally based on their context.

Environmental Education: Is a discipline that deals with how human beings engage with their environment, its conservation, and preservation.

Pedagogies: Are approaches to teaching.

Pre-Service Teachers: Are student teachers under training.

Teacher Education: Is a program intended for students who after graduation take up teaching as a profession.

Chapter 4
Importance of Gender Inclusiveness in Sustainable Climate Change Education

Falak Shad Memon
Institute of Business Management, Karachi, Pakistan

ABSTRACT

This study investigates the importance of gender-inclusive climate change education programs and the challenges involved with designing such curriculum in rural Sindh, Pakistan. The study addresses three research areas: (1) perspectives and consequences of gender inclusivity in climate change education, (2) gender inclusivity's effect the efficacy and success of Sindh's rural climate change education initiatives, and (3) the hurdles to attaining gender equality in sustainable climate change education and its effect on the overall effectiveness of such initiatives. The qualitative research method was applied, and interviews were conducted from eight leading educationist in Pakistan. The findings show that women face various hurdles in climate change education, including restricted access to education, societal norms, and cultural barriers. However, research identifies various tactics and recommendations to improve gender inclusion, such as the promotion of female role models from best practices, the incorporation of gender viewpoints in curricula, as well as programs to empower women.

DOI: 10.4018/978-1-6684-9099-0.ch004

INTRODUCTION

Despite the growing acknowledgment of the crucial role that education plays in addressing the challenges presented by climate change, the education sector remains underutilized as a strategic resource for effectively tackling and adapting to climate change. Many education stakeholders in various countries have yet to establish a unified and comprehensive framework for climate change education (Anderson, 2012). In order to advance climate change education within the broader context of Education for Sustainable Development (ESD) or Climate Change Education for Sustainable Development (CCESD), it becomes imperative to enhance learners' comprehension of the causes and consequences of climate change, as well as foster their willingness to take proactive measures to mitigate its impacts (Gerrard, 2008). Education is widely acknowledged as a vital instrument for achieving sustainability-related goals. As a result, the concepts of Education for Sustainable Development (ESD) and climate change education have gained significant prominence in academic and policy discourse.

Over the past decade, there has been a substantial surge in interest in climate education. This noteworthy growth can be credited to various factors, including the spreading of increased knowledge and support by leaders involved in climate education programs, as well as the growing awareness of the far-reaching global impacts of climate change across diverse regions. However, climate change education presents a complex and challenging domain, leading researchers to engage in ongoing debates regarding its effectiveness within different operational contexts and populations (Denton et al., 2019).

Despite the growing interest in climate education, misconceptions about the causes and impacts of climate change continue to persist. The level of climate change awareness and understanding varies and is influenced by several factors, such as socio-economic conditions, gender, age, and educational background. According to Chigwanda (2016), education emerges as the most influential predictor of climate change awareness. Therefore, educators play a pivotal role in cultivating this knowledge and promoting climate literacy.

However, it is crucial to recognize that many primary and secondary level educators tend to prioritize the delivery of information rather than focusing on the development of critical thinking and problem-solving skills, which are vital for individuals to effectively, address climate change. This neglect of crucial skill-building aspects postures a significant interference to meaningful action and engagement in climate change mitigation and adaptation efforts (Brownlee, Robert, & Jeffery, 2013).

Scholars agree that climate education should be approached differently from teaching techniques employed in other life sciences. This emphasizes the necessity for careful consideration and tailored approaches to ensure the effectiveness of climate education (Brownlee, Robert, & Jeffery, 2013). It is vital to adopt innovative and contextually relevant teaching methods that encourage critical thinking, problem-solving, and a deep understanding of the complexities associated with climate change, thereby empowering individuals to actively address this pressing global issue.

When designing and developing climate change education programs, it is essential to give careful consideration to the influence of cultural perspectives on individuals' perception and awareness of climate change (Guy, Yoshihisa, Iain, & Neill, 2014). Furthermore, an additional challenge in climate change education arises from the need to engage diverse audiences effectively, which embrace different genders and age groups. Altering teaching methods and mediums becomes crucial in this regard, as youth can be educated within the formal settings of schools and educational institutions, while adults may require suitable opportunities in community centers and similar community-based settings. Moreover, the knowledge required and its specific relevance to the vulnerabilities associated with each audience or population may vary significantly.

Recognizing the unique characteristics and needs of different stakeholders is crucial in developing an effective climate change education programs. Educators must possess the competence to deliver accurate knowledge and facilitate the development of relevant skills (Rainard, Smith & Pachauri, 2023; Monroe, Oxarart, & Richard, 2013). They should be well-equipped with the necessary expertise to adapt their teaching approaches and materials to suit the specific requirements of diverse groups. This includes addressing cultural perspectives, understanding different learning styles, and recognizing the varying levels of climate change awareness and understanding among different audiences. In addition, a comprehensive focus should be given to the availability of resources to all stakeholders.

It is important to note that Pakistan, as a country vulnerable to climate change, faces a wide range of risks and hazards associated with this global phenomenon. Hence, the need for effective climate change education programs becomes even more pronounced in order to enhance preparedness, resilience, and adaptive capacities within the population. By equipping individuals with knowledge, skills, and awareness, they can actively contribute to climate change mitigation and adaptation efforts, both at the individual and collective levels, and minimize the potential negative impacts on their communities and the environment (Memon et al., 2022).

Over the past two decades, Pakistan has witnessed a notable escalation in the intensity of climate-related disasters, including devastating floods, prolonged droughts, scorching heatwaves, declining agricultural productivity, increased health risks, and alterations in rainfall patterns. These occurrences have brought to light the urgent need for climate education. Acknowledging the significance of climate education, the Pakistani government has implemented various initiatives aimed at enhancing climate resilience and raising awareness through diverse channels, both formal and informal (Memon and Amjad, 2020).

A pivotal aspect of Pakistan's climate change policy is the emphasis on developing curricula for climate education and integrating it into the formal education system (NCCP, 2012). This integration aims to equip the younger generation with essential knowledge and skills to understand and respond effectively to climate change challenges. To gauge the effectiveness of these climate change education initiatives, it is crucial to comprehend the perceptions and viewpoints of the individuals involved.

Evaluating policy initiatives and assessing their impact in specific regions necessitates familiarity with the programs conducted within the domain of climate change awareness. Moreover, it is essential to ensure inclusivity and consider the diverse members of society to address the unique needs and concerns of all individuals (Martha et al., 2019). In Pakistan, where the out-of-school population ranks second globally, with a significant majority being females, it becomes imperative to conduct gender-oriented perception research. This research aims to identify the extent of inclusivity of women in climate change education (Memon & Amjad, 2020). Understanding women's understanding and perspectives on climate change education initiatives assumes great importance as it enables the incorporation of gender inclusivity into climate policies and strategies.

Research Question

The study explores the answers to two main research questions:

Q1) What are the perceptions and implications of gender inclusiveness in sustainable climate change education, particularly in the context of rural areas in Sindh, Pakistan?

Q2) How does gender inclusiveness impact the effectiveness and outcomes of sustainable climate change education initiatives in rural areas of Sindh, Pakistan?

Q3) What are the barriers in achieving gender inclusiveness in sustainable climate change education in rural areas of Sindh, Pakistan, and how do they influence the overall success of such initiatives?

LITERATURE REVIEW

Women and Climate Change Education

Climate change impacts and reactions are not gender neutral. Women and men are affected differently by climate change. Natural events are not discriminating but individual resistance and resilience make some more vulnerable than others. The varied effects of climate change on men and women are determined by social norms and gender inequality in society (Omoeva, Hatch & Moussa, 2018; Cannon, 2002; Denton et al., 2019; Blum et al., 2013). Imbalanced power relations, both formal (as in institutions) and informal (as in communities and the private sphere), are at the foundation of women's disproportionate vulnerability relative to males (Cannon, 2002; Shackleton, Cobban & Cundill, 2014); UN Women, 2020). Studies have shown that women often face higher risks and greater challenges in coping with climate change impacts due to existing gender inequalities (Wen et al., 2023; Denton et al., 2019; Enarson, 2012; UN Women, 2020).

However, most gender and climate change research to date has focused on women and their specific vulnerabilities (Rainard, Smith & Pachauri, 2023; Ikeda, 2009; O'Brien, 2012). While understanding women's experiences and vulnerabilities is essential, it is equally important to examine how inequalities contribute to vulnerabilities and, as a result, how gender relations contribute to the differentiated effects of climate change on women (Denton et al., 2019; Enarson, 2012). Addressing the broader gender dimensions of climate change requires considering the intersecting factors such as age, socioeconomic status, ethnicity, and geographical location that shape women's experiences and capacities to adapt to and mitigate climate change impacts (Enarson, 2012; O'Brien, 2012; Shackleton, Cobban & Cundill, 2014).

Incorporating gender analysis into climate change research and policy frameworks is crucial for understanding the gendered dimensions of vulnerability and resilience (Denton et al., 2019). It enables the identification of gender-specific needs, priorities, and capacities, and informs the development of gender-responsive and equitable climate change adaptation and mitigation strategies (Enarson, 2012; Shackleton, Cobban & Cundill, 2014; UN Women, 2020). Furthermore, acknowledging and addressing gender inequalities within climate change policies and interventions can contribute to more effective and sustainable outcomes, as it recognizes the importance of gender equality in building resilience and achieving sustainable development goals (Wen et al., 2023; Denton et al., 2019; UN Women, 2020).

Therefore, it is crucial to expand gender and climate change research beyond women's vulnerabilities to also explore the ways in which gender inequalities intersect with other social factors to shape the impacts of climate change on individuals and communities (Denton et al., 2019). By adopting an intersectional and gender-responsive approach, policymakers and practitioners can develop more inclusive and effective strategies to address climate change impacts and promote gender equality (Enarson, 2012; UN Women, 2020).

According to the Pakistan Bureau of Statistics (PBS, 2017), women make up approximately 50% of Pakistan's population. This demographic statistic highlights the significance of considering women's experiences and needs in climate change initiatives. In many developing countries, including Pakistan, women are disproportionately affected by climate change impacts due to pre-existing gender inequalities and socio-cultural norms (Denton et al., 2019; Enarson, 2012). They often have limited access to resources, including education, healthcare, and economic opportunities, which hinders their ability to cope with climate-related challenges.

Additionally, women's representation in decision-making processes related to climate change is often insufficient, further disregarding their voices and limiting their influence in shaping policies and programs (UN Women, 2020). This underrepresentation restricts their ability to advocate for their specific needs and priorities in climate change adaptation and mitigation strategies (Shackleton, Cobban & Cundill, 2014). In Pakistan, women face additional barriers to engagement in climate education programs and initiatives. Societal norms and cultural practices restrict their mobility and access to information, limiting their capacity to participate in community knowledge programs and decision-making processes (Ilahi & Grimard, 2000). Scholars suggests that gender disparities in access to education result in limited opportunities for female children, particularly in rural areas, to receive basic education (Blum et al., 2013).

It is essential to recognize the importance of gender-responsive climate education policies and programs in addressing the specific needs and vulnerabilities of women. Assessing the effectiveness of climate education initiatives in meeting the local women's requirements is crucial for ensuring their meaningful engagement and empowerment (O'Brien, 2012). The vulnerability of Pakistan to climate change is well-documented, with the country consistently categorized as a region highly vulnerable to climate impacts (Khan, Gao & Abid 2020). This vulnerability further emphasizes the urgency of understanding and addressing the unique challenges faced by women in the context of climate change.

This emphasis on addressing climate change has placed increasing pressure on traditional disciplinary areas like education and teaching to develop knowledge and awareness in this field. To ensure a better future, interdisciplinary curricula focused on climate change are being developed and implemented (Hamilton, 2009). While many of these programs are currently offered at the university level, there is a need for educational institutions at all levels to incorporate climate education in order to prepare the future workforce. The primary objective of climate change education is to collectively combat the impacts of climate change. This can be achieved through education, training, public awareness, and active participation (UNESCO, 2017). The term "Action for Climate Empowerment" (ACE) is used by organizations such as UNESCO and UNFCCC to emphasize the importance of education, awareness, and training of local populations in priority areas, enabling them to develop the skills necessary to address the challenges posed by global climate change. ACE encourages governments to build resilience by implementing educational campaigns and promoting public awareness through both formal and informal education (UNESCO, 2017).

According to the Intergovernmental Panel on Climate Change (IPCC), climate change education involves educating people about the potential risks, uncertainties, and changes associated with climate variations, while also making them aware of measures that can be taken to mitigate these impacts (IPCC, 2014). One of the main approaches to climate education is incorporating it into existing curricula and formal education systems. This ensures that the information reaches a wider audience, including communities through informal channels (Rainard, Smith & Pachauri, 2023 ; Mochizuki & Bryan, 2015). Many authors argue that climate change education is essential for preparing the current generation to critically understand and adapt to the potential risks and uncertainties associated with climate change (Stevenson, Nicholls, & Whitehouse, 2017). By promoting education and awareness, particularly among women, it is possible to enhance their perception of climate change and their role in promoting education initiatives.

Pakistan has been ranked as the second-to-last country from 2012 to 2014 in the gender gap index (GGI), indicating significant gender disparities. The GGI is based on multiple dimensions, including education. Gender gaps in education are particularly evident in Sindh, where 58% of girls are reported to be out of school (Sustainable Development Policy Institute [SDPI], 2019). Factors contributing to girls' dropout rates include inadequate facilities, especially sanitation for girls, long distances to schools, household responsibilities, and cultural barriers (Ali & Cheema, 2018). The lack of inclusiveness in addressing the gender dimension in educational planning can hinder overall effectiveness.

Safe transportation for girls living in remote areas and the availability of female teachers pose additional barriers to the dissemination of climate change education (Ali & Cheema, 2018). The Government of Pakistan recognizes that women are particularly vulnerable to climate change, as anticipated disasters such as droughts, heatwaves, and floods are expected to increase (Dankelman, 2010). It is crucial for strategies and policies to be developed to capture and mitigate the damages arising from climate change.

The National Climate Change Policy in Pakistan places emphasis on enhancing women's resilience through government and community efforts, with a focus on training, capacity building, education, awareness promotion, and gender mainstreaming (National Climate Change Policy [NCCP], 2012; Afriyie, Ganle, & Santos, 2018). Women have demonstrated commendable roles during climate change-induced disasters, showcasing their potential as change agents when provided with timely information and knowledge (Dankelman, 2010). Their deep understanding of the changing environmental patterns, rooted in their experiences and social responsibilities, particularly in rural areas, contributes to their survival (Khan, Gao & Abid 2020).

While current programs in Pakistan primarily concentrate on sanitation and hygiene knowledge, the incorporation of basic climate change understanding remains unclear (Sustainable Development Policy Institute [SDPI], 2019). Molthan-Hill et al., (2019) suggest two approaches for integrating climate change education into existing programs: mainstreaming climate change education within the current educational curriculum and developing specialized programs that exclusively focus on climate change in the context of specific populations. It is crucial to utilize existing knowledge and experiences to inform future strategies, making them more accurate and viable.

METHODOLOGY

This study utilized qualitative methods of research, which are particularly valuable for uncovering in-depth insights in small case study-based research. The qualitative approach allowed us to explore the perspectives, experiences, and opinions of the participants in a rich and nuanced manner (Srikanthan & Dalrymple, 2003). A qualitative technique makes it possible to collect information that cannot be measured or represented mathematically, such as sensitive data or descriptive discourse. It is especially useful when attempting to understand variables affecting marginalized groups, because quantitative research may need a large number of participants, which may be difficult to recruit, or the meanings connected to descriptions may not be well reflected in quantitative data. The Institute of Business Management's ethical committee approved the study's aims and methods. The data collection process involved gathering information from all relevant stakeholders who had valuable knowledge and experiences related to the research topic. In order to ensure ethical standards, informed consent for participation in this research study was obtained from each participant. Before the interviews commenced, participants were provided with consent forms to review and sign. The consent forms clearly outlined the purpose of the study, assured participants of the anonymity of their names and designations (therefore are referred as respondent one, respondent two and so on), and informed them of their rights to withdraw from the study at any point without consequence.

Data Collection and Analysis

A total of eight in-depth interviews of senior members of the education sector were conducted, each lasting approximately 1 hour. The interviews followed a semi-structured format and specifically addressed the research questions of this study. The purpose of these interviews was to gather rich and detailed insights from knowledgeable professionals in the field. To ensure a rigorous analysis, the interviews were transcribed and manually coded for themes that were directly related to the research questions. The coding process involved identifying key ideas and patterns in the interview data, which were then grouped into thematic categories. In order to minimize bias, an independent coder, who was not involved in the interview process, analyzed the coded data separately (Srikanthan & Dalrymple, 2003). This approach helped to ensure objectivity and enhance the reliability of the findings.

RESULTS AND DISCUSSION

After analysis following main themes were identified. Themes under each questions are mentioned section wise.

1) Themes for research question one

Upon asking regarding the perception and implication of gender inclusiveness in Pakistan, following main themes were identified.

a) Awareness and Recognition of Gender Inequalities in Climate Change Education

Respondent one explained that

It is a high time that we should examine the existing literature on gender disparities in climate change education and then take measures accordingly as perception matters a lot so is women current awareness level.

Examining the current research on gender discrepancies in climate change education is critical to properly tackling the issue. Several academic studies stress the significance of identifying and correcting gender inequities in this environment. Policymakers, educators, and stakeholders can acquire insights into the particular issues and limitations experienced by women in climate change education by analyzing the current research. Respondent three explained same thing and he mentioned,

The role of awareness in recognizing and addressing these inequalities is important for sustainable education program development.

Several studies have found gender differences in climate change education. These inequities can emerge in a variety of ways, including unequal access to educational resources, a lack of female representation in decision-making processes, and cultural norms that reinforce gender stereotypes (Afriyie, Ganle, & Santos, 2018). Women, for example, may encounter impediments to education owing to cultural norms, societal expectations, and economic restraints.

The respondent five highlighted,

The importance of integrating gender perspectives into educational programs is undeniable and to do so we need to understand the current awareness level in rural areas.

National Climate Change Policy emphasizes on how climate change education can be key to resolve the gender inequality in climate change awareness. The current awareness and formation of dedicated teaching and information material can bridge the gap and can result in sustainable development (Afriyie, Ganle, & Santos, 2018). Respondent six said during interview highlighted the same thing and mentioned,

Identifying current attitude and perceptions we can also explore potential barriers or resistance that can hinder the implementation process of gender-inclusive approaches.

To close the gender gap in climate change education, it is critical to adopt steps that meet the unique demands and obstacles that women confront. This might include building inclusive educational settings, supporting gender-responsive policies and curriculum, ensuring equitable participation and leadership opportunities for women, and addressing cultural norms that promote gender inequities.

b) Enhances learning outcomes and participation

Second theme focuses that if we want women to actively participate and become climate-conscious members of society, it is crucial to increase their involvement in climate change initiatives. Numerous scholarly articles emphasize the importance of women's participation in addressing climate change and achieving sustainable outcomes. Respondent two explained during interview

See gender inclusiveness can empower individuals, especially women and marginalized groups, to actively participate in climate change initiatives if we continue to ignore them during curriculum development there will be less participation from them.

Respondent three said the same thing in his words:

If we want women to participate and be the part of active and climate conscious member of society we need to increase their participation that will results in better outcomes.

Research indicates that women's participation leads to better outcomes in climate change mitigation and adaptation efforts (Afriyie, Ganle, & Santos, 2018). For example, a study by Jabeen et al. (2020) found that involving women in community-based climate change adaptation projects resulted in more effective and sustainable solutions (Wen et al., 2023). Women bring unique perspectives, knowledge, and skills to the table, which enriches the decision-making process and leads to more comprehensive and inclusive strategies.

Respondent eight mentioned,

Sustainable development goal is also focused of climate and social justice and we in Pakistan need to work towards this in order to increase participation of rural women.

Increasing women's participation in climate change education and decision-making processes has several benefits. It promotes gender equality and women's empowerment, contributing to broader social justice goals. Women's active engagement also enhances the effectiveness of climate change policies and programs by incorporating diverse viewpoints and priorities (Jabeen et al., 2020).

Moreover, involving women in climate change initiatives can nurture resilience within communities. Women often play critical roles in resource management, food security, and disaster preparedness, making their participation vital for building climate resilience at the community level (Dankelman, 2010). To increase women's participation, it is essential to address the barriers and challenges they face. These can include gender stereotypes, limited access to resources and education, cultural norms, and discriminatory practices. Scholarly literature underscores the need for targeted interventions, such as gender-responsive policies, capacity-building programs, and inclusive decision-making processes, to overcome these barriers and promote women's active involvement (Rainard, Smith & Pachauri, 2023).

Promoting women's participation in climate change education and action requires a multi-faceted approach. It involves creating supportive environments that encourage women's engagement, providing educational opportunities and skills development, fostering mentorship and leadership programs, and ensuring equal access to resources and decision-making platforms (UN women, 2020).

2) Themes for research question 2

The following main themes were identified with reference to research question two:

a) Empowerment and livelihood

Gender-inclusive climate change education is critical to the establishment of sustainable livelihoods, especially in rural regions.

Respondent seven mentioned,

Gender-inclusive climate change education supports the development of sustainable livelihoods in rural areas we have success stories from Bangladesh we can do same here.

Respondent five said that:

Actually by addressing the gender specific needs during climate education we just don't reach to the marginalized population but also foster innovative solutions and promote sustainable practices that financially benefit individuals.

According to research, gender-inclusive climate change education promotes the development of information and skills required for sustainable livelihoods. Education efforts that target the distinct needs and perspectives of both men and women can provide individuals with the tools they need to adapt successfully to climate-related changes and encourage sustainable practices in their everyday lives (Wodon et al., 2018). Scholarly publications show the good influence of such education in empowering individuals and communities to adapt to the difficulties of climate change while supporting sustainable practices. Successful examples from Bangladesh show the possibility for similar ideas to be used in other situations (Jabeen et al, 2020). Bangladesh is one example of gender-inclusive climate change education success. The government has launched creative programs to empower women and increase their engagement in climate change adaptation and mitigation activities. For example, the Grameen Shakti program has trained rural women in renewable energy technology, allowing them to become entrepreneurs and contribute to sustainable development while tackling energy poverty (Amin & Langendoen, 2012).

Gender-inclusive climate change education can provide favorable results if comparable techniques are used in other locations. Women's participation in sustainable livelihood programs can have a variety of advantages, including poverty reduction, greater food security, and increased community resilience (Wodon et al., 2018). Women's active participation in income-generating climate change adaptation activities such as sustainable agriculture, eco-tourism, and natural resource management can generate economic possibilities while also promoting environmental sustainability (Afriyie, Ganle, & Santos, 2018; Haque, Kumar & Bhullar, 2023).

To replicate Bangladesh's success stories, it is critical to evaluate context-specific characteristics and modify interventions accordingly. Understanding local socio-cultural dynamics, recognizing unique needs and constraints encountered by rural communities, and incorporating key stakeholders in the design and execution of gender-inclusive climate change education programs are all part of this.

In addition, efficient implementation of such projects requires coordination among government agencies, non-governmental organizations, educational institutions, and local communities. Strong collaborations may allow information exchange, capacity building, and resource mobilization, hence enhancing the sustainability and scalability of gender-inclusive climate change education programs (Ashrafuzzaman, 2023).

b) Sustainable development

Respondent three during interview highlighted and explained how inclusiveness in climate change education equips rural communities with the knowledge and skills needed to build resilience and adapt to climate-related challenges.

In his own words, "we need to build climate change resilience in rural areas that is key to sustainable development and in my opinion climate change education can be the main area that is needed to be focused on."

Respondent five mentioned the same thing and said "Gender-specific vulnerabilities and coping strategies can be addressed by climate education, ensuring that women in rural areas are as equipped as men to respond effectively to climate change impacts."

Building resistance to climate change in rural regions is critical for attaining sustainable development. Past studies frequently emphasize the importance of strengthening rural communities' resilience in order to overcome the difficulties posed by climate change effectively. Climate change education emerges as a critical area of concentration and an important tool for improving resilience in this respect (Dankelman, 2010). Climate change education is crucial in providing individuals and communities with the information, skills, and attitudes required to identify and respond to climate-related risks and vulnerabilities. Education may allow rural people to take proactive actions and make educated decisions to develop resilience by boosting understanding about climate change consequences, adaption options, and sustainable practices (Rainard, Smith & Pachauri, 2023; UNESCO, 2017).

Individuals get a better knowledge of the interdependence between their livelihoods, the environment, and climate change via climate change education. They become more aware of the specific dangers and problems that their communities face, such as shifting rainfall patterns, a rise in the frequency of extreme weather events, or the impact on agricultural output (Guenther et al., 2023, Haque, Kumar & Bhullar, 2023). Rural communities can gain the capacity to identify and execute suitable adaptation and mitigation strategies if they have access to climate change education (Afriyie, Ganle, & Santos, 2018). Reddy et al. (2012) define this as strategies for sustainable agriculture, water management, land-use planning, and natural resource conservation. Climate change education also promotes critical thinking, problem solving, and creativity, allowing communities to adopt context-specific solutions that increase their resistance to the effects of climate change.

Climate change education promotes participation and commitment of all rural community members, including women, youth, and marginalized groups. By guaranteeing inclusion and diversity in educational programs, it is feasible to harness the expertise and experiences of many stakeholders, supporting more comprehensive and long-term approaches to resilience-building (Djoudi et al., 2016). It is critical to recognize that climate change education extends beyond traditional educational settings. Informal and community-based learning activities, such as workshops, community dialogues, and indigenous knowledge systems, help to improve resilience by encouraging local

c) Policy and community transformation

Policy and community development was the last theme identified as benefit of sustainable climate education.

Respondent seven said,

Gender-inclusive education initiatives can drive policy and community-level transformations in rural areas that will be more sustainable and on long run will be more effective.

Gender-inclusive education efforts in rural regions have the ability to generate policy and community-level improvements, resulting in more durable and successful outcomes. Scholarly publications emphasize the significance of incorporating gender views into education programs and policies in order to ensure long-term sustainability and good transformation. Respondent three mentioned,

By highlighting the importance of gender equality and inclusiveness in climate action, we can influence policy frameworks and community practices, promoting more equitable and effective approaches to sustainable development.

Gender-inclusive education goes beyond offering equal opportunity for both genders; it entails questioning gender conventions, addressing power relations, and encouraging all persons to participate actively. Rural communities can get a better knowledge of the social, economic, and environmental components of sustainability by including gender views into school curriculum (Dankelman, 2010). Establishing gender-inclusive educational programs has the potential to improve people's lives in a variety of ways. Policymakers may create and execute gender-responsive education policies, including gender views into processes for making decisions. This can lead to the creation of more effective and inclusive policies that meet the particular demands and issues that rural communities confront (Rainard, Smith & Pachauri, 2023).

Past case studies support the transformational potential of gender-inclusive education efforts. According to Ethiopian study, incorporating gender views into agricultural education programs can increase women's engagement, improve agricultural practices, and help to food security and poverty reduction (Mangheni et al., 2019). Gender-inclusive education projects at the community level can promote changes in attitudes, beliefs, and behaviors. These efforts encourage individuals to participate in sustainable practices actively and decision-making processes by defiance of gender stereotypes and fostering equality. This can lead to the adoption of more sustainable livelihood options, improved resource management, and increased community resilience to the effects of climate change (Bhadwal et al., 2019). Gender-inclusive education projects can also help to empower women and marginalized populations in rural regions. These efforts assist individuals to break free from gender-based barriers, get access to resources, and engage in economic activities by offering possibilities for education and skill development. Individuals that are empowered may then promote good change in their communities and push for more inclusive and long-term policies and practices (Kwauku & Braga, 2017).

3) Themes for research question 3

During analysis following main themes were identified in relation to research question three.

a) Sociocultural barrier

The current study indicates that traditional gender roles and expectations limit opportunities for women and girls to engage in climate change education initiatives in some contexts. Respondent six during interview explained,

Sociocultural norms and stereotypes play a significant role in hindering gender inclusiveness in climate change education especially in rural context.

Gender norms and stereotypes have a substantial impact on gender inclusion in climate change education, particularly in rural environments. Scholarly studies frequently highlight the significance of cultural expectations and gender biases in sustaining gender equality hurdles in environmental and climate change education (Kwauku & Braga, 2017). Respondent four highlighted that

Societal perceptions of gender roles in Pakistan rural community may perpetuate the notion that environmental issues are primarily the responsibility of men, excluding women from participating fully which act as one of the major barrier.

Sociocultural norms influence attitudes and expectations regarding gender roles, which can hinder women's participation in climate change education. These norms frequently perpetuate conventional gender boundaries, with women assigned to domestic and caregiving responsibilities and males positioned as key decision-makers and providers. These sociocultural norms might be more established in rural regions, making gender inclusivity in climate change education difficult to attain. Traditional gender norms and expectations are strongly embedded in many rural areas, restricting women's access to education, resources, and decision-making platforms, according to research (Haque, Kumar & Bhullar, 2023). As a result, women may face additional challenges when it comes to engaging in climate change education activities.

These gender biases and stereotypes can manifest in various ways, such as limited educational opportunities for girls, unequal distribution of household and caregiving responsibilities, and cultural beliefs that prioritize men's perspectives and contributions over women's (Foss & Ko, 2019; Guenther et al., 2023). Consequently, women may be excluded from discussions and decision-making processes related to climate change adaptation and mitigation, leading to a lack of diverse perspectives and a missed opportunity to harness their knowledge and experiences. It is critical to address sociocultural norms and preconceptions in order to promote gender inclusivity in climate change education. Scholarly literature emphasizes the significance of questioning and changing social norms through targeted interventions and public awareness campaigns (Kwauku & Braga, 2017). This involves working with local communities, religious and community leaders, and other important stakeholders to combat gender prejudices, empower women, and create welcoming conditions for gender equality (Jabeen et al., 2020).

It is possible to question established gender roles, raise awareness about the importance of gender equality in climate change education, and encourage women's active involvement in decision-making processes through fostering conversation, education, and community engagement. Furthermore, efforts should be taken to ensure that climate change education activities are culturally responsive, respect the local context, and promote inclusivity and gender equality (Alston, 2014).

b) Lack of resources and infrastructure

In rural regions, limited access to education promotes gender inequities. It promotes conventional gender norms and expectations, in which women are frequently allocated home tasks and caregiving duties while males are given preferential access to school and other resources. Respondent two mentioned,

Here women have limited access to resources and infrastructure, such as education facilities, in rural areas can create barriers to gender inclusiveness in climate change education.

As a result, women are marginalized in climate change education activities, as are their views and viewpoints (Foss & Ko, 2019; Guenther et al., 2023). Respondent seven also said on same notations that:

Unequal distribution of resources between genders may disproportionately affect girls' and women's access to education and opportunities for involvement in climate change initiatives.

Women in rural communities sometimes lack access to crucial resources and infrastructure, including educational institutions. This lack of access raises enormous obstacles to attaining gender equality in climate change education. Scholarly studies frequently emphasize the difficulties that women experience in rural communities, where gender inequities in education and resource allocation impede their involvement and engagement in climate change education projects. Education is a critical resource that enables people to comprehend and respond to environmental concerns such as climate change. However, women in many rural communities have restricted access to educational resources such as schools, training centers, and instructional materials (Foss & Ko, 2019; Kwauku & Braga, 2017). Women's capacity to gain the information and skills required to actively engage in climate change education is hampered by a lack of educational opportunities.

Women's capacity to engage in climate change education is hampered by a lack of education facilities in rural regions, which also inhibits their overall empowerment and assistance. Education is a vital driver of social and economic success since it enables women to access better career possibilities, make educated decisions, and actively participate in community development (Dankelman, 2010). Rural areas lose out by denying women access to education. Infrastructure, such as transportation, energy, and internet access, is also critical for increasing gender inclusivity in climate change education. Women face extra challenges in many rural regions due to limited infrastructure, making it difficult for them to access educational materials, participate in training programs, or engage in online learning platforms. Women are further marginalized and excluded from climate change education activities in the absence of adequate infrastructure (Rainard, Smith & Pachauri, 2023).

Addressing these limitations needs specific initiatives to increase rural women's access to resources and infrastructure. This involves investing in the development and enhancement of educational facilities, advocating for gender-responsive educational policy, and offering scholarships and financial assistance to encourage females to attend school. Furthermore, boosting infrastructural development, such as expanding access to power and internet connectivity can dramatically increase women's engagement in climate change education (Jabeen et al., 2020).

c) Gender Bias and discrimination

The study highlights instances of gender-based discrimination, such as unequal treatment, stereotyping, and limited representation of women in decision-making roles, which hinder the active participation and leadership of women in climate change initiatives. Respondent four identified that,

Gender bias and discrimination within educational institutions and the broader society can impede gender inclusiveness in climate change education.

Respondent eight said during his interview,

Even if we somehow bring women to education center but then there will be discrimination on every step and we need to overcome that in order to achieve our sustainable development targets.

Numerous studies have shed light on the prevalence of gender-based discrimination in climate change programs, emphasizing cases of uneven treatment, stereotyping, and inadequate representation of women. These discriminatory practices serve as substantial impediments to women's active engagement and leadership in climate change activities. Recognizing and resolving these issues allows for a more inclusive and effective response to climate change. Gender inequality presents itself in a variety of ways in the context of climate change activities (Omoeva, Hatch & Moussa, 2018). One critical component is the uneven treatment of women in comparison to men. According to research, women in climate change initiatives frequently experience limited access to resources, financing, and decision-making possibilities (Wisner, 2004). This uneven treatment affects women's agency and engagement, making it difficult for them to successfully participate to climate change mitigation and adaptation activities.

Stereotyping also contributes significantly to the persistence of gender-based discrimination. Due to preconceived notions and preconceptions about gender roles and capacities, society frequently undervalues women's efforts and expertise in climate change initiatives. Women's expertise and experiences in agriculture, natural resource management, and community resilience are frequently ignored or marginalized (Alston, 2014, Haque, Kumar & Bhullar, 2023). These misconceptions perpetuate the notion that women are less capable or informed, resulting in less opportunities for women to participate in and lead climate change projects.

Furthermore, the underrepresentation of women in decision-making roles is a significant barrier to attaining gender inclusivity in climate change programs. Women are routinely excluded from crucial decision-making processes and positions of leadership, where significant policy and strategy decisions are made (Djoudi et al., 2016; Biesbroek et al., 2018). This lack of representation makes it difficult to include varied viewpoints and impedes the creation of comprehensive and equitable climate change policies and programs. Addressing gender-based discrimination in climate change programs necessitates taking proactive steps to question and reform current conventions and practices (Foss & Ko, 2019). It is critical to develop gender-responsive policies and practices that recognize and respect women's contributions and enable equitable participation possibilities for women. This involves providing platforms and venues for women's views to be heard, including women in decision-making processes, and offering training and capacity-building opportunities to help them become more effective leaders (Rainard, Smith & Pachauri, 2023).

In addition, creating gender-transformative methods is critical for overcoming preconceptions and promoting gender equality in climate change projects. Recognizing and addressing the underlying power dynamics and cultural norms that perpetuate discrimination, as well as actively trying to change these dynamics through education, awareness raising, and the promotion of gender equality principles (Agarwal, 2009), is part of this process.

d) Institutional and Policy Barriers:

Institutional and policy-level barriers play a significant role in hindering the integration of gender inclusiveness in climate change education. These barriers arise from the absence or inadequacy of gender-responsive policies and frameworks that address the specific needs and challenges faced by women in the context of climate change. Governments and relevant institutions need to recognize and address these barriers to ensure effective and equitable climate change education. Respondent one highlighted

Institutional and policy-level barriers, such as the lack of gender-responsive policies and frameworks, can hinder the integration of gender inclusiveness in climate change education government should look into this.

Respondent six explained and said,

See inadequate funding, absence of supportive policies, and insufficient institutional commitment to gender equality may impede the implementation and sustainability of gender-inclusive initiatives we have been talking about this but so far result are not shown.

One of the primary challenges is the lack of gender-responsive policies and strategies within the education sector. Many existing policies and frameworks fail to explicitly consider the gender dimensions of climate change and how they intersect with education. This oversight hampers the integration of gender perspectives, resulting in a limited understanding of the unique experiences, roles, and needs of women in relation to climate change (Djoudi et al., 2016). Additionally, the absence of gender mainstreaming in policy development and implementation processes is a significant barrier to gender inclusiveness in climate change education. Gender mainstreaming involves systematically integrating gender perspectives into all stages of policy formulation, implementation, and evaluation. When this approach is lacking, the specific needs and concerns of women are often overlooked, leading to ineffective and inequitable education initiatives (Agarwal, 2009).

Institutional barriers also contribute to the lack of gender inclusiveness in climate change education. Limited coordination and collaboration among government departments, education institutions, and relevant stakeholders can impede the design and implementation of comprehensive and coherent gender-responsive strategies. The absence of institutional structures dedicated to gender equality and climate change education further exacerbates the challenges (Djoudi et al., 2016; Aryal et al., 2020). To overcome these barriers, governments should prioritize the development and implementation of gender-responsive policies and frameworks in the realm of climate change education. This includes conducting gender analysis and impact assessments to understand the differential impacts of climate change on women and men, as well as their respective educational needs and opportunities (Wisner, 2004).

It is high time that government should promote inter-sectoral collaboration and coordination among relevant stakeholders to ensure a holistic and integrated approach to gender-inclusive climate change education. This involves establishing partnerships between education institutions, government agencies, civil society organizations, and local communities to jointly develop and implement gender-responsive strategies and initiatives (Alston, 2014). Moreover, capacity-building efforts are necessary to enhance the understanding and skills of policymakers, educators, and practitioners in integrating gender perspectives into climate change education. Training programs, workshops, and awareness campaigns can contribute to building the capacity of stakeholders to address gender-related challenges and promote gender equality in climate change education (UNESCO, 2017).

CONCLUSION

Finally, the analysis of the data and conversations showed many major issues about gender inclusivity in climate change education in Pakistan.

The first topic found in terms of the perception and implications of gender inclusivity was the understanding and identification of gender inequities in climate change education. Respondents emphasized the significance of reviewing current research on gender gaps in order to properly address the issue. Uneven access to educational resources, a lack of female representation in decision-making processes, and cultural practices that reinforce gender stereotypes all contribute to these discrepancies. Building inclusive educational settings, promoting gender-responsive policies and curricula, and addressing cultural norms were mentioned as ways to overcome the gender gap.

The second highlighted topic was improving learning outcomes and involvement through gender inclusivity. Gender inclusivity, according to respondents, encourages individuals, particularly women and marginalized groups, to actively engage in climate change projects. According to research, women's engagement in climate change mitigation and adaptation activities leads to better outcomes because they bring distinct views, expertise, and skills to the board.

The third theme identified was the empowerment and livelihood benefits of gender-inclusive climate change education. Respondents highlighted that such education supports the development of sustainable livelihoods in rural areas. Successful examples from Bangladesh were cited, demonstrating the positive impact of empowering women and promoting their engagement in climate change adaptation and mitigation activities. Gender-inclusive climate change education can provide individuals with the information and skills necessary for sustainable livelihoods, contributing to poverty reduction, food security, and community resilience.

The fourth theme identified was the contribution of gender-inclusive climate change education to sustainable development. It was emphasized that climate change education equips rural communities with the knowledge and skills needed to build resilience and adapt to climate-related challenges. By understanding the interdependence between livelihoods, the environment, and climate change, individuals become more aware of the specific risks their communities face and can identify and execute suitable adaptation and mitigation strategies. Climate change education also promotes involvement and engagement of all community members, fosters critical thinking and problem-solving, and encourages context-specific solutions.

The fifth theme identified was the role of gender-inclusive climate change education in policy and community transformation. Respondents highlighted that such education can drive policy and community-level transformations, promoting more equitable and effective approaches to sustainable development. Gender-inclusive education goes beyond providing equal opportunities and entails questioning gender conventions, addressing power relations, and encouraging active participation. It can lead to the creation of more inclusive policies, changes in attitudes and behaviors, and the empowerment of women and marginalized populations.

Lastly, the analysis revealed two main themes related to the barriers and challenges faced in achieving gender inclusiveness in climate change education. The first theme was sociocultural barriers, where traditional gender roles and expectations limit opportunities for women and girls to engage in climate change education initiatives. Sociocultural norms and gender biases perpetuate gender equality hurdles, and addressing these norms through targeted interventions and public awareness campaigns is crucial. The second theme was the lack of resources and infrastructure, with limited access to education, resources, and infrastructure, particularly for women in rural areas. Efforts should be made to provide equal access to resources and create supportive conditions for gender inclusiveness.

In conclusion, the findings highlight the importance of gender inclusiveness in climate change education in Pakistan. Addressing gender disparities, enhancing participation, empowering individuals, and transforming policies and communities are key factors in achieving sustainable development and resilience to climate change. Overcoming sociocultural barriers and improving access to resources and infrastructure are essential steps towards creating an inclusive and equitable climate change education system.

RECOMMENDATIONS

Following are the study recommendation to enhance gender inclusivity in climate change education in Pakistan. This, in turn, can contribute to gender equality, sustainable development, and climate resilience at the individual, community, and national levels.

1. Encourage Gender-Responsive Policies and Curriculum: Policymakers, educators, and stakeholders should make gender perspectives in climate change education a priority. This involves creating gender-responsive policies and curricula that meet the unique demands and problems that women and marginalized groups confront. Education programs may become more inclusive and successful in promoting gender equality and sustainable development by recognizing and resolving gender inequities.
2. Increase Participation of Women: Efforts should be made to actively incorporate women in climate change projects and decision-making processes. This may be accomplished through building inclusive educational environments, offering equitable access to resources and opportunities, and encouraging women to take up leadership responsibilities. Diverse views and experiences may be enriched by boosting women's engagement in the creation and implementation of climate change policies and programs.
3. Build Sustainable Livelihoods: Gender-inclusive climate change education should focus on enabling people and communities to create long-term livelihoods. This may be accomplished by offering education and skill training tailored to the unique requirements and problems of rural areas. Learning from

successful examples, such as Bangladesh's Grameen Shakti program, can aid in the development of initiatives that encourage entrepreneurship, renewable energy, sustainable agriculture, and other income-generating activities.

4. Strengthen Climate Change Resilience: Climate change education should provide rural people with the information and skills they need to build resilience and adapt to climate-related difficulties. This involves encouraging environmentally friendly practices, improving resource management, and cultivating critical thinking and problem-solving abilities. Communities may improve their resilience and long-term sustainability by expanding their awareness of climate change consequences and adaptation alternatives.
5. Overcome Sociocultural Barriers: Attempts should be made to address sociocultural conventions and preconceptions that stymie gender inclusion in climate change education. This necessitates collaboration with local populations, religious and community leaders, and other stakeholders in order to foster conversation, awareness, and behavior change. It is feasible to establish more inclusive and inviting conditions for women's active engagement in climate change education by addressing gender prejudices and misconceptions.
6. Improve Access to Resources and Infrastructure: Action should be done to alleviate the scarcity of resources and infrastructure, which disproportionately affects women's access to education and involvement in climate change activities. This involves expanding rural access to high-quality education, training, and technical resources. Barriers to education and active participation can be eliminated by investing in infrastructural development and guaranteeing equal opportunities for women.
7. Foster Collaboration and Partnerships: Collaboration among government agencies, non-governmental organizations, educational institutions, and local communities is required for the effective implementation of gender-inclusive climate change education. Partnerships that are strong can allow information exchange, capacity building, and resource mobilization, resulting in more sustainable and scalable programs. Stakeholders may utilize their skills and resources to make long-term effects and achieve policy and community-level reforms by working together.

REFERENCES

Afriyie, K., Ganle, J.K., & Santos, E. (2018). The floods came and we lost everything': Weather extremes and households' asset vulnerability and adaptation in rural Ghana. *Climate and Development, 10*(3), 259–274. doi:10.1080/17565529.2017.1291403

Agarwal, B. (2009). Gender and forest conservation: The impact of women's participation in community forest governance. *Ecological Economics, 69*(11), 2785–2799. doi:10.1016/j.ecolecon.2009.04.025

Ali, S., & Cheema, A. R. (2018). *Exploring the many barriers to a girl's education in Sindh.* LSE South Asia Center. Retrieved from http://blogs.lse.ac.uk/ southasia/2018/05/16/exploring-the-many-barriers-to-a-girls-education-in-sindhpakistan/

Alston, M. (2014). Gender mainstreaming and climate change. *Women's Studies International Forum, 47*(B), 287-294.

Anderson, A. (2012). Climate change education for mitigation and adaptation. *Journal of Education for Sustainable Development, 6*(2), 191–206. doi:10.1177/0973408212475199

Aryal, J. P., Sapkota, T. B., Rahut, D. B., Krupnik, T. J., Shahrin, S., Jat, M. L., & Stirling, C. M. (2020). Major climate risks and adaptation strategies of smallholder farmers in coastal Bangladesh. *Environmental Management, 66*(1), 105–120. doi:10.100700267-020-01291-8 PMID:32388655

Ashrafuzzaman, M. (2023). Local Context of Climate Change Adaptation in the South-Western Coastal Region of Bangladesh. *Sustainability, 15*(8), 6664. doi:10.3390/su15086664

Bhadwal, S., Sharma, G., Gorti, G., & Sen, S. M. (2019). Livelihoods, gender and climate change in the Eastern Himalayas. *Environmental Development, 31*, 68–77. doi:10.1016/j.envdev.2019.04.008

Biesbroek, R., Ford, L., Ford, J., Tanabe, A., Austin, S., & Lesnikowski, A. (2018, November). Data, concepts and methods for large-n comparative climate change adaptation policy research: A systematic literature review. *Wiley Interdisciplinary Reviews: Climate Change, John Wiley & Sons, 9*(6). Advance online publication. doi:10.1002/wcc.548

Blum, N., Nazir, J., Breiting, S., Chuan Goh, K., & Pedretti, E. (2013). Balancing the Tensions and Meeting the Conceptual Challenges of Education for Sustainable Development and Climate Change. *Environmental Education Research, 19*(2), 206–217. doi:10.1080/13504622.2013.780588

Brownlee, M. T. J., Robert, B. P., & Jeffery, C. H. (2013). A review of the foundational processes that influence beliefs in climate change: Opportunities for environmental education research. *Environmental Education Research, 19*(1), 1–20. doi:10.1080/13504622.2012.683389

Cannon, T. (2002). Gender and climate hazards in Bangladesh. Climate Change and Gender Justice. *Gender and Development, 10*(2), 45–50. doi:10.1080/13552070215906

Chigwanda, E. (2016). A framework for building resilience to climate change through girls' education programming. Brookings.

Dankelman, I. (2010). *Gender and climate change: Women as agents of change*. Zed Books.

Denton, F., Wilbanks, T. J., Abeysinghe, A. C., Burton, I., Gao, Q., Lemos, M. C., . . . Rahman, A. (2019). Chapter 11: Strengthening and implementing the global response. In V. Masson-Delmotte, H.-O. Pörtner, D. Roberts, J. Skea, P.R. Shukla, A. Pirani, W. Moufouma-Okia, C. Péan, R. Pidcock, S. Connors, J.B.R. Matthews, Y. Chen, X. Zhou, M.I. Gomis, E. Lonnoy, T. Maycock, M. Tignor, & T. Waterfield (Eds.), Global warming of 1.5°C. An IPCC Special Report on the impacts of global warming of 1.5°C above pre-industrial levels and related global greenhouse gas emission pathways, in the context of strengthening the global response to the threat of climate change, sustainable development, and efforts to eradicate poverty (pp. 641-775). World Meteorological Organization.

Djoudi, H., Locatelli, B., Vaast, C., Asher, K., Brockhaus, M., & Basnett Sijapati, B. (2016). Beyond dichotomies: Gender and intersecting inequalities in climate change studies. *Ambio, 45*(S3, Suppl 3), 248–262. doi:10.100713280-016-0825-2 PMID:27878531

Enarson, E. (2012). *Women confronting natural disaster: From vulnerability to resilience*. Lynne Rienner Publishers. doi:10.1515/9781588269560

Foss, A. W., & Ko, Y. (2019). Barriers and opportunities for climate change education: The case of Dallas-Fort Worth in Texas. *The Journal of Environmental Education, 50*(3), 145–159. doi:10.1080/00958964.2019.1604479

Gerrard, E. (2008). Climate Change and human rights: Issues and opportunities for indigenous peoples. *The University of New South Wales Law Journal, 31*(3), 941–952.

Guenther, L., Jörges, S., Mahl, D., & Brüggemann, M. (2023). Framing as a Bridging Concept for Climate Change Communication: A Systematic Review Based on 25 Years of Literature. *Communication Research, 0*(0). Advance online publication. doi:10.1177/00936502221137165

Guy, S., Yoshihisa, K., Iain, W., & O'Neill, S. (2014). Investigating the effects of knowledge and ideology on climate change beliefs. *European Journal of Social Psychology, 44*(5), 421–429. doi:10.1002/ejsp.2039

Hamilton, L. (2009). Education, politics and opinions about climate change: Evidence for interaction effects. *Climatic Change*, *104*(2), 231–242. doi:10.100710584-010-9957-8

Haque, A. T. M. S., Kumar, L., & Bhullar, N. (2023, February 20). Kumar. L & Bhullar, N. (2023) Gendered perceptions of climate change and agricultural adaptation practices: A systematic review. *Climate and Development*, 1–18. Advance online publication. doi:10.1080/17565529.2023.2176185

Ikeda, K. (2009). How women's concerns are shaped in community-based disaster risk management in Bangladesh. *Contemporary South Asia*, *17*(1), 65–78. doi:10.1080/09584930802624679

Ilahi, N., & Grimard, F. (2000). Public infrastructure and private costs: Water supply and time allocation of women in rural Pakistan. *Economic Development and Cultural Change*, *48*(3), 793–813. doi:10.1086/452490

IPCC. (2014). *Climate change 2014: Impacts, adaptation, and vulnerability*. Cambridge University Press.

Jabeen, S., Haq, S., Jameel, A., Hussain, A., Asif, M., Hwang, J., & Jabeen, A. (2020). Impacts of Rural Women's Traditional Economic Activities on Household Economy: Changing Economic Contributions through Empowered Women in Rural Pakistan. *Sustainability*, *12*(7), 2731. doi:10.3390/su12072731

Khan, N. A., Gao, Q., & Abid, M. (2020). Public institutions' capacities regarding climate change adaptation and risk management support in agriculture: The case of Punjab Province, Pakistan. *Scientific Reports*, *10*(1), 14111. doi:10.103841598-020-71011-z PMID:32839499

Kwauku, C., & Braga, A. (2017). *Three platforms for girls' education in climate strategies. Brooke Shearer Series, number 6*. Brookings Institution.

Mangheni, M. N., Tufan, H. A., Boonabana, B., Musiimenta, P., Miiro, R., & Njuki, J. (2019). Building gender research capacity for non-specialists: lessons and best practices from gender short courses for agricultural researchers in subSaharan Africa. In T. S. Marcia, K. Kristy, & D. Vasilikie (Eds.), *Gender and practice: knowledge, policy, organizations* (pp. 99–118). Emerald Publishing. doi:10.1108/S1529-212620190000028006

Martha, C. M., Richard, R. P., Oxarart, A., Bowers, A., & Willandia, A. C. (2019). Identifying effective climate change education strategies: A systematic review of the research. *Environmental Education Research*, *25*(6), 791–812. doi:10.1080/13504622.2017.1360842

Memon, F., Husain, I., Abdullah, F., Iqbal, R., Ahmad, S., & Abdullah, M. (2022). Addressing women's climate change awareness in Sindh, Pakistan: An empirical study of rural and urban women. *Climate and Development, 14*, 1–13. doi:10.1080/17565529.2022.2125784

Memon, F. S & Amjad, S. (2022). Understanding Women's Perceptions of Promoting Education and Policy Initiatives about Climate Change in Rural Areas of Sindh, Pakistan. *Journal of Education and Educational Development, 7*(1), 140-156. doi:10.22555/joeed.v7i1.3223

Mochizuki, Y., & Bryan, A. (2015). Climate Change Education in the Context of Education for Sustainable Development: Rationale and Principles. *Journal of Education for Sustainable Development, 9*(1), 4–26. doi:10.1177/0973408215569109

Molthan-Hill, P., Worsfold, N., Nagy, G. J., Leal-Filho, W., & Mifsud, M. (2019). Climate change education for universities: A conceptual framework. *Journal of Cleaner Production, 226*, 1092-1101.

Monroe, M. C., Oxarart, A., & Richard, R. P. (2013). A role for environmental education in climate change for secondary science educators. *Applied Environmental Education and Communication, 12*(1), 4–18. doi:10.1080/1533015X.2013.795827

National Climate Change Policy. (2012). *Government of Pakistan.* Retrieved from http://www.climatechange.gov.pk/en/policy

O'Brien, K. (2012). Global environmental change II: From adaptation to deliberate transformation. *Progress in Human Geography, 36*(5), 667–676. doi:10.1177/0309132511425767

Omoeva, C., Hatch, R., & Moussa, W. (2018). *The Effects of Armed Conflict on Educational Attainment and Inequality.* Education Policy and Data Center Paper No. 18-03. FHI 360.

Rainard, M., Smith, C. J., & Pachauri, S. (2023). Gender equality and climate change mitigation: Are women a secret weapon? *Front. Clim., 5*, 946712. doi:10.3389/fclim.2023.946712

Reddy, K. V., Paramesh, V., Arunachalam, V., Das, B., Ramasundaram, P., Pramanik, M., & Sridhara, S. (2022). Farmers' Perception and Efficacy of Adaptation Decisions to Climate Change. *Agronomy, 12*(5), 1023. doi:10.3390/agronomy12051023

Shackleton, S., Cobban, L., & Cundill, G. (2014). A gendered perspective of vulnerability to multiple stressors, including climate change, in the rural Eastern Cape, South Africa. *Agenda (Durban, South Africa), 28*(3), 73–89. doi:10.1080/10130950.2014.932560

Srikanthan, G., & Dalrymple, J. (2003). Developing alternative perspectives for quality in higher education. *International Journal of Educational Management, 17*(3), 126–136. doi:10.1108/09513540310467804

Stevenson, R. B., Nicholls, L., & Whitehouse, H. (2017). What is climate change education. *Curriculum Perspectives, 37*(1), 67–71. doi:10.100741297-017-0015-9

Sustainable Development Policy Institute. (2019). *Gender equity in education: Fact sheet Sindh*. Retrieved from https://www.sdpi.org/publications/files/FA-85.pdf

UNESCO. (2017). *Gender equality in and through education: Policy guidelines*. Retrieved from https://unesdoc.unesco.org/ark:/48223/pf0000245964

Wen, J., Wan, C., Ye, Q., Yan, J., & Li, W. (2023). Disaster Risk Reduction, Climate Change Adaptation and Their Linkages with Sustainable Development over the Past 30 Years: A Review. *Int J Disaster Risk Sci, 14*(1), 1–13. doi:10.100713753-023-00472-3

Wisner, B., Blaikie, P., Cannon, T., & Davis, I. (2004). *At risk, natural hazards. People's Vulnerability and Disasters*. Routledge.

Wodon, Q., Montenegro, C., Nguyen, H., & Onagoruwa, A. (2018). *Missed Opportunities: The High Cost of Not Educating Girls. The Cost of Not Educating Girls Notes Series*. The World Bank. doi:10.1596/29956

Women, U. N. (2020). *Making gender equality central to addressing climate change*. Retrieved from https://www.unwomen.org/-/media/headquarters/attachments/sections/library/publications/2020/issue-brief-making-gender-equality-central-to-addressing-climate-change-en.pdf?la=en&vs=5651

KEY TERMS AND DEFINITIONS

Adaptation: The process of adjusting and modifying in response to change or new circumstances.

Barriers: Obstacles or hindrances that impede progress or restrict access to desired outcomes.

Climate Change: Long-term alteration of Earth's climate patterns, primarily caused by human activities.

Community: A social group sharing common characteristics or interests, fostering a sense of belonging and mutual support.

Education: The process of acquiring knowledge, skills, and values through formal or informal means to facilitate personal growth and development.

Empowerment: The process of enabling individuals or groups to gain control over their lives, make informed decisions, and act to improve their circumstances.

Gender: The social and cultural roles, behaviors, and expectations associated with being male or female, which go beyond biological differences.

Gender Discrepancies: Differences or inequalities between genders in terms of opportunities, resources, rights, or treatment within a society.

Livelihood: The means by which individuals or communities obtain the necessary resources, such as income, assets, and employment, to sustain their basic needs and well-being.

Mitigation: Actions or measures taken to reduce or prevent the severity, impact, or extent of a problem or challenge, often used in the context of addressing climate change.

Recourses: The available means, assets, or materials that can be utilized to achieve specific goals, meet needs, or support various activities.

Resilience: The ability of individuals, communities, or systems to withstand, adapt to, and recover from shocks, stresses, or disruptions while maintaining functionality and well-being.

Rural: Referring to areas, communities, or regions located outside urban centers, typically characterized by agricultural activities, natural landscapes, and lower population densities.

Socio Cultural: Relating to the social and cultural aspects of a society, including norms, values, beliefs, traditions, customs, and interpersonal relationships that shape human behavior and interactions.

Stake Holder: Individuals, groups, organizations, or entities that have a vested interest or are affected by a particular issue, project, or decision, often possessing the ability to influence outcomes or be impacted by them.

Sustainable Development: The concept of meeting present needs without compromising the ability of future generations to meet their own needs, considering environmental, social, and economic factors to ensure long-term well-being and resource stewardship.

Chapter 5
Ethical Dimensions of Sustainable Development:
The Philosophy, Logistics, and Climate Change Education

Ebua Jarvis Ebua
University of Bamenda, Cameroon

ABSTRACT

This chapter examines the ethical dimensions of sustainable development within the context of climate change education. It explores the intersection of philosophy, logistics, and climate change education. The chapter sheds light on ethical implications and considerations arising in pursuing sustainable development. It seeks to provide valuable insights for climate change analysts, policymakers, curriculum and instructional designers, faculty, administrators, teacher educators, and other stakeholders interested in averting the impact of climate change and fostering sustainable development.

INTRODUCTION

This chapter examines the ethical dimensions of sustainable development within the context of climate change education. It explores the intersection of philosophy, logistics, and climate change education. The chapter sheds light on ethical implications and considerations arising in pursuing sustainable development. It seeks to provide valuable insights for climate change analysts, policymakers, curriculum and instructional designers, faculty, administrators, teacher educators, and other

DOI: 10.4018/978-1-6684-9099-0.ch005

stakeholders interested in averting the impact of climate change and fostering sustainable development.

Sustainable development and climate change education are critical components in addressing global environmental challenges. In order to promote responsible actions and decision-making, it is essential to apply ethical theories and frameworks. We explore various ethical perspectives relevant to sustainable development and climate change education, citing academic sources to support the analysis.

Utilitarianism, an ethical theory advocated by Jeremy Bentham and John Stuart Mill, centers on the principle of maximizing overall happiness or utility. In the context of sustainable development and climate change education, this theory promotes actions and policies that lead to the greatest good for the greatest number of people, taking into account environmental considerations. By incorporating utilitarian principles into educational programs, educators can emphasize the importance of collective efforts to address climate change and promote sustainable practices (Mill, 1863, p. 15).

Deontology, as proposed by Immanuel Kant, focuses on the inherent rightness or wrongness of actions rather than their consequences. In the context of climate change education, this ethical framework emphasizes the duty of individuals and societies to protect the environment and act in ways that respect the intrinsic value of nature. Kantian ethics can guide educators in cultivating a sense of responsibility and moral duty among students to take active roles in sustainability efforts (Kant, 1785, p. 30).

Virtue ethics, attributed to philosophers like Aristotle, centers on the development of moral character and virtues. In the context of sustainable development and climate change education, this framework emphasizes cultivating virtues like environmental consciousness, empathy, and responsibility. By promoting virtuous behavior, educators can instill a deep-rooted commitment to sustainable practices and environmental preservation among students (Aristotle, 350 BCE, p. 112).

Environmental ethics is a specific branch of ethics that focuses on the moral relationship between humans and the environment. This framework encourages viewing nature as having intrinsic value beyond its utility to human beings. Educators can apply environmental ethics in climate change education to foster a sense of ecological interconnectedness, emphasizing the need to protect and respect all living beings and ecosystems (Callicott, 1989, p. 54).

Intergenerational ethics is concerned with the moral obligations of the current generation towards future generations. In the context of sustainable development, this framework emphasizes the importance of considering the long-term impacts of our actions and policies on future inhabitants of the planet. By incorporating intergenerational ethics into climate change education, educators can instill a sense of responsibility for preserving natural resources and ensuring a sustainable world for future generations (O'Neill, 1993, p. 87).

Ethical theories and frameworks offer valuable insights into promoting sustainable development and climate change education. By integrating utilitarianism, deontology, virtue ethics, environmental ethics, and intergenerational ethics into educational initiatives, we can cultivate a generation of environmentally conscious individuals who understand the importance of making ethically responsible choices to safeguard our planet for the well-being of all living beings. As we navigate the complexities of climate change and sustainability, these ethical perspectives provide guidance for fostering a more harmonious and sustainable relationship between humanity and the environment.

PRECAUTIONARY PRINCIPLE AND INTERGENERATIONAL JUSTICE, IN CLIMATE CHANGE EDUCATION

The application of ethical frameworks, such as the precautionary principle and intergenerational justice, in climate change education is essential to instill a sense of responsibility and urgency in tackling the climate crisis. By promoting proactive measures and considering the interests of future generations, climate change education becomes a powerful tool to inspire action and foster a more sustainable and just future for all.

The precautionary principle is an ethical guideline that advocates taking preventive action in the face of uncertain risks to protect human health and the environment (O'Riordan & Cameron, 1994, p. 32). In climate change education, this principle underscores the need for proactive measures to address climate change and its potential consequences. Climate change education can raise awareness about the risks associated with climate change, such as extreme weather events, rising sea levels, and loss of biodiversity (IPCC, 2014, p. 117). Understanding these risks can motivate individuals to support and advocate for climate change mitigation.

The precautionary principle encourages the adoption of mitigation strategies even in the absence of complete scientific certainty (Gardiner, 2006, p. 481). Climate change education can emphasize the importance of reducing greenhouse gas emissions and transitioning to renewable energy sources as precautionary measures. Educating about adaptation measures is crucial, as climate change's effects are already evident in many regions (Adger et al., 2007, p. 876). By preparing communities and individuals to adapt to changing conditions, the precautionary principle can help minimize the adverse impacts on vulnerable populations.

Intergenerational justice is an ethical concept that emphasizes the moral responsibility of the present generation to protect the well-being and rights of future generations (Gosseries, 2008, p. 68). Climate change, with its long-term consequences, exemplifies an intergenerational justice issue. Climate change education can expand the time horizon of decision-making, encouraging individuals to consider the impacts of their actions on future generations (Gardiner, 2011, p. 145). It fosters a sense of responsibility to preserve the environment for the benefit of those yet to come. By incorporating intergenerational justice principles, climate change education can promote sustainable practices that balance current needs with the needs of future generations (Barry, 2012, p. 215). This includes advocating for sustainable agriculture, responsible consumption, and conservation efforts.

Understanding intergenerational justice can motivate individuals to advocate for stronger climate change policies that prioritize long-term sustainability over short-term gains (Page, 2008, p. 390). This can lead to increased political will to address climate change effectively.

THE INTERSECTION OF LOGISTICS, SUSTAINABILITY, AND PHILOSOPHY IN PROMOTING ENVIRONMENTALLY RESPONSIBLE SUPPLY CHAINS

In recent years, there has been growing recognition of the crucial role that logistics, sustainability, and philosophy play in promoting environmentally responsible supply chains. Supply chain logistics involves the coordination of various activities, including transportation, warehousing, and inventory management, which significantly impact the environmental footprint of businesses (Smith & Johnson, 2020, p. 560). By adopting sustainable practices in logistics operations, such as implementing eco-friendly transportation modes and optimizing routes to reduce emissions, companies can make substantial contributions to environmental conservation.

The concept of sustainability further reinforces the importance of environmentally responsible supply chains. Sustainability emphasizes meeting the needs of the present without compromising the ability of future generations to meet their needs (Brown, 2018, p. 73). By integrating sustainable practices into supply chain management, such as sourcing materials ethically and reducing waste generation, companies can contribute to preserving natural resources and minimizing their overall environmental impact (Jones, 2019, p. 112).

Philosophy also plays a pivotal role in shaping attitudes and values within the business community (Johnson, 2017, p. 85). Adopting a philosophical approach that prioritizes ecological integrity and responsible decision-making can lead to profound changes in how supply chains are structured and operated. For instance, incorporating environmental ethics into the decision-making process may result in choosing suppliers with strong sustainability commitments or favoring locally sourced materials to reduce carbon emissions associated with long-distance transportation (Roberts, 2016, p. 501).

The intersection of logistics, sustainability, and philosophy provides a fertile ground for promoting environmentally responsible supply chains. By acknowledging the significance of each component and integrating them harmoniously, businesses can drive positive change and work towards a more sustainable and eco-friendly future (Miller et al., 2021, p. 70).

Logistics plays a vital role in reducing greenhouse gas emissions, optimizing resource utilization, and promoting circular economy principles. By strategically managing the movement of goods, information, and resources across the supply chain, logistics can contribute significantly to sustainability efforts and mitigate environmental impacts. Here's a discussion of how logistics achieves these goals.

One of the major contributors to greenhouse gas emissions is the transportation of goods. Logistics can help reduce emissions by optimizing transportation routes, using fuel-efficient vehicles, and promoting alternative transportation methods, such as electric or hybrid vehicles (Abe, 2018). By consolidating shipments and adopting intermodal transportation, logistics can further minimize emissions per unit of cargo transported (Simpson et al., 2019).

Last-mile delivery is a crucial phase of logistics where emissions can be high due to frequent stops and short distances. Implementing innovative delivery methods like crowd shipping, bike couriers, or drone delivery can significantly reduce emissions in urban areas (Zhang et al., 2020).

Efficient logistics planning and inventory management can reduce the need for expedited shipments, thereby reducing emissions associated with rush deliveries (Rogers & Tibben-Lembke, 2016).

Effective logistics helps in managing inventory levels efficiently, reducing excess stock and minimizing waste. By using advanced technologies like RFID, IoT, and analytics, logistics can enable real-time monitoring and optimize inventory holding levels (Kim et al., 2018).

Logistics also plays a role in handling product returns and waste disposal. Implementing effective reverse logistics processes allows for the recovery and recycling of materials, reducing the consumption of raw resources and promoting resource circularity (Genovese et al., 2017). Logistics facilitates the design and implementation of closed-loop supply chains, where products are refurbished, remanufactured, or recycled after use. This approach reduces waste generation and extends the lifecycle of products, aligning with circular economy principles (Govindan et al., 2015).

Logistics encourages collaboration among different stakeholders, such as manufacturers, retailers, and waste management companies. Through these collaborative networks, surplus materials and by-products can be exchanged, promoting resource sharing and reducing waste (Russo & Comi, 2019). Logistics can support the transition from a traditional ownership model to a PaaS model, where consumers lease or rent products rather than purchasing them outright. This shift incentivizes manufacturers to design products with longevity and recyclability in mind, leading to reduced resource consumption and waste generation (Telenko et al., 2019).

Logistics plays a pivotal role in reducing greenhouse gas emissions, optimizing resource utilization, and promoting circular economy principles. By embracing sustainable practices and leveraging innovative technologies, the logistics industry can drive positive environmental change while meeting the demands of an increasingly environmentally conscious world.

THE ROLE OF PHILOSOPHY IN SHAPING SUSTAINABLE LOGISTICS PRACTICES AND POLICIES

The development and implementation of sustainable logistics practices are crucial in addressing the environmental and social impacts of global supply chains. Philosophical theories such as environmental ethics, sustainability ethics, and social responsibility play significant roles in shaping and guiding sustainable logistics initiatives.

Environmental ethics promotes the sustainable use of resources in logistics operations (DesJardins, 2013, p. 17). Companies may adopt strategies like reducing waste, optimizing transportation routes, and using eco-friendly packaging materials to minimize their environmental footprint. Sustainable logistics practices can be influenced by the need to protect biodiversity and ecosystems (Taylor, 1981, p. 135). Companies may choose suppliers and transportation modes that have lower impacts on sensitive ecosystems. Sustainability ethics encourages businesses to consider the entire lifecycle of products and services in logistics (Savaskan et al., 2004, p. 865). Companies may conduct life cycle assessments to identify areas where sustainability improvements can be made.

Adopting circular economy principles, where materials are recycled and reused, aligns with sustainability ethics (Kirchherr et al., 2017, p. 757). Companies can implement reverse logistics processes to recover, refurbish, and recycle products. Social responsibility emphasizes fair labor practices in logistics operations (Carter & Rogers, 2008, p. 56). Companies may promote safe working conditions, fair wages, and e Community Engagement: Socially responsible logistics practices involve engaging with local communities to address their concerns and contribute positively to their development (Cruz & Marques, 2011, p. 100). Companies may participate in community projects or support local initiatives. Employee well-being within their supply chains.

INTEGRATION OF PHILOSOPHICAL PRINCIPLES INTO GREEN SUPPLY CHAIN MANAGEMENT, REVERSE LOGISTICS, AND SUSTAINABLE TRANSPORTATION STRATEGIES

The integration of philosophical principles into green supply chain management, reverse logistics, and sustainable transportation strategies is crucial in promoting environmentally and socially responsible practices within these domains.

Green supply chain management emphasizes the moral responsibility of businesses to protect the environment and minimize their ecological footprint. By incorporating environmental ethics, companies can prioritize eco-friendly practices, such as sourcing materials from sustainable suppliers, using energy-efficient manufacturing processes, and reducing waste generation (Ciliberti et al., 2008, p. 97). This approach aligns with the principle of valuing nature's intrinsic worth and respecting the interconnectedness of all living beings.

Sustainability ethics encourages businesses to adopt long-term thinking and consider the impacts of their actions on future generations. In green supply chain management, this translates into implementing strategies that ensure the continued availability of resources, minimize negative environmental impacts, and promote social and economic well-being (Walker et al., 2019, p. 250). Emphasizing sustainability ethics in supply chain decision-making can lead to more resilient and responsible operations.

Social responsibility in green supply chain management involves recognizing and addressing the welfare of workers, communities, and other stakeholders affected by supply chain activities. Businesses can integrate social responsibility by promoting fair labour practices, supporting local communities, and engaging in philanthropic efforts (Seuring & Müller, 2008, p. 114). Considering the well-being of all stakeholders aligns with the ethical principle of acting in the best interests of society.

REVERSE LOGISTICS

The application of environmental ethics in reverse logistics involves responsible handling and disposal of end-of-life products or materials. By adopting practices like recycling, refurbishing, and remanufacturing, businesses can minimize waste and prevent harmful materials from polluting the environment (Fleischmann et al., 2001, p. 88). This approach reflects the ethical value of preserving nature's integrity and reducing human impacts on ecosystems.

Embracing sustainability ethics in reverse logistics entails designing products with recyclability and reusability in mind. Companies can implement circular economy principles, where products are designed for easy disassembly and recovery of valuable materials (Guide & Van Wassenhove, 2001, p. 201). This approach supports the principle of sustainable development by ensuring the efficient use of resources throughout a product's life cycle.

Social responsibility in reverse logistics involves considering the social implications of end-of-life product management. Businesses can engage in programs to responsibly dispose of hazardous materials, support recycling initiatives in local communities, and create job opportunities in the recycling industry (Braunscheidel et al., 2018, p. 77). This aligns with the ethical principle of promoting the well-being of individuals and communities affected by reverse logistics activities.

SUSTAINABLE TRANSPORTATION STRATEGIES

Sustainable transportation strategies rooted in environmental ethics seek to reduce greenhouse gas emissions and mitigate the environmental impacts of transportation. Companies can invest in low-carbon transportation modes such as electric vehicles, fuel-efficient fleets, and alternative fuels (Button, 2017, p. 24). By doing so, they demonstrate a commitment to environmental protection and reducing their carbon footprint.

Sustainability ethics in transportation strategies involves considering the broader ecological and social consequences of transportation decisions. Businesses can adopt a multi-modal approach, favoring transportation modes with lower environmental impacts and assessing the social benefits of transportation initiatives (Cruz et al., 2019, p. 3). This aligns with the ethical principle of achieving a balance between economic, social, and environmental interests.

Socially responsible transportation strategies involve ensuring that transportation operations prioritize safety, accessibility, and the well-being of communities along transport routes (Barla et al., 2017, p. 82). Companies can engage with local communities to address their transportation needs and concerns, contributing positively to their development. This approach reflects the ethical value of promoting social welfare and inclusivity.

The integration of philosophical principles into green supply chain management, reverse logistics, and sustainable transportation strategies is vital in promoting responsible and sustainable practices within these domains. By aligning business decisions with environmental ethics, sustainability ethics, and social responsibility, companies can contribute to a more sustainable and ethical future, where ecological, social, and economic interests are harmonized for the well-being of present and future generations.

ETHICAL CONSIDERATIONS IN RESOURCE ALLOCATION AND WASTE MANAGEMENT

Resource allocation and waste management are critical aspects of sustainable development and environmental stewardship. Ethical considerations play a pivotal role in guiding decision-making processes to ensure equitable, responsible, and sustainable use of resources, as well as effective waste management practices.

Environmental ethics emphasizes the moral responsibility of individuals and societies to protect the environment and respect the intrinsic value of nature (DesJardins, 2013, p. 26). In resource allocation, this ethical framework calls for sustainable practices that prioritize the long-term health of ecosystems and biodiversity. It raises questions about the ethical implications of over-extracting finite resources and the need to promote renewable and regenerative resource use.

Ethical considerations in resource allocation include social justice concerns, particularly regarding access to resources and their distribution. Addressing disparities in resource access and ensuring equitable allocation align with the principles of fairness and social responsibility (Rawls, 1971, p. 7). Ethical resource allocation aims to minimize environmental and social inequalities and provide equal opportunities for all individuals and communities.

Environmental ethics informs waste management practices by emphasizing the responsibility to minimize environmental harm and pollution caused by waste (Feinberg, 1970, p. 341). Ethical waste management involves adopting strategies like recycling, composting, and waste-to-energy conversion to reduce the environmental impact of waste disposal.

Ethical waste management recognizes the social implications of waste disposal, particularly the potential health hazards and environmental justice concerns for marginalized communities (Gupta & Noe, 2016, p. 42). Responsible waste management involves considering the well-being of communities affected by waste disposal facilities and ensuring proper waste treatment and disposal to protect public health. Ethical considerations in waste management also encompass the impact of waste on future generations (Sandler, 2010, p. 5). Ethical waste practices prioritize minimizing waste generation and ensuring that current disposal methods do not burden future generations with long-lasting environmental and health consequences.

Prioritizing resource efficiency, waste reduction, and responsible disposal practices is essential in promoting sustainable development and environmental conservation. However, the pursuit of these goals raises ethical considerations that warrant careful examination.

CARBON FOOTPRINT REDUCTION AND ETHICAL ACCOUNTABILITY

Carbon footprint reduction is a critical aspect of mitigating climate change and promoting environmental sustainability. As individuals, organizations, and governments strive to lower their carbon emissions, ethical accountability becomes central in recognizing the moral responsibility to address the climate crisis.

Carbon footprint reduction aligns with this ethical framework, as it aims to minimize greenhouse gas emissions, mitigating the impacts of climate change on ecosystems and future generations (DesJardins, 2013, p. 26). Ethical considerations prompt individuals and entities to take action to safeguard the planet's well-being. The choices made today regarding carbon emissions will have far-reaching consequences for the well-being of those yet to come (Sandler, 2010, p. 5). Ethical intergenerational considerations necessitate carbon reduction efforts that protect the interests of future inhabitants and preserve a sustainable planet.

Corporate Social Responsibility (CSR) and Carbon Footprint Reduction involve a corporation's commitment to consider the interests of stakeholders, including employees, communities, and the environment. Carbon footprint reduction is a vital component of CSR, as corporations are ethically responsible for minimizing their environmental impact (Crane et al., 2019, p. 8). Ethical accountability drives corporations to adopt sustainable practices and reduce their carbon emissions to protect both the planet and their stakeholders. Ethical accountability in carbon footprint reduction is demonstrated through ethical leadership within organizations. Ethical leaders prioritize sustainability and inspire their teams to implement carbon reduction strategies (Lambert, 2019, p. 35). Ethical leadership fosters a culture of responsibility and environmental consciousness, driving the organization's commitment to reducing its carbon footprint. Individuals have an ethical obligation to take personal responsibility for their carbon footprint. Making conscious choices, such as reducing energy consumption, using public transportation, and adopting eco-friendly habits, reflects ethical accountability towards the environment and society (Johnson, 2015, p. 52).

Ethical accountability extends beyond individual actions to collective efforts in reducing carbon footprints. Individuals have a moral responsibility to advocate for policy changes, support sustainable practices, and engage in collective action to address climate change (Gardiner, 2011, p. 163). Ethical considerations prompt individuals to recognize their role in a larger global community striving for carbon reduction.

ETHICAL DILEMMAS AND CHALLENGES FACED BY NGOS IN THE CONTEXT OF CLIMATE CHANGE EDUCATION AND SUSTAINABLE DEVELOPMENT

Non-Governmental Organizations (NGOs) play a crucial role in advocating for climate change education and promoting sustainable development initiatives. However, in their pursuit of these noble goals, NGOs encounter various ethical dilemmas and challenges. Ethical Dilemmas in Climate Change Education.

NGOs involved in climate change education face ethical dilemmas regarding the accuracy and transparency of information presented to the public. Maintaining scientific integrity while communicating complex climate science can be challenging (Oreskes, 2004, p. 372). Striking a balance between engaging the public and avoiding oversimplification can be ethically demanding. NGOs often encounter ethical dilemmas in framing and messaging climate change issues. The choice of language and imagery can influence public perceptions and responses (Nisbet & Myers, 2007, p. 359). Ethical considerations involve avoiding fear-mongering or underplaying the severity of climate change impacts.

They may face ethical challenges in deciding where to allocate limited resources. Prioritizing certain projects over others can lead to trade-offs between social, economic, and environmental goals (De Kadt, 1979, p. 82). Balancing competing interests while ensuring maximum positive impact is an ethical dilemma. Sustainable development efforts must involve the affected communities and respect their rights and interests (Kaplinsky, 2000, p. 26). NGOs may encounter ethical challenges in ensuring inclusivity and meaningful participation of all stakeholders in decision-making processes.

NGOs face ethical dilemmas when receiving funding from donors with conflicting interests or agendas (Miller, 2004, p. 196). Balancing financial sustainability while maintaining independence and prioritizing the organization's mission requires careful ethical considerations. Ethical challenges arise in maintaining transparency and accountability in the use of funds and project outcomes (Amaeshi & Amao, 2012, p. 170). NGOs must ensure that donors' trust is upheld by demonstrating their commitment to ethical practices and responsible use of resources.

NGOs face the ethical challenge of involving all stakeholders, including local communities, in decision-making processes. The preferences and needs of different groups may diverge, making it difficult to satisfy all parties (Reed, 2008, p. 57). Inclusive and participatory decision-making processes are essential to address this challenge. Ensuring the representation of diverse perspectives and empowering local communities to have a meaningful say in development initiatives can foster ethical and sustainable outcomes (Reed, 2008, p. 60).

NGOs often encounter ethical dilemmas when balancing the need for environmental conservation with economic development goals. Economic interests may promote projects that have adverse environmental impacts (Bartelmus, 1999, p. 7). A comprehensive assessment of the environmental and social impacts of development projects is crucial. Adopting sustainability frameworks that incorporate environmental protection and social welfare can guide decision-making in ways that promote both development and conservation goals (Bartelmus, 1999, p. 12).

NGOs face ethical dilemmas in allocating resources among various projects and initiatives. Prioritizing one community or environmental cause over others may lead to inequitable outcomes (Rogerson, 2006, p. 478). Transparent resource allocation processes, guided by ethical principles, can address this challenge. NGOs should prioritize projects that have the greatest positive impact on local communities and the environment, considering both immediate and long-term benefits (Rogerson, 2006, p. 483).

NGOs face ethical challenges when collaborating with corporations and governments. These entities may have different interests, including economic gain or political considerations, which can affect the NGO's independence and mission (Wapner, 1996, p. 159). Maintaining integrity and independence in collaborations is essential. NGOs should establish clear boundaries and principles for engagement to ensure that their actions align with their ethical values and goals (Wapner, 1996, p. 163).

CHALLENGES OF PRIORITIZING CLIMATE CHANGE EDUCATION IN RESOURCE-CONSTRAINED SETTINGS AND ADDRESSING THE DIVERSE NEEDS OF DIFFERENT COMMUNITIES

Resource-constrained settings may lack sufficient infrastructure and access to technology, hindering the dissemination of climate change information through modern educational channels (Kruse et al., 2020, p. 8). Adopting low-tech and community-based educational approaches, such as workshops, community meetings, and radio programs, can be effective in reaching diverse populations in areas with limited technology access (Kruse et al., 2020, p. 11).

Diverse communities in resource-constrained settings may have different languages, cultural norms, and beliefs. Standardized educational materials may not resonate with all groups, leading to reduced efficacy of climate change education initiatives (Norgaard, 2011, p. 33). Climate change education materials should be culturally and linguistically sensitive. Engaging local community leaders and educators can ensure that educational content is tailored to specific cultural contexts, improving its relevance and impact (Norgaard, 2011, p. 35).

In resource-constrained settings, communities may have low awareness of climate change and its impacts, leading to a lack of interest in climate change education (Hountondji et al., 2020, p. 8). Conducting awareness campaigns to highlight the relevance of climate change to local livelihoods can generate interest and support for climate change education. Demonstrating how sustainable practices can benefit communities can motivate active participation (Hountondji et al., 2020, p. 12).

Ensuring inclusivity and gender equality in climate change education is essential. However, women and marginalized groups may face barriers to accessing education and participating in decision-making processes (Dossou-Yovo et al., 2019, p. 219). Implementing gender-sensitive and inclusive educational approaches can address this challenge. Engaging women and marginalized groups in the design and implementation of climate change education initiatives can promote equity and social empowerment (Dossou-Yovo et al., 2019, p. 223).

Resource allocation processes often face disparities in resource distribution among different regions and communities. Historically marginalized groups may have limited access to resources compared to more privileged communities (Kaplan, 2019, p. 54).

Promoting social equity in resource allocation requires overcoming systemic biases and ensuring fair and just distribution of resources. Identifying and addressing the needs of vulnerable and disadvantaged populations is essential (Leichenko & O'Brien, 2008, p. 124). Resource allocation processes may be susceptible to undue influence from powerful entities, such as corporations or governments. These influences can skew resource allocation in favor of specific interests, undermining social equity (Ebrahim & Rangan, 2010, p. 15).

Political and policy biases may influence resource allocation processes. Partisan interests can impact decisions, potentially exacerbating social inequities (Harvey, 2005, p. 22). Ensuring public participation and representation in resource allocation processes can be challenging. A lack of inclusive decision-making can lead to decisions that do not reflect the diverse needs of the population (Reed, 2008, p. 2417).

POTENTIAL CONFLICTS OF INTEREST, POWER DYNAMICS, AND ACCOUNTABILITY ISSUES IN PARTNERSHIPS

Partnerships play a crucial role in implementing climate change education initiatives, particularly in resource-constrained settings. While collaborations can bring valuable expertise and resources, they can also lead to potential conflicts of interest, power imbalances, and accountability challenges.

Climate change education partnerships may involve stakeholders with different objectives and interests. For instance, corporate partners may prioritize their brand reputation, while NGOs focus on achieving sustainable outcomes (Yaro, 2018, p. 9). Conflicts of interest can hinder decision-making and compromise the integrity of educational content. There may be pressures to downplay certain environmental concerns or present information in a biased manner (Yaro, 2018, p. 12). Climate change education partnerships often involve organizations with unequal power and influence. Larger entities, such as governments or multinational corporations, may dominate decision-making processes (Reed, 2008, p. 57). Unequal power dynamics can marginalize the voices of smaller stakeholders, such as local communities, leading to initiatives that may not adequately address their needs and priorities (Reed, 2008, p. 60).

In partnerships, establishing clear lines of accountability can be challenging. Different stakeholders may have varying levels of responsibility for project outcomes (Leal Filho et al., 2018, p. 1756). Unclear accountability can lead to finger-pointing in case of project failures or mismanagement. Lack of accountability may also result in suboptimal use of resources and inefficient project implementation (Leal Filho et al., 2018, p. 1759). Corporate partners may engage in greenwashing - portraying themselves as environmentally responsible without implementing substantive changes (Brown & Kasser, 2005, p. 517). Greenwashing can undermine the credibility of climate change education initiatives and mislead the public. It may divert attention from the urgent need for genuine sustainability efforts (Brown & Kasser, 2005, p. 522).

NGOs working in climate change education partnerships may face the risk of being co-opted by powerful entities. They might align their agendas with those of corporate or governmental partners at the expense of their original mission (Ebrahim & Rangan, 2010, p. 123). Co-optation compromises the independence of NGOs, affecting their ability to advocate for sustainable practices and hold partners accountable. It can erode public trust and diminish the effectiveness of climate change education (Ebrahim & Rangan, 2010, p. 129).

DECISION-MAKING PROCESSES IN SUSTAINABLE DEVELOPMENT PROJECTS

Examining the ethical implications of decision-making processes in sustainable development projects is of paramount importance (Gupta & Bleischwitz, 2018, p. 367). Sustainable development projects aim to balance economic, social, and environmental objectives to meet the needs of the present generation without compromising the ability of future generations to meet their own needs (Thompson & Tansey, 2017, p. 511). However, these projects often involve complex trade-offs and decision-making dilemmas that can have significant ethical ramifications (Wiek et al., 2014, p. 7865). Here are some key ethical implications to consider:

Sustainable development projects must address issues of equity and justice, ensuring that benefits and burdens are distributed fairly among various stakeholders, including marginalized communities (Gupta & Bleischwitz, 2018, p. 368). Decision-makers must be vigilant in avoiding environmental injustice and not perpetuating social inequalities while implementing projects (Gupta & Bleischwitz, 2018, p. 368).

Many sustainable development projects involve interactions with local communities and indigenous groups. Respecting the principle of informed consent is crucial, ensuring that affected communities have access to accurate information, are adequately consulted, and have the right to participate in the decision-making process (Florini, 2016, p. 1).

Sustainable development inherently involves long-term thinking. Decision-makers must consider the interests and well-being of future generations and not sacrifice their needs for short-term gains or convenience (Thompson & Tansey, 2017, p. 512). Ethical decision-making requires safeguarding the integrity of the natural environment. Projects must not compromise vital ecosystems or lead to irreversible damage to biodiversity and ecological systems (Wapner, 2016, p. 3). Decision-makers must be transparent in their actions and accountable for the consequences of their choices. This involves open communication with stakeholders, sharing information, and being willing to adjust strategies based on feedback (Wiek et al., 2014, p. 7872).

Ethical dilemmas can arise when striking a balance between sustainable practices and economic development. It's crucial to ensure that development activities do not undermine sustainability efforts or lead to negative externalities that harm vulnerable communities or the environment (Jamali et al., 2018, p. 357).

Many sustainable development projects rely on innovative technologies. Ethical considerations include ensuring that these technologies are safe, not harmful to human health or the environment and that they are introduced responsibly (Jasanoff & Kim, 2015, p. 1). Corporate Social Responsibility: Involvement of private corporations in sustainable development projects raises ethical issues related to their motivations, potential conflicts of interest, and the sincerity of their commitment to social and environmental objectives (Kolk, 2016, p. 23).

Decision-making in sustainable development should embrace the precautionary principle, being cautious when evidence is uncertain and avoiding actions that may cause severe or irreversible harm (Holling, 2018, p. 138). Ethical considerations are paramount when making decisions in sustainable development projects (Wapner, 2016, p. 5). By recognizing and addressing these implications, decision-makers can ensure that projects align with principles of justice, intergenerational equity, environmental responsibility, and human well-being, fostering a more sustainable and inclusive future.

CONFLICTS BETWEEN ECONOMIC INTERESTS, SOCIAL JUSTICE, AND ENVIRONMENTAL PRESERVATION IN THE CONTEXT OF SUSTAINABLE DEVELOPMENT

Analyzing the potential conflicts between economic interests, social justice, and environmental preservation in the context of sustainable development is essential for achieving a balanced and inclusive approach to progress (Gupta & Bleischwitz, 2018, p. 367). Sustainable development seeks to meet the needs of the present without compromising the ability of future generations to meet their own needs (Thompson & Tansey, 2017, p. 511). However, pursuing economic growth, social equity, and environmental protection simultaneously can lead to inherent conflicts (Wiek et al., 2014, p. 7865). Here are some key points to consider when examining these potential conflicts:

Economic activities often require the extraction of natural resources, leading to environmental degradation, habitat loss, and pollution (Dresner, 2008, p. 29). Industrial development and infrastructure projects may encroach on ecologically sensitive areas, putting ecosystems at risk (Elkington, 1997, p. 1). Economic development can exacerbate social inequalities by benefiting certain groups while marginalizing others (Banerjee & Bell, 2007, p. 1742). Profit-driven endeavors might neglect the well-being of vulnerable communities, particularly in areas with limited resources and infrastructure (Chambers, 1983, p. 629). Conservation efforts may lead to the displacement and disenfranchisement of indigenous and local communities that rely on natural resources for their livelihood (West et al., 2006, p. 251). Strict environmental regulations might hinder economic opportunities and limit access to resources for disadvantaged populations (Agrawal & Gibson, 1999, p. 629).

Integrative Approaches: Sustainable development requires an integrated approach that recognizes the interconnectedness of economic, social, and environmental aspects (WCED, 1987, p. 43). Inclusive Decision-Making: Involving diverse stakeholders, including local communities, in the decision-making process helps address conflicts and fosters equitable solutions (Sen, 1999, p. 1753). Sustainable Business Practices: Encouraging businesses to adopt environmentally friendly and socially responsible practices contributes to reducing conflicts (Eccles & Krzus, 2010, p. 1).

Implementing well-designed policies and regulations can create incentives for businesses to align their economic interests with social and environmental goals (Porter & van der Linde, 1995, p. 97). Navigating the potential conflicts between economic interests, social justice, and environmental preservation is crucial in sustainable development. By adopting integrated approaches, engaging diverse stakeholders, promoting sustainable business practices, and implementing effective policies, societies can work towards a more harmonious and sustainable future that respects the principles of social equity, economic growth, and environmental protection.

TENSIONS AND TRADE-OFFS THAT ARISE WHEN ECONOMIC INTERESTS CLASH WITH SOCIAL JUSTICE AND ENVIRONMENTAL CONCERNS

When economic interests clash with social justice and environmental concerns, tensions and trade-offs arise that can be challenging to resolve (Gupta & Bleischwitz, 2018, p. 367). Balancing the pursuit of economic growth and development with the objectives of social equity and environmental sustainability requires careful consideration of the trade-offs involved (Thompson & Tansey, 2017, p. 511). Economic activities that generate jobs and livelihoods may also have adverse environmental consequences, such as pollution and habitat destruction (Dresner, 2008, p. 29). Striking a balance between economic development and environmental preservation is crucial (Elkington, 1997, p. 1).

Extractive industries, such as mining and logging, often involve the exploitation of natural resources found on indigenous or ancestral lands. This raises ethical concerns regarding the rights and well-being of indigenous communities (Anderson et al., 2017, p. 66). Economic growth may not necessarily result in equitable distribution of wealth, leading to income disparities and social injustices (Banerjee & Bell, 2007, p. 1742).

Building infrastructure, such as roads and dams, can be essential for economic progress, but it may also lead to the destruction of critical ecosystems and habitats (Nascimento & Fearnside, 2020, p. 5790). Pursuing short-term economic gains without considering long-term environmental and social impacts can lead to the depletion of natural resources and the degradation of ecosystems (Kallis, 2011, p. 873). Implementing strong economic regulations to address social and environmental concerns may be perceived as impeding market freedom and hindering economic growth (Braithwaite, 2017, p. 1). Providing essential services, such as electricity and water, to underserved communities may involve the use of energy sources and water management practices with negative environmental consequences (Daly, 2015, p. 447). Maximizing profits in the private sector may not always align with achieving social welfare and addressing the needs of vulnerable populations (Hart, 2005, p. 1).

Addressing these tensions and trade-offs requires a comprehensive and multidimensional approach that takes into account the interests of all stakeholders and considers the long-term impacts of decisions. Engaging in transparent and inclusive decision-making processes, incorporating the principles of sustainable development, and adopting responsible business practices can help mitigate conflicts and create a more harmonious path toward economic, social, and environmental well-being.

THE ROLE OF POWER DYNAMICS AND DECISION-MAKING IN SUSTAINABLE DEVELOPMENT

Power dynamics and decision-making play a critical role in shaping the outcomes of sustainable development initiatives (Gupta & Bleischwitz, 2018, p. 367). Sustainable development aims to address the complex interplay between economic, social, and environmental dimensions, but achieving this balance requires navigating various power structures and ensuring equitable participation in decision-making processes (Thompson & Tansey, 2017, p. 511).

1. **Distribution of Power**: Power imbalances, whether at the local, national, or global level, can influence the prioritization and implementation of sustainable development projects (Ribot, 2004, p. 32). Decision-making processes may favor the interests of powerful stakeholders, leading to the marginalization of vulnerable groups and reinforcing existing inequalities.
2. **Inclusive Participation**: Meaningful and inclusive participation of all stakeholders is essential for sustainable development (Cooke & Kothari, 2001, p. 102). Involving local communities, indigenous peoples, civil society organizations, and marginalized groups in decision-making empowers them to influence the direction and outcomes of development initiatives.
3. **Access to Information**: Access to relevant information is a crucial element of informed decision-making (Leach et al., 2019, p. 287). Ensuring transparency and providing accessible information empowers communities and stakeholders to engage effectively in sustainable development processes.
4. **Accountability and Governance**: Effective governance structures are necessary to hold decision-makers accountable for their actions and promote responsible decision-making (Gaventa & Barrett, 2010, p. 15). Transparent and accountable governance fosters trust and legitimacy in sustainable development initiatives.
5. **Conflicting Interests**: Sustainable development often involves reconciling conflicting interests among various stakeholders (Pattberg & Stripple, 2008, p. 367). Balancing economic growth, social justice, and environmental protection requires navigating divergent priorities and finding common ground.
6. **Global Power Relations**: At the global level, power dynamics influence the allocation of resources, technology transfer, and policy-making in sustainable development (Biermann et al., 2009, p. 16). Addressing global challenges like climate change requires cooperation and collaboration among nations with varying levels of power and influence.
7. **Capacity Building**: Enhancing the capacity of local communities and institutions to participate effectively in decision-making is crucial for sustainable development (Kuehnast et al., 2017, p. 8). Empowering individuals and

organizations with knowledge and skills enables them to influence development processes positively.

8. **Institutional and Policy Design**: The design of institutions and policies can either reinforce or challenge existing power structures (Keck & Sikkink, 1998, p. 367). Institutional arrangements that promote inclusive decision-making and social justice are more likely to foster sustainable development outcomes.

Addressing these tensions and trade-offs requires a comprehensive and multidimensional approach that takes into account the interests of all stakeholders and considers the long-term impacts of decisions. Engaging in transparent and inclusive decision-making processes, incorporating the principles of sustainable development, and adopting responsible business practices can help mitigate conflicts and create a more harmonious path toward economic, social, and environmental well-being.

ENGAGING WITH PHILOSOPHICAL PERSPECTIVES TO ADDRESS CULTURAL, SOCIAL, AND ECONOMIC ASPECTS OF CLIMATE CHANGE EDUCATION

Engaging with philosophical perspectives is essential for addressing the cultural, social, and economic aspects of climate change education in a comprehensive and meaningful manner (Gardiner, 2006, p. 397). Philosophical insights can enrich climate change education by providing critical analysis, ethical frameworks, and holistic perspectives on the complex challenges posed by climate change (Jamieson, 2014, p. 1).

Philosophical perspectives on environmental ethics delve into the moral principles and values that guide human interactions with the natural world (Jamieson, 2014, p. 1). By exploring ethical responsibilities towards the environment and future generations, climate change education can foster a sense of environmental stewardship and a commitment to sustainable practices. Philosophical inquiry into epistemology helps learners critically assess the sources of climate change information and distinguish between reliable scientific evidence and misinformation (Hulme, 2010, p. 1). Encouraging critical thinking skills can empower individuals to make informed decisions and engage in constructive climate action.

A philosophical exploration of cultural perspectives on climate change can lead to a deeper understanding of diverse worldviews and belief systems regarding nature and the environment (Escobar, 1996, p. 325). Recognizing and respecting different cultural values can enhance the effectiveness of climate change education across various communities. Philosophical discussions on social justice and climate change highlight the disproportionate impacts of climate change on vulnerable communities

(Gardiner, 2006, p. 397). Climate change education can address climate justice, equity, and inclusion issues, fostering a commitment to address socioeconomic disparities.

Examining economic philosophies and their implications for climate change can illuminate how economic systems influence environmental behaviors (Sandel, 2012, p. 1). Climate change education can explore sustainable economic models that prioritize environmental protection and social well-being. Philosophical reflection on human values and aspirations can help individuals reassess their consumerist and materialistic tendencies, encouraging a shift towards more sustainable lifestyles and consumption patterns (Norton, 1987, p. 1).

Philosophical inquiry into environmental aesthetics explores the relationship between humans and nature in terms of beauty, art, and aesthetic experience (Carlson, 2000, p. 1). Integrating aesthetics into climate change education can evoke emotional connections with nature, motivating environmental conservation efforts. Philosophical contemplation of existential questions related to climate change can prompt individuals to confront the scale and urgency of environmental challenges (O'Brien & Hochachka, 2010, p. 1). Climate change education can engage learners in exploring personal and collective responsibility in the face of global crises.

By engaging with these philosophical perspectives, climate change education can move beyond merely presenting scientific facts and data (Gardiner, 2006, p. 397). It can foster a deeper understanding of climate change's ethical, cultural, social, and economic dimensions, encouraging individuals to become informed, empathetic, and responsible stewards of the planet.

CONSIDERATIONS IN THE USE OF ICT FRAMEWORKS AND LEARNING RESOURCES FOR SUSTAINABLE EDUCATION

Ethical considerations in the use of Information and Communication Technology (ICT) frameworks and learning resources for sustainable education are crucial for fostering responsible and effective teaching practices (Gupta & Bleischwitz, 2018, p. 367). As educators and learners increasingly rely on digital tools and resources, ethical awareness becomes essential to ensure that the use of ICT aligns with the principles of sustainability and promotes positive societal and environmental impacts (Thompson & Tansey, 2017, p. 511).

Ensuring equitable access to ICT resources and learning opportunities is an ethical imperative (Hargittai, 2002, p. 1). Addressing the digital divide and providing inclusive access to technology for disadvantaged communities can empower individuals to participate in sustainable education initiatives. ICT infrastructure, including data centers and electronic devices, has an ecological footprint. Ethical considerations involve minimizing energy consumption, using renewable energy sources, and responsibly disposing of e-waste (Pires et al., 2020, p. 1). The collection and storage of personal data in ICT systems require strict adherence to privacy and data protection regulations (Milberg et al., 2000, p. 35). Educators must be mindful of privacy concerns and safeguard learners' sensitive information.

Ethical use of ICT resources involves respecting copyright laws and promoting open access to educational materials whenever possible (UNESCO, 2019, p. 1). Properly attributing and acknowledging content creators is essential in sustainable education practices. Promoting digital literacy and critical thinking skills is vital for empowering learners to navigate digital spaces responsibly and discern reliable information from misinformation (Purcell et al., 2012, p. 1).

ICT resources and learning materials should respect diverse cultural perspectives and avoid stereotypes or misrepresentations (Squires, 2011, p. 245). Ethical considerations involve ensuring content is culturally sensitive and inclusive. Ethical practices entail securing ICT systems against cyber threats and educating learners about online safety and responsible online behavior (Kumaraguru & Cranor, 2010, p. 13). Opting for sustainable and eco-friendly ICT solutions, such as energy-efficient devices and cloud-based resources, aligns with the ethical principles of environmental stewardship (Eriksson & Hilty, 2016, p. 1). Considering the broader social implications of ICT use in education is essential. Ethical educators should assess how technology affects social interactions, relationships, and well-being (Koivusilta et al., 2021, p. 212). Ethical considerations involve finding a balance between online and in-person learning to ensure the holistic development of learners while minimizing the potential negative consequences of excessive screen time (Rideout et al., 2017, p. S137).

By incorporating these ethical considerations into the use of ICT frameworks and learning resources, sustainable education practices can be fostered, benefiting learners, educators, and the broader society. Promoting ethical use of technology helps create a more inclusive, environmentally responsible, and socially just learning environment, aligning with the principles of sustainability.

CONCLUSION

Our discussions have delved deep into the crucial interplay between ethics, logistics, and climate change education in the pursuit of sustainable development. Throughout this chapter, we have examined the pressing need for a multidisciplinary approach to address the complex challenges posed by climate change and the ethical considerations that underpin sustainable development initiatives.

The integration of philosophy in sustainable development endeavors has brought forth a deeper understanding of the moral implications associated with our actions and decisions. By questioning our values and beliefs, we can better align our goals with the well-being of both current and future generations, fostering a more inclusive and equitable approach to development.

Furthermore, the role of logistics in sustainable development cannot be overstated. Efficient resource management, supply chain optimization, and innovative technologies have the potential to drastically reduce environmental footprints and contribute to a more sustainable world. However, we must ensure that ethical principles guide logistical practices to prevent unintended negative consequences and prioritize the long-term health of our planet.

Climate change education emerges as a critical component in the broader context of sustainable development. By empowering individuals with knowledge, awareness, and a sense of responsibility, we can foster a global movement towards more sustainable lifestyles and decision-making. Ethical considerations are vital in designing educational programs that inspire and motivate people to become environmental stewards and advocates for positive change.

It is evident that the intersection of philosophy, logistics, and climate change education plays a pivotal role in shaping the future of sustainable development. This chapter has highlighted the importance of adopting an integrated and ethical approach that acknowledges the interconnectedness of environmental, social, and economic issues.

As we move forward, it is imperative for policymakers, educators, businesses, and individuals to collaborate and embrace the ethical dimensions of sustainable development. By doing so, we can forge a path towards a more harmonious relationship between humanity and the environment, where our actions are guided by a profound commitment to the well-being of our planet and all its inhabitants. Only through collective responsibility, mindful decision-making, and a commitment to learning and evolving, can we hope to create a sustainable and thriving future for generations to come.

REFERENCES

Abe, Y. (2018). Logistics Emissions and Energy Consumption Models. In A. Nagurney (Ed.), *Sustainable Supply Chains* (pp. 121–146). Springer.

Adams, W. M., & Hutton, J. (2007). People, parks, and poverty: Political ecology and biodiversity conservation. *Conservation & Society*, *5*(2), 147–183.

Adger, W. N., Agrawala, S., Mirza, M. M. Q., Conde, C., O'Brien, K., Pulhin, J., Pulwarty, R., Smit, B., & Takahashi, K. (2007). Assessment of adaptation practices, options, constraints, and capacity. In M. L. Parry, O. F. Canziani, J. P. Palutikof, P. J. van der Linden, & C. E. Hanson (Eds.), *Climate Change 2007: Impacts, Adaptation, and Vulnerability. Contribution of Working Group II to the Fourth Assessment Report of the Intergovernmental Panel on Climate Change* (pp. 717–743). Cambridge University Press.

Agrawal, A., & Gibson, C. C. (1999). Enchantment and disenchantment: The role of community in natural resource conservation. *World Development, 27*(4), 629–649. doi:10.1016/S0305-750X(98)00161-2

Agyeman, J., Bullard, R. D., & Evans, B. (Eds.). (2003). *Just Sustainability's: Development in an Unequal World*. MIT Press.

Amaeshi, K., & Amao, O. O. (2012). Corporate Accountability and the Politics of CSR in a Developing Country. *Journal of Business Ethics, 117*(1), 173–190.

Anderson, K., Stahl, C., & Utterwulghe, A. (2017). Resource Extraction and Indigenous Livelihoods: A Case Study in Papua New Guinea. *Ecological Economics, 138*, 66–73.

Aristotle. (350 BCE). *Nicomachean ethics* (W. D. Ross, Trans.). Oxford University Press.

Banerjee, A. V., & Bell, C. R. (2007). Why Does Poverty Persist? *The American Economic Review, 97*(5), 1742–1759.

Bansal, P., & Song, H. C. (2017). Similar but not the same: Differentiating corporate sustainability from corporate responsibility. *The Academy of Management Annals, 11*(1), 105–149. doi:10.5465/annals.2015.0095

Barry, C. (2012). *The Ethics of Educational Leadership*. Routledge.

Bartelmus, P. (1999). *Environment, Growth and Development: The Concepts and Strategies of Sustainability*. Routledge.

Bebbington, A., Bury, J., & Warnaars, X. (2019). Hidden voices: From adaptation as a neoliberal burden to the collective potential of social practice theories. *Environmental Science & Policy, 100*, 29–36.

Bell, S., Tyrväinen, L., Sievänen, T., & Pröbstl-Haider, U. (2018). Connecting the social and the ecological in European peri-urban spaces. *Landscape and Urban Planning, 170*, 300–308.

Biermann, F., Pattberg, P., van Asselt, H., & Zelli, F. (2009). The Fragmentation of Global Governance Architectures: A Framework for Analysis. *Global Environmental Politics*, *9*(4), 14–40. doi:10.1162/glep.2009.9.4.14

Braithwaite, J. (2017). *Economic Analysis and the Design of Policy: Lessons from the New Institutional Economics*. Routledge.

Brechin, S. R., Wilshusen, P. R., Fortwangler, C. L., & West, P. C. (2003). Beyond the square wheel: Toward a more comprehensive understanding of biodiversity conservation as social and political process. *Society & Natural Resources*, *16*(10), 905–920.

Brown, A. (2018). *Sustainability and Business Ethics*. Routledge.

Brown, K., & Kasser, T. (2005). Are Psychological and Ecological Well-being Compatible? The Role of Values, Mindfulness, and Lifestyle. *Social Indicators Research*, *74*(2), 349–368. doi:10.100711205-004-8207-8

Büscher, B., & Davidov, V. (2013). States of exception: Conservation, militarization, and 'democracy' in the forested borderlands of Poland and Belarus. *Conservation & Society*, *11*(1), 31–43.

Callicott, J. B. (1989). Intrinsic Value, Quantum Theory, and Environmental Ethics. *Environmental Ethics*, *11*(1), 53–82.

Camagni, R., & Capello, R. (2013). Regional competitiveness and territorial capital: A conceptual approach and empirical evidence from the European Union. *Regional Studies*, *37*(9), 891–904. doi:10.1080/00343404.2012.681640

Carlson, A. (2000). *Aesthetics and the Environment: The Appreciation of Nature, Art, and Architecture*. Routledge.

Carter, C. R., & Rogers, D. S. (2008). A Framework of Sustainable Supply Chain Management: Moving Toward New Theory. *International Journal of Physical Distribution & Logistics Management*, *38*(5), 360–387. doi:10.1108/09600030810882816

Chambers, R. (1983). *Rural Development: Putting the Last First*. Longman.

Cooke, B., & Kothari, U. (2001). *Participation: The New Tyranny?* Zed Books.

Cruz, J., & Marques, P. (2011). Sustainable Transportation and Corporate Social Responsibility. *Procedia: Social and Behavioral Sciences*, *20*, 100–110.

Daly, H. E. (2015). *The Unraveling of the Growth Economy and the Transition to a Sustainable Era*. University of Chicago Press.

De Kadt, E. (1979). Poverty: The Ethical Dilemma. *World Development*, *7*(1-2), 81–87.

DesJardins, J. R. (2013). *Environmental Ethics: An Introduction to Environmental Philosophy* (5th ed.). Wadsworth Publishing.

Dossou-Yovo, E. R., Zinsou, C. B., & Dovonou, F. (2019). Women's Roles in Climate Change Adaptation in Sub-Saharan Africa: A Systematic Review. *Gender and Development*, *27*(2), 217–235.

Dresner, S. (2002). *The principles of sustainability*. Earthscan.

Ebrahim, A., & Rangan, V. K. (2010). The Limits of Nonprofit Impact: A Contingency Framework for Measuring Social Performance. *Harvard Business School Working Paper, 10*(99), 1–36.

Eccles, R. G., & Krzus, M. P. (2010). *One Report: Integrated Reporting for a Sustainable Strategy*. John Wiley & Sons.

Elkington, J. (1997). *Cannibals with Forks: The Triple Bottom Line of 21st Century Business*. Capstone.

Eriksson, E., & Hilty, L. M. (2016). ICT for Sustainability: An Emerging Research Field. *Environmental Impact Assessment Review*, *59*, 1–16.

Escobar, A. (1996). Construction Nature: Elements for a Post-structuralist Political Ecology. *Futures*, *28*(4), 325–343. doi:10.1016/0016-3287(96)00011-0

Florini, A. (2016). *The Right to Know: Transparency for an Age of Environmental Uncertainty*. Columbia University Press.

Gardiner, S. M. (2006). A Perfect Moral Storm: Climate Change, Intergenerational Ethics, and the Problem of Moral Corruption. *Environmental Values*, *15*(3), 397–413. doi:10.3197/096327106778226293

Gardiner, S. M. (2011). Ethics and Global Climate Change. In D. G. Arnold (Ed.), *The Ethics of Global Climate Change* (pp. 143–169). Cambridge University Press.

Gasper, D. (2014). Understanding the diversity of conceptions of well-being and quality of life. *Journal of Socio-Economics*, *40*(5), 559–570.

Gasper, D. (2014). Understanding the diversity of conceptions of well-being and quality of life. *Journal of Socio-Economics*, *40*(5), 559–570.

Gaventa, J., & Barrett, G. (2010). *So What Difference Does it Make? Mapping the Outcomes of Citizen Engagement.* IDS Working Paper, 347.

Genovese, A., Acquaye, A. A., Figueroa, A., & Koh, S. C. L. (2017). Sustainable supply chain management and the transition towards a circular economy: Evidence and some applications. *Omega, 66*(Part B), 344-357.

Gosseries, A. (2008). Intergenerational Justice. *Stanford Encyclopedia of Philosophy*. Retrieved from https://plato.stanford.edu/archives/win2008/entries/justice-intergenerational/

Govindan, K., Soleimani, H., & Kannan, D. (2015). Reverse logistics and closed-loop supply chain: A comprehensive review to explore the future. *European Journal of Operational Research, 240*(3), 603–626. doi:10.1016/j.ejor.2014.07.012

Gupta, J., & Bleischwitz, R. (2018). Public Choices and Ethical Dilemmas in Managing a Limited Natural Resource: The Case of Water. *Environmental Values, 27*(4), 367–389.

Hargittai, E. (2002). Second-Level Digital Divide: Differences in People's Online Skills. *First Monday, 7*(4). Advance online publication. doi:10.5210/fm.v7i4.942

Hart, S. L. (2005). *Capitalism at the Crossroads: Aligning Business, Earth, and Humanity*. Wharton School Publishing.

Harvey, D. (2005). *A Brief History of Neoliberalism*. Oxford University Press. doi:10.1093/oso/9780199283262.001.0001

Holling, C. S. (2018). Ethics in sustainable development. In M. Redclift & G. Woodgate (Eds.), *The International Handbook of Environmental Sociology* (pp. 127–138). Edward Elgar Publishing.

Hountondji, Y.-C., Xu, J., & Yang, C. (2020). Exploring the Relationship between Social Capital and Climate Change Awareness in Rural Communities. *International Journal of Environmental Research and Public Health, 17*(21), 7851. PMID:33120876

Hulme, M. (2010). *Why We Disagree About Climate Change: Understanding Controversy, Inaction, and Opportunity*. Cambridge University Press.

IPCC. (2014). *Climate Change 2014: Synthesis Report. Contribution of Working Groups I, II, and III to the Fifth Assessment Report of the Intergovernmental Panel on Climate Change. Core Writing Team* (R. K. Pachauri & L. A. Meyer, Eds.). IPCC.

Jamali, D., Zanhour, M., & Keshishian, T. (2018). Peculiar strengths and relational attributes of SMEs in the context of CSR. *Journal of Business Ethics, 82*(4), 355–363.

Jamieson, D. (2014). *Reason in a Dark Time: Why the Struggle Against Climate Change Failed—and What It Means for Our Future*. Oxford University Press. doi:10.1093/acprof:oso/9780199337668.001.0001

Jasanoff, S., & Kim, S. H. (2015). *Dreamscapes of Modernity: Sociotechnical Imaginaries and the Fabrication of Power*. University of Chicago Press. doi:10.7208/chicago/9780226276663.001.0001

Johnson, M. (2017). *Environmental Philosophy: From Theory to Practice*. Bloomsbury Academic.

Jones, P. (2019). *Sustainable Supply Chain Management: Principles, Strategies, and Practices*. Routledge.

Kallis, G. (2011). In Defence of Degrowth. *Ecological Economics*, *70*(5), 873–880. doi:10.1016/j.ecolecon.2010.12.007

Kant, I. (1785). *Groundwork of the Metaphysics of Morals* (M. Gregor, Trans.). Cambridge University Press.

Kaplan, S. (2019). Climate Change and Social Equity. *Annual Review of Environment and Resources*, *44*, 53–80.

Kaplinsky, R. (2000). Globalisation and Unequalisation: What Can Be Learned from Value Chain Analysis? *The Journal of Development Studies*, *37*(2), 117–146. doi:10.1080/713600071

Keck, M. E., & Sikkink, K. (1998). *Activists Beyond Borders: Advocacy Networks in International Politics*. Cornell University Press.

Kim, K., Suh, E., & Kim, S. (2018). *Sustainability in logistics and supply chain management: Selected papers from the 5th International Conference on Logistics and Maritime Systems*. Springer.

Kirchherr, J., Reike, D., & Hekkert, M. (2017). Conceptualizing the Circular Economy: An Analysis of 114 Definitions. *Resources, Conservation and Recycling*, *127*, 221–232. doi:10.1016/j.resconrec.2017.09.005

Koivusilta, L., Lintonen, T., & Rimpelä, A. (2021). Digital Divide and Digital Literacy in the Context of Finnish Schoolchildren. *Nordic Journal of Digital Literacy*, *16*(4), 212–230.

Kolk, A. (2016). The social responsibility of international business: From ethics and the environment to CSR and sustainable development. *Journal of World Business*, *51*(1), 23–34. doi:10.1016/j.jwb.2015.08.010

Kruse, E., Read, D., Negin, J., Hunter, E., & Rosewell, A. (2020). Responding to the Needs of Rural and Remote Populations: Insights from a Rapid Global Response to Climate Change Education in Vanuatu. *Journal of Geography in Higher Education, 44*(1), 1–19.

Kuehnast, K., Naujoks, J., Smith, D., & Wimmer, A. (2017). *Building Local Capacity for Peace and Development*. United States Institute of Peace.

Kumaraguru, P., & Cranor, L. F. (2010). Privacy and Security in Online Social Networking Sites. *IEEE Security and Privacy, 8*(3), 13–15.

Leach, M., Raworth, K., & Rockström, J. (2019). Between Social and Planetary Boundaries: Navigating Pathways in the Safe and Just Space for Humanity. *World Development*, 283–295.

Leal Filho, W., Azeiteiro, U. M., Alves, F., Molthan-Hill, P., Pace, P., Mifsud, M., & Caeiro, S. (2018). Reinvigorating the Sustainability Agenda: Education for Sustainable Development (ESD) for Responsible and Transformational Leadership. *Sustainability, 10*(10), 3415.

Leichenko, R. M., & O'Brien, K. L. (2008). *Environmental Change and Globalization: Double Exposures*. Oxford University Press. doi:10.1093/acprof:oso/9780195177329.001.0001

Milberg, S. J., Smith, J. H., & Burke, S. J. (2000). Information Privacy: Corporate Management and National Regulation. *Organization Science, 11*(1), 35–57. doi:10.1287/orsc.11.1.35.12567

Mill, J. S. (1863). *Utilitarianism* (5th ed.). Longmans, Green, and Co.

Miller, C., Jackson, T., & Thompson, A. (2021). Sustainable Logistics: Past, Present, and Future. *Journal of Sustainable Supply Chain Management, 10*(2), 65–78.

Miller, T. (2004). The Well-Tempered Self. *The Philosophical Forum, 35*(3), 195–210.

Miteva, D. A., Pattanayak, S. K., & Ferraro, P. J. (2015). Evaluation of biodiversity policy instruments: What works and what doesn't? *Oxford Review of Economic Policy, 31*(1-2), 1–34.

Nascimento, J. D. D., & Fearnside, P. M. (2020). Mega-Dams in Brazilian Amazonia: Cost-Benefit Analysis and the Persistence of Policy Stereotypes. *Sustainability, 12*(14), 5790.

Newig, J., & Fritsch, O. (2009). Environmental governance: Participatory, multi-level – and effective? *Environmental Policy and Governance, 19*(3), 197–214. doi:10.1002/eet.509

Nisbet, M. C., & Myers, T. (2007). The Polls - Trends: Twenty Years of Public Opinion about Global Warming. *Public Opinion Quarterly, 71*(3), 444–470. doi:10.1093/poq/nfm031

Norgaard, K. M. (2011). *Living in Denial: Climate Change, Emotions, and Everyday Life*. The MIT Press. doi:10.7551/mitpress/9780262015448.001.0001

Norton, B. G. (1987). *Why Preserve Natural Variety?* Princeton University Press.

O'Brien, K., & Hochachka, G. (2010). *The Human Relationship with Nature: Development and Culture*. MIT Press.

O'Neill, J. (1993). *Ecology, Policy, and Politics: Human Well-Being and the Natural World*. Routledge.

O'Riordan, T., & Cameron, J. (Eds.). (1994). *Interpreting the Precautionary Principle*. Earthscan Publications.

Oreskes, N. (2004). The Scientific Consensus on Climate Change. *Science, 306*(5702), 1686. doi:10.1126cience.1103618 PMID:15576594

Page, E. A. (2008). *Climate Change, Justice and Future Generations*. Edward Elgar Publishing.

Pattberg, P., & Stripple, J. (2008). Beyond the Public and Private Divide: Remapping Transnational Climate Governance in the 21st Century. *International Environmental Agreement: Politics, Law and Economics, 8*(4), 367–388. doi:10.100710784-008-9085-3

Pires, I. M., Martinho, A. P., & Gonçalves, R. (2020). The Ecological Footprint of Data Centers. *Journal of Cleaner*.

Porter, M. E., & van der Linde, C. (1995). Toward a New Conception of the Environment-Competitiveness Relationship. *The Journal of Economic Perspectives, 9*(4), 97–118.

Reed, M. S. (2008). Stakeholder Participation for Environmental Management: A Literature Review. *Biological Conservation, 141*(10), 2417–2431. doi:10.1016/j.biocon.2008.07.014

Ribot, J. C. (2004). *Waiting for Democracy: The Politics of Choice in Natural Resource Decentralization*. World Resources Institute.

Roberts, E. (2016). Environmental Ethics and Supply Chain Management. *Business Ethics Quarterly*, *26*(4), 489–512.

Rogers, D. S., & Tibben-Lembke, R. S. (2016). Going Backwards: Reverse Logistics Trends and Practices. *Reverse Logistics Magazine*, *1*(1), 20–34.

Rogerson, C. M. (2006). Ethical Aspects of Tourism: Progress and Challenges. *Tourism Management*, *27*(2), 171–183.

Russo, I., & Comi, A. (2019). Circular Economy and Collaborative Networks. In M. Peruzzini & R. Raffaeli (Eds.), *Collaboration in a Hyperconnected World* (pp. 297–308). Springer.

Sachs, W. (2015). Sustainable Development Goals and Human Rights. *Ethics & International Affairs*, *29*(1), 103–116.

Sandel, M. J. (2012). *What Money Can't Buy: The Moral Limits of Markets*. Farrar, Straus and Giroux.

Savaskan, R. C., Bhattacharya, S., & Van Wassenhove, L. N. (2004). Closed-Loop Supply Chain Models with Product Remanufacturing. *Management Science*, *50*(2), 239–252. doi:10.1287/mnsc.1030.0186

Sen, A. (1999). *Development as Freedom*. Oxford University Press.

Simpson, D., Wells, P., & Mason, D. (2019). Green Logistics. In R. C. Hoyle & P. W. Daniels (Eds.), *Transport and Logistics: Past, Present and Predictions* (pp. 297–324). Elsevier.

Smith, R., & Johnson, L. (2020). Green Logistics: A Pathway to Sustainable Supply Chain Management. *International Journal of Logistics Management*, *31*(3), 555–574.

Sovacool, B. K., Tan-Mullins, M., & Ren, X. (2017). Sustainable, pro-poor, and ethical: Exploring moral underpinnings of low-carbon energy. *Wiley Interdisciplinary Reviews. Energy and Environment*, *6*(2), e214.

Taylor, P. W. (1981). The Ethics of Respect for Nature. *Environmental Ethics*, *3*(3), 135–145. doi:10.5840/enviroethics19813321

Telenko, C., Seifert, R., Amankwah-Amoah, J., & D'Amico, K. (2019). Circular Economy in Reverse Supply Chain Management: A Comprehensive Review and Framework Development. *Sustainability*, *11*(22), 6303.

Thompson, P. B., & Tansey, G. (2017). *The ethical implications of global environmental change. The Oxford Handbook of Food Ethics*. Oxford University Press.

Wapner, P. (1996). Environmental Activism and World Civic Politics. *World Politics*, *49*(1), 62–89.

Wapner, P. (2016). Is sustainability possible? *Environmental Sociology*, *2*(1), 1–8.

WCED (World Commission on Environment and Development). (1987). *Our Common Future*. Oxford University Press.

West, P., Igoe, J., & Brockington, D. (2006). Parks and Peoples: The Social Impact of Protected Areas. *Annual Review of Anthropology*, *35*(1), 251–277. doi:10.1146/annurev.anthro.35.081705.123308

Wiek, A., Ness, B., Brand, F. S., & Schweizer-Ries, P. (2014). Fostering sustainability transitions in the energy domain through ethical leadership. *Sustainability*, *6*(12), 7865–7886.

Yaro, S. A. (2018). Conflicts of Interest and Climate Change Communication in the News Media. *Science Communication*, *39*(1), 8–34.

Zhang, H., Lai, K. K., & Zhang, M. (2020). Green and Reverse Logistics. In R. R. Negenborn & Z. Lukszo (Eds.), *Smart Port Perspectives* (pp. 187–211). Springer.

KEY TERMS AND DEFINITIONS

Biodiversity Conservation: The active protection, preservation, and sustainable management of the variety and variability of life forms, species, and ecosystems on Earth to maintain ecological balance and ensure the continued existence of diverse species.

Carbon Footprint: The total amount of greenhouse gases, specifically carbon dioxide (CO2) and other carbon compounds, emitted directly or indirectly by an individual, organization, event, or product over a given period. It is used to measure the environmental impact in terms of carbon emissions.

Climate Adaptation: The process of adjusting to the changing climate and its impacts to reduce vulnerability and enhance resilience. It involves strategies and actions to cope with the effects of climate change.

Climate Change Education: The process of imparting knowledge, skills, and awareness about climate change, its causes, impacts, and potential solutions to foster informed and responsible action.

Climate Justice: A movement and concept that emphasizes the fair distribution of the benefits and burdens of addressing climate change. It seeks to ensure that vulnerable communities, often disproportionately affected by climate change, are treated equitably and have their rights respected.

Corporate Social Responsibility (CSR): A business approach that involves a company's commitment to operate in an ethical and sustainable manner by considering the impact of its activities on society, the environment, and stakeholders beyond its financial bottom line. CSR initiatives may encompass social, environmental, and philanthropic efforts aimed at making a positive contribution to the community and minimizing negative impacts.

Deontology: A moral theory that emphasizes the importance of following moral rules and principles, regardless of the potential outcomes. Deontological ethics focuses on duty, obligations, and the inherent rightness or wrongness of actions.

Eco-Consciousness: A heightened awareness and consideration of environmental issues in everyday actions and decision-making, with a focus on reducing harm to the environment and promoting sustainability.

Eco-Philosophy: A branch of philosophy that explores the relationships between humans and the natural world, often emphasizing ethical and moral considerations in how humans interact with and impact the environment.

Eco-System Thinking: A holistic approach to understanding and analyzing complex systems, such as ecosystems, by considering the interactions and interdependencies among various components and factors.

Environmental Ethics: A branch of philosophy that examines the moral relationship between humans and the natural environment, addressing questions about the inherent value of nature and our ethical responsibilities toward it.

Environmental Stewardship: The responsible and ethical management, protection, and care of natural resources and the environment to ensure their sustainability and well-being for present and future generations.

Ethical Decision-Making: The process of making choices and taking actions based on moral principles, values, and ethical considerations, with the aim of achieving the most morally justifiable outcome.

Green Logistics: Also known as sustainable logistics or eco-logistics, it refers to the integration of environmentally friendly practices into the process of managing the flow of goods and services, with the goal of reducing environmental impact.

Green Supply Chain Management: A strategic approach to supply chain management that integrates environmentally friendly practices and principles to minimize the environmental impact of the entire supply chain. Green supply chain management encompasses sustainable sourcing, production, distribution, and end-of-life management of products, with a focus on reducing resource consumption, waste, and emissions.

Intergenerational Ethics: A perspective that addresses the ethical obligations and responsibilities of current generations toward future generations. It emphasizes considering the long-term impacts of decisions on the well-being and sustainability of future societies and environments.

Intersectionality: A concept that highlights how various social identities (such as gender, race, class, etc.) intersect and interact, leading to unique experiences of privilege and oppression.

Logistics: The planning, implementation, and management of the flow of goods, services, and information from the point of origin to the point of consumption. It involves coordination, transportation, warehousing, and distribution.

Renewable Energy: Energy derived from naturally occurring sources that are replenished on a human timescale, such as sunlight, wind, water (hydropower), and biomass. It is considered environmentally sustainable and less harmful compared to fossil fuels.

Reverse Logistics: The process of managing the return, recycling, refurbishment, or disposal of products and materials after they have been used or consumed. Reverse logistics involves the handling of items from the end consumer back to the manufacturer or other points in the supply chain to recapture value, reduce waste, and promote sustainability.

Sustainable Development: The process of achieving societal progress and economic growth while preserving and enhancing the environment for present and future generations. It involves balancing economic, social, and environmental factors.

Sustainable Practices: Actions and behaviors that aim to meet present needs without compromising the ability of future generations to meet their own needs. These practices prioritize long-term environmental, social, and economic well-being.

Utilitarianism: A moral theory that suggests that the right action is the one that maximizes overall happiness or well-being. It focuses on the consequences of actions and aims to achieve the greatest good for the greatest number of people.

Virtue Ethics: A moral theory that centers on the development of virtuous character traits and emphasizes the importance of cultivating virtues such as honesty, compassion, and courage. Virtue ethics focuses on becoming a morally excellent person rather than solely evaluating actions.

Chapter 6

Climate Change Reporting and the Role of Media in Shaping Public Discourse in Nepal:

Evidence From a Literature Review

Sudeep Uprety
https://orcid.org/0000-0002-9650-8321
Prince of Songkla University, Thailand

Arkom Palamanit
Prince of Songkla University, Thailand

Kuaanan Techato
Prince of Songkla University, Thailand

ABSTRACT

Though the role of media in climate change messaging has been understood in academic and public spheres, limited studies have been conducted and published, especially in the context of Nepal. A review of 65 readings highlights increased focus on behavior change communication than media advocacy, lack of innovation scope in funded media projects, absence of empirical research. The study recommends comprehensive study climate change reporting through framing and factor analysis to identify reasons and incentives.

DOI: 10.4018/978-1-6684-9099-0.ch006

INTRODUCTION

The Climate Risk Index statistics shows that Nepal falls under the ten most affected countries in the world, which is a major concern (Eckstein, Kunzel, & Schafer, 2021). Majority of Nepali people (about 80 percent) are vulnerable and likely to extreme heat stress, flooding, and air pollution (Ministry of Home Affairs, 2018). The risk has aggravated even more due to increasing trends of occurrence of storms, erosion, and landslides, resulting in the loss of life and livelihoods (ADB, 2021). Specifically, when we look at the rise in temperature, the projection is increment by 0.92–1.070C in between 2016–2045; 1.3–1.80C in between 2036–2065 taking the reference of 20-year period between 1981 and 2010 (MoFE, 2019). Predictions also show Nepali people annually affected by river flooding caused by climate change could more than double to around 350,000 in 2030 compared to 157,000 in 2010 (ADB, 2021). Consequently, there is likelihood of great financial loss, due to the flood as the estimates are that the loss could be threefold (ibid).

Climate change has already considerably affected lives of people during the last four decades in Nepal. As 647 people lose their lives each year because of climate-induced disasters (MoFE, 2021). The 'Climate Change Scenarios for Nepal' report highlights that in general, climate in Nepal will be significantly warmer and wetter in the coming years. The extremities in terms of temperature and precipitation will directly affect social sectors such as health, water, energy, biodiversity, agriculture, urban planning, and livelihoods (MoFE, 2019). When we see the increasing trend of climate-induced disasters in Nepal, we also need to take into context the current land and forest management. There is only one hydroelectric plant which is run-of-river. The productivity of this plant is very much dependent on river runoff volume and sedimentation, which has been affected by poor land and forest management. All this leads to potential vulnerability in the context of uncertain future precipitation regimes, glacial melting, and potential increases in the risks of landslides and glacial melting.

Media and Development Journalism

Media is a very powerful tool for raising voices against the wrong and empowering lives. In this context, responsible journalism has even greater significance. When journalists act responsibly, there can raise the right kind of questions to initiate public debates, create discourses around the issue, advocate with the people in power and authority to make amendments or changes, and therefore, bring about the change or transformation, which the society is hoping for (Mittal, 2012).

A journalist can play a critical role in laying out facts, figures, and scenarios regarding a matter of public importance. While journalists report about an issue, it automatically leads to an open discussion or triggers a series of thoughts, ideas, and discourses, whether implicitly or explicitly (Boykoff & Smith, 2010).

It is therefore understandable that journalists have been using frames to craft appealing news reports, and policymakers apply frames to contribute broad solutions to the problem while audiences need frames to understand and discuss an issue (Nisbet, 2009).

Discussing about nature of journalism, journalists in the Global South also adopt development journalism which emphasizes on the need for social change through development. Development journalists have been categorized into various types such as: Statists, Watchdogs, Interventionists, to Guardians of Transparency (Chua, 2019). All in all, this form of journalism relies on the ideologies of diversity and social inclusion.

Prioritization of issues and stories by the media makes a significant difference on how the incidents are perceived and re-prioritized. Agenda setting is considered a very important tool for shaping public perception as well as perception of the academic scholars and the practitioners. Prioritization can take place in various forms such as getting the news published on the front page, using big-size images and cartoons, using provocative headlines, among others. When media agencies deliberately prioritize a particular incident or other incidents on a similar theme, then it results in agenda-setting – or otherwise public debates and discourses will be centered around only particular issue, viewpoint, and perspective (Hornig, 1992). Framing of any issue by the media happens in an episodic manner (comprehensive discussion regarding an incident) or thematically, having a comprehensive view on the issue-at-rise. Therefore, various principles, ideologies, and culture of conducting journalism influences the attention to the issues as well as themes and dimensions while doing climate change coverage.

In a nutshell, this chapter investigates the evolution of climate change discourses through the lens of media and how media has played a role contributing to the discourse, particularly in the context of Nepal.

Media Reporting on Climate Change

Discussion and coverage around climate change is perceived to be a recent phenomenon, however, the coverage dates to the 1930s (Boykoff & Roberts, 2007). Since then, many journalists and communicators have covered climate change, with the aim of informing and educating the audiences about climate change and its glaring effects (Christians, Glasser, McQuail, Nordenstreng, & White, 2009). It is often a tricky situation where issues surrounding environment and climate are very much technical and scientific, which requires rigorous academic understanding and training for the journalists to understand, comprehend and report climate change. Whether it is lack of thematic understanding or the intention (prioritization or de-prioritization), the notion of accuracy, balance, and credibility (which are considered the three pillars of media) in climate change coverage has been considered problematic. Despite, growing level of public awareness regarding climate change, it is a general scenario worldwide that coverage on politics, economy, celebrities, sports, and crime are still prioritized more than climate change issue (Swain, 2021).

Coverage Trends and Patterns

When we investigate the general scale and frequency of coverage, it is observed that the frequency of climate change coverage is higher by the Western/Global North media than those of Global South. On the other hand, in terms of the nature of coverage, Global South journalists have emphasized on the socio-economic, cultural dimensions and prioritized more often issues related to scarcity of water, unpredictable weather conditions, agriculture and livestock, and threats to animal habitat (Althor, Watson, & Fuller, 2016). Global South journalists are more interested to discuss on the social change aspects and less interested to play the role of detached observers like in the case of Global North journalists (Hase, Mahl, Schäfer, & Keller, 2021).

There are also various factors that has led journalists to prioritizing the climate change issues such as preference of journalists at personal level, at institutional level (news organization) and at comprehensive level interlinking society, politics, and economy (Shoemaker & Resse, 1991). For instance, at the macro level, some studies have shown that journalists' education affects editorial decision-making (Vu, 2013). Biasness in the news is also consequence of the journalism norms and expectation of incentives on the part of the journalists (Hackett, 1984).

Another factor that has led to the differences in prioritization is the economy. There is a tendency by resourceful countries in the West to picturize climate change from the discipline of science or a technical subject, given they have adequate financial resources. On the other hand, media from poorer Global South countries are likely to categorize it as a transnational environmental phenomenon, which needs to be dealt with geo-strategically and as part of environmental diplomacy. Therefore, a country's economy and development status, political and social environment also has a major influence on the media framing (Hase, Mahl, Schäfer, & Keller, 2021).

Differences exist within the countries of similar economic and development status. In the US press, the media has framed climate change to emphasize scientific controversy (Boykoff & Boykoff, 2004) while newspapers in Germany have given a frame of international relations or international politics (Brossard, Shanahan, & McComas, 2004). For instance, the Southern Nigerian media has used the scenario (blame) frame to blame the developed countries for climate change, which indicates that various contexts (country's geopolitics, religion, culture) could affect and influence reporting 'tone' thereby questioning journalistic norms (Chetty, Devadas, & Fleming, 2015).

Thus, as compared to earlier trends and patterns, in the recent years, there has been a shift from denial that climate change is a direct consequence of human acts to the notion that climate change is a result of negligence or undermining of actions and policies that were developed particularly to address climate change (Schmid-Petri, Adam, Schmucki, & Häussler, 2017).

Evolution of Climate Change Prioritization in the Media

Since the 1990s, there have been six distinct phases in climate change prioritization by the media as outlined below:

1. **Denial Phase (Early 1990s):** A group of skeptical spokesmen and scientists denied that climate change is led by human contributions due to their beliefs and partly also because they were receiving funding from carbon-based industry.
2. **Scientific Consensus Phase (Mid-1990s):** A massive group of more than 2000 researchers who came together with the help of their scientific research and publications, establishing the notion that humans influence the global climate.
3. **Awareness Phase (End of 1990s):** During this period, media agencies started disseminating warnings of the climate change impact which helped in raising public awareness regarding climate change at some level. At this stage, people had started showing concern but not motivated enough to take actions.
4. **Climate Change Adaptation Phase (Between 2003 and 2006):** During this phase, media coverage of climate change adaptation increased substantially. In developing countries and rural areas, climate change news was communicated primarily through community radio.
5. **Great Recession and Climate Change Deprioritization Phase (2007-2009):** Though this phase sparked a surge in public interest about the environment, energy, and pollution, the Great Recession and the loss of journalism jobs led to limited print space and airtime. As a result, editors were often disinterested and, in some cases, forced to cover issues other than climate change and the environment.
6. **Go-Green Phase (Early 2010s):** Since the 2010s, there has been wider interest in the green economy also due to corporate pressures and advocacy by civil society, interest groups.

METHODS

Various readings – journal articles, media articles, research reports, policies and guidelines were reviewed while conducting this review. Various search engines such as Google Scholar, Science Direct, Research Gate, Scopus were used to identify the relevant readings. The key words or search terms used for the review included: 'Climate Change and Media', 'Climate Change and Media in Nepal', 'Media Discourses on Climate Change', 'Climate Change Communication', among others. Altogether 65 readings have been discussed and mentioned in this review. The basis of selection of these papers are mainly: 1) relevance to the context (related to Nepal); 2) relevance to the issue (media discourse); and 3) relevance to the theme (climate change).

Media Theories and Models

After careful reading and re-reading of various literature, the prominent theories, and models particularly relevant to climate change are discussed below. While there are several other media, theories, and frameworks, only the following have been discussed for the purpose of this study.

Encoding and Decoding: There is encoding and decoding process of the media content (Hall, 1980). The content creators or the producers encode the content based on their preference or prioritization. However, the audience or the readers decode the content and assign their own meaning to the content, depending upon their political, socio-cultural context and background. Hall (1980) further categorizes the media content into: dominant, negotiated, or oppositional. Dominant reading refers to the audience acceptance of the producer's interpretation; negotiated reading refers to acceptance of certain elements in media content and interpretation based on own judgement; and oppositional reading refers to total rejection of producer's interpretation and having a total opposite view.

Uses and Gratification Theory: The Uses and Gratification Theory (Blumler & Katz, 1974) especially contributes to elaborate how audiences conceive and interpret a media content. The proponents of this theory argue that audience, as a media user has choice of media content and therefore alternatives to choose the suitable content to be gratified or satisfied from. Particularly while discussing about climate, a study was conducted to understand whether and how exposure to "nature 2.0" – a nature-based concept similar to web 2.0, would help in pro-environmental behavior. The findings indeed suggested functional, relational, and emotional gratifications of the users or the audiences (Ng, 2022).

Two-Step Flow Theory: Two-Step Flow theory is also closely associated with interpretation of the messages conveyed through media. The first step involves the dissemination of information from mass media to a select group of individuals or influencers (who are basically opinion makers, political party leaders) while the second step is the message dissemination to the public (Weimann, 2015).

Climate Change and Media Discourses

Discourses are considered to both - instrument as well as manifestation of power, shaping institutions, behaviors, and practices (Foucault, 1975). Language, discourses, and narratives are instrumental in picturizing a problem or issue related to environment and climate change – ultimately shaping how that issue is highlighted as an agenda of wider geopolitical debate or domestic politics (Hajer, 1995). Likewise, economic, and political interests also influence, in a major way, lead to knowledge production and use (Forsyth, 2003) In the context of climate change too, there is a tendency to discredit or exclude voices and perspectives that are not conforming to dominant discourses (Swyngedouw, 2010).

To prove an argument or a point, the extent to which the scholars share their perspectives demonstrates the nature and extent of politicization or depoliticization. Methmann and Rothe (2012) discuss the role of dominant framing in establishing climate change as an issue that poses risks globally. Consequently, such categorization has led to managing risks stemming from climate change and interruptions in the environment – either through adaptation or mitigation rather than seeking alternatives. In the similar vein, economic factors, journalistic norms, political context, ideological cultures, and citizen decoding also shape the identities, worldviews, and belief systems that are implicitly or explicitly present in mediated public discourse about climate change (Pepermans & Maeseele, 2014).

The journalists' attempt and desire to be portrayed as objective and impartial could also be a reason to avoid cover climate change if it threatens interests of the establishment. This is particularly evident when the changes in particular governments' positions on climate change have resulted in changes in editorial stances (Carvalho & Burgess, 2005). Media discursions around climate change – considering it as a closed matter of fact or a controversial matter of concern depends on who owns the issue (politicians, scientists, journalists, or NGOs) and whether contrarian voices achieve legitimate standing outside the media arena (Aykut, Comby, & Guillemot, 2012). Here in comes the notion that "matters of concern", explaining that climate change cannot be directly observed and therefore, allowing the alternative space for discourse or discussion (Maeseele & Pepermans, 2017).

There is also tendency of journalists to personalize stories related to climate which have been initiated or instigated by individuals than groups. As a result, the considerations of power, context and vested interests would then have limited space for any direct influence (Pepermans & Maeselle, 2017). Therefore, from a political standpoint, climate change becomes a broad issue which could be beyond the interest of the media which prefers small tangible stories and only could be prioritized when there is a dramatic incident such as natural disaster.

Climate Change and Media Framing

The news media, under the umbrella of the information and dissemination ecosystem help in spreading specific messages across from various channels and medium to the general public. The way discourses are shaped for the public and general audiences is also very much determined by the framing that takes place. Framing is defined as the process of selecting "some aspects of a perceived reality and make them more salient in a communicating text, in such a way as to promote a particular problem definition, causal interpretation, moral evaluation, and/or treatment recommendation of the item described" (Entman, 1993).

The news media, a primary source of public knowledge, can strongly influence the peoples' understanding of any issue and might encourage them to engage in actions that ultimately influence decision making processes when reporters employ framing in their articles. The media, thus, has a significant role in raising public awareness to understand and interpret climate change, and has an important influence on actions to engage individuals in any decision-making process. As a means as well as an art of communication, skillful use of media framing can help to reach the audiences with the intended message, conveying an argument which you strongly feel about in the right manner. This is even more critical when issues such as climate change, which often affects our lives in a critical manner, would be viewed from various perspectives and schools of thought. The extent to which you optimize the art of framing shapes your audiences' understanding, attitude and behavior on the issue (Druckman, 2001).

Various frames related to climate change have been identified as part of analysis by media agencies such as financial loss or advantages, uncertainties in the environment, among others. These frames of climate communication have helped in shaping the public attitude and behavior (Stecula & Merkley, 2019). Scholars such as Boykoff and Roberts (2007) opine that the coverage of climate change is often event oriented, which is also labeled as 'episodic' framing. Likewise, another study has compared framing of US media based on political ideologies when Republicans considered global warming as "scientific uncertainty" or "unfair economic burden" while for the Democrats, they labeled global warming as a "Pandora's box" of catastrophe (Nisbet & Mooney, 2007).

It is observed that media frames climate change as a concern, developed as a byproduct of capitalism, interrelationship between nature and culture, and how the timely response to climate change is possible through social transformation in order to change the course of action. Word play and choices of words also matter to create the urgency or realization among the public/audiences. Some scholars also opine that audiences get more detached when terms such as "climate change" are used rather than "global warming" (Shove, 2010).

When prioritizing climate change at the public level, an assessment of media frames on climate change in four climate change conferences, the most common media frames included: 'global warming victims,' 'political dispute,' 'sustainable energy,' and 'common sense'. Despite the commonality, the contextual diversity across different countries have led to multiple interpretations (Lück, Wessler, Wozniak, & Lycarião, 2016). There is a good development in terms of the enhanced focus on the risks posed by climate inaction, which will ultimately contribute to greater understanding and awareness (through media) among the stakeholders and the public regarding the need for climate change interventions (Schafer, 2015).

Consequences of Media Framing

Discussing about the results of how a media story will be interpreted and responded, at the public level, there could be wholehearted support for climate action such as voting for environmentally friendly candidates, carbon reduction, political activism. On the other hand, public might also be skeptic based on the framing as they might feel climate change mitigation as costly, idealistic, or too late for any effective response (Fielding & Hornsey, 2016). When it comes to nuances of various frames, the responsibility frame discusses blame for the problem and the solution. The conflict frame refers to conflict and disagreement among parties. The economic consequences frame refers to the economic costs and benefits of an issue. The human-interest frame considers the personal, emotional side of a problem. The morality frame offers a moral judgment, usually by quoting others giving a moral perspective (Valkenburg, Semetko, & Vreese, 1999). When it comes to nature of media framing, the risk frame is also seen to be very prominent. Based on a media analysis study, risk frame was predominantly present in the Intergovernmental Panel on Climate Change (IPCC) reports, headlines, or first few lines of the media articles. Likewise, other prominent frames included Uncertainty and Opportunity frame. Very few articles mentioned about transitioning into low-carbon economy (Painter, 2013).

Role of Media in Agenda Setting

Scholars such as Nisbet (2009), Carvalho & Burgess (2005) though have differed in terms of their perspectives have come to a common consensus that climate change discourses are political. This also been reiterated by Hulme (2009) that climate have been used (and misused) by various social and civic movements to promote their own ideologies and agendas. Therefore, the has role of media in agenda setting is very much a reality.

When the solutions to climate change cannot be exclusively technocratic, climate change communication through media also needs to be reframed, moving away from exclusive scientific orientation or explanation but more of social engagement and understanding at a broader and comprehensive level (Pidgeon & Fischhoff, 2011). Concepts such as environmental risks are often difficult messages to convey to the public as journalists need to be conscious of balanced reporting. It is also to be taken into context that various interest groups would interpret a particular angle or perspective to a story, which may not be the intended take or perspective according to the journalist. The journalists are also pressurized by these groups to either intensify or downplay the technical details on the environment and climate, which as a consequence, could lead to lack of in-depth media coverage (Painter, 2013).

Journalists and scientists, given the professionalism they need to maintain provided the impact they would have on society, often are reluctant to play the advocacy role. While on the other hand, journalists and scientists use their professional autonomy against change and intrusion from the outside. This created space thus gives the journalists and scientists with vested interests to stick to their comfort zone. There is resistance to new forms of journalism where climate is emphasized and lack of innovation and freshness in climate reporting (Brüggemann, 2017).

The problems in the environment and the surrounding policies and actions that have led to climate change. However, the political economy of media also creates that hurdle. The media prioritization on climate change is also dependent on advertisement pressure, editorial policy, reporting culture, tight deadlines, limitations of space, among others. The newsworthiness of all climate change related news has been a major concern. Moreover, the newsworthiness is an issue which is often contested within newsrooms and editorial rooms, and often dependent upon editor's take, his/her own socio-economic context and funding environment.

The ideal role of media and communication channels is to promote the culture of authoritative, accurate storytelling which will inform and educate the public and policymakers regarding climate change. This process will involve the presentation of accurate and latest data (facts and figures), scenarios and projections based on current scientific assessments, anticipated risks, short-term risk management and mid to long term adaptation strategies (Swain, 2021).

Climate Change Coverage/Reporting Scenario in Nepali Media

The major issues related to climate change reporting in Nepali media are discussed below:

Journalists' technical knowledge and understanding: Based on the literature search and review of the available literature, it is observed that climate change coverage in Nepal is not very comprehensive. One of the prime factors for this is the lack of technical knowledge and understanding on the issue. Though there is some level of awareness these days due to increasing coverage worldwide, the level of understanding and knowledge to be able to comprehend complex climate terms is limited (Dunn, 2022).

Enabling environment and resource availability: It is also observed that particularly in developing countries like Nepal, there is distinct resource constraint to do comprehensive stories on the climate and the environment. Often it takes days even to reach the rural areas where there are real impacts of climate change, and it takes good amount of prior research, coordination, interviews with the locals, translators, camera crew (for visual stories) and most importantly the patience and the go-ahead from the editors to do such kind of 'dry' reporting. This issue leaves climate change reporting more of a 'luxury'. Another very important aspect is the lack of reliable data to see the tangible changes brought about by the climate crisis, in order to project an impactful story (Hase, Mahl, Schäfer, & Keller, 2021).

Nature of discourses on environment, sustainable energy, and climate change in Nepali media: The growth of private Nepali media, especially since the 2000s have increased the access to information among the general public. As a result, there is also increased public trust in the media. On the other hand, there are also questions regarding the conflict of interest and independence of these media houses as they are likely to be under the influence of government and private interests (Onta, 2006).

While environment and climate change issues have been covered by the Nepali media since 1990s, in terms of prioritization of news related to conservation, sustainable energy or development are only mentioned in the feature pages. Furthermore, there is no in-depth reporting that draws the links regarding flaws in the policy, over exploitation of natural resources and the need to think of alternatives to current energy demands and consumption. Rarely are links drawn between policy failures and the over-exploitation of natural resources or the need for changing Nepal's energy mix (Bhattarai, 2021).

There is a tendency in the Nepali media to focus at the surficial level, lamenting on the poor air quality. However, the underlying causes of these environmental hazards are not discussed adequately. Consequently, it is observed that there is a fatalistic tone to the media coverage regarding environmental issues in Nepal, which doesn't serve the purpose well – in terms of awareness generation among the public and stakeholders. Due to this lack of awareness on the solutions, the public and stakeholders are leaning towards taking the environmental problems lightly and a 'natural' phenomenon due to urbanization, population growth, among others (NDI, 2022).

In terms of timeline of the prioritization by the media, during the 2000s, coverage was mostly on how agriculture was affected by the climate change (especially irrigation due to lack of water availability), and how governments and the communities adapt and mitigate to climate change impacts in 2010s. Mostly media coverage related to environment, conservation and disasters is event-based and does not link well with the climate crisis. Despite the boom in terms of community radios and the television networks, there are hardly any dedicated shows or airtime on climate change, let alone sustainable energy (Bhattarai, 2021).

Due to Nepal's geographical context and glacial melting, most of the climate change reporting in the media is dominated by the coverage on impacts on water scarcity and agricultural productivity. There is also greater attention towards adaptation rather than mitigation, considering that Nepal with less industrial dominance falls under the country with nominal emission.

A general tendency regarding the media reporting is that the global events and processes regarding environment and climate change affects the nature and scale of media coverage by Nepali agencies. However, with the media agencies now spread in every nook and corner of the country, attention to national and local issues has been increasing considerably.

Role of Media in Shaping Public Discourses on Environment and Climate Change

Nepali media has been instrumental in shaping various policy agenda in Nepal. While there was massive deforestation of Sal Forest ongoing in Southern Nepal, Nepali media aggressively covered the news and mentioned the direct involvement of community leaders, government officials and political leaders. As a result, a parliamentary task force was commissioned to carry out an investigation. The report led to the ban on imposed a ban on the harvesting and trade of timber for about six months and filing of corruption cases (Khatri, Bhusal, Paudel, & Gurung, 2012).

Various non-government organizations and community radios have also come together to aware the communities regarding the impacts of climate change. A project, 'Communities challenging climate change' was initiated by Communities Radio Support Center (CRSC) which was implemented in Taplejung, Khotang, Kavre, Baglung and Kailali districts of Nepal covering mountain, hills, and plain ecological belt (Koirala & Bhatta, 2010). Nepal Forum for Environmental Journalists (NEFEJ) through Asian Development Bank's support also implemented a project to raise public and media awareness on climate change and environmental management (NEFEJ, 2011).

CONCLUSION

The analysis of the readings demonstrate that discussion of the interventions related to climate change communication. Majority of the interventions which wanted to focus on the improvement in the coverage of climate change focused on the behavior change communication among the audiences. However, it is equally imperative to focus on the advocating the decision makers at the government, the Ministry of Forests and Environment to bring out conducive policies that address long term impacts on climate change. So, media advocacy also plays a vital role.

Most of the projects focusing on the improvement in climate change communication are funded through some aid agencies such as United States Agency for International Development (USAID), European Union, Department of Foreign Affairs and Trade (DFAT), Foreign, Commonwealth and Development Office (FCDO), among others. These aid agencies would have their own scheme of values, norms that would sometimes be an obstacle for innovation. This aspect of how aid affects the media intervention has not been discussed adequately.

Empirical research on processes of depoliticization regarding climate change in public and media discourses is currently absent: There is a need for a broad systematic research agenda revealing the extent to which depoliticizing discourses are found to influence public and media discourses, either geographically or historically. This would allow communication researchers to draw conclusions on the nature and extent of democratic debate and citizenship about climate change either in specific locales or time-periods.

There's also a lack of proper distinction on journalistic take on climate change as incidental event or structural development. Media thrives on big news and therefore mainstream journalism focuses on the big incidents (often negative) that makes a good headline and gathers audience attention. Climate change, thus, is modelled more on the change perspective – which is represented as a sudden, exceptional, and negative disruption. However, climate does not change every day. It is a series of episodic events that lead towards a change. Therefore, proper media analysis is required to distinguish between daily incident-based coverage and investigative critical coverage.

Media framing plays a vital role in how a climate change incident is perceived by the media agency and then by the audiences, it is imperative to conduct an in-depth framing and factor analysis to find out that factors affecting the framing of climate change, and the socio-economic, cultural, and domestic and international politics surrounding it.

While some media research is available that describes the media narratives on climate change in Nepal, there is a distinct lack of national surveys and studies that meticulously analyzes the trend of climate change discourses in Nepal. It is particularly relevant to understand the sub-national coverage scenario where dimensions in coverage of climate change are likely to differ given the diversity in socio-cultural, political context.

REFERENCES

ADB. (2021). *Climate Risk Country Profile: Nepal.* Manila: Asian Development Bank. Retrieved from https://www.adb.org/sites/default/files/publication/677231/climate-risk-country-profile-nepal.pdf

Althor, G., Watson, J. E., & Fuller, R. A. (2016). Global mismatch between greenhouse gas emissions and the burden of climate change. *Scientific Reports*, *6*(1), 20281. Advance online publication. doi:10.1038rep20281 PMID:26848052

Aykut, S. C., Comby, J.-B., & Guillemot, H. (2012). Climate Change Controversies in French Mass Media 1990-2010. *Journalism Studies*, *13*(2), 154–174. doi:10.1080/1461670X.2011.646395

Balzacq, T. (2005). The Three Faces of Securitization: Political Agency, Audience and Context. *European Journal of International Relations, 11*(2), 171-201. https://journals.sagepub.com/doi/10.1177/1354066105052960

Bhattarai, S. (2021, January 29). *Underreporting the climate emergency.* Retrieved from Nepali Times: https://www.nepalitimes.com/latest/underreporting-the-climate-emergency/

Blumler, J. G., & Katz, E. (1974). The Uses of Mass Communications: Current Perspectives on Gratifications Research. *American Journal of Sociology*, 3.

Boykoff, M., & Smith, J. (2010). Media representations of climate change. In *Handbook of Climate Change and Society* (pp. 210–218). Routledge.

Boykoff, M. T., & Boykoff, J. M. (2004). Balance as bias: Global warming and the US prestige press. *Global Environmental Change*, *14*(2), 125–136. doi:10.1016/j.gloenvcha.2003.10.001

Boykoff, M. T., & Roberts, J. T. (2007). *Media Coverage of Climate Change: Current Trends, Strengths, Weaknesses.* UNDP.

Brossard, D., Shanahan, J., & McComas, K. (2004). Are Issue-Cycles Culturally Constructed? A Comparison of Global Climate Change. *Mass Communication & Society*, *7*(3), 359–377. doi:10.120715327825mcs0703_6

Brüggemann, M. (2017). Post-normal Journalism: Climate Journalism and Its Changing Contribution to an Unsustainable Debate. In P. Berglez, U. Olausson, & M. Ots (Eds.), *What is Sustainable Journalism? Integrating the Environmental, Social, and Economic Challenges of Journalism* (pp. 57–73). Peter Lang.

Carvalho, A., & Burgess, J. (2005). Cultural Circuits of Climate Change in U.K. Broadsheet Newspapers, 1985–2003. *Risk Analysis, 25*(6), 1457–1469. doi:10.1111/j.1539-6924.2005.00692.x PMID:16506975

Chetty, K., Devadas, V., & Fleming, J. (2015). The framing of climate change in New Zealand newspapers from June 2009 to June 2010. *Journal of the Royal Society of New Zealand, 45*(1), 1-20. https://www.tandfonline.com/doi/full/10.1080/03036758.2014.996234

Christians, C. G., Glasser, T. L., McQuail, D., Nordenstreng, K., & White, R. A. (2009). *Normative Theories of the Media*. University of Illinois Press.

Chua, Y. T. (2019, June 25). *Communication*. Retrieved from Oxford Research Encyclopedias: https://oxfordre.com/communication/view/10.1093/acrefore/9780190228613.001.0001/acrefore-9780190228613-e-799

Ciano, T., Ferrara, M., Gangemi, M., Merenda, D. S., & Pansera, B. A. (2021). Climate Change Sustainability: From Bargaining to Cooperative Balanced Approach. *Games, 12*(2), 45. doi:10.3390/g12020045

Druckman, J. N. (2001). Framing: Toward Clarification of a Fractured Paradigm. *Political Behavior, 23*(3), 225–256. doi:10.1023/A:1015006907312

Dunn, K. (2022, March 30). *The impact of climate change in the Himalayas matters. It's a global and regional issue*. Retrieved from Reuters Institute: https://reutersinstitute.politics.ox.ac.uk/news/impact-climate-change-himalayas-matters-its-global-and-regional-issue

Eckstein, D., Kunzel, V., & Schafer, L. (2021). *Global Climate Risk Index 2021*. Berlin: Germanwatch. Retrieved from https://www.germanwatch.org/sites/default/files/Global%20Climate%20Risk%20Index%202021_2.pdf

Entman, R. M. (1993). Framing: Toward Clarification of a Fractured Paradigm. *Journal of Communication, 43*(4), 51–58. doi:10.1111/j.1460-2466.1993.tb01304.x

Fielding, K. S., & Hornsey, M. J. (2016). A Social Identity Analysis of Climate Change and Environmental Attitudes and Behaviors: Insights and Opportunities. *Frontiers in Psychology, 7*. Advance online publication. doi:10.3389/fpsyg.2016.00121 PMID:26903924

Forsyth, T. (2003). *Critical Political Ecology: The politics of environmental science*. Routledge.

Foucault, M. (1975). *The birth of the clinic: An archeology of medical perception*. Vintage/Random House.

Hackett, R. A. (1984). Decline of a paradigm? Bias and objectivity in news media studies. *Critical Studies in Mass Communication, 1*(3), 229-259. https://www.tandfonline.com/doi/abs/10.1080/15295038409360036

Hajer, M. A. (1995). *The Politics of Environmental Discourse: Ecological Modernization and the Policy Process*. Oxford University Press.

Hall, S. (1980). Encoding/decoding. In *Culture, Media, Language*. Hutchinson.

Hase, V., Mahl, D., Schäfer, M. S., & Keller, T. R. (2021). Climate change in news media across the globe: An automated analysis of issue attention and themes in climate change coverage in 10 countries (2006–2018). *Global Environmental Change, 70*, 102353. Advance online publication. doi:10.1016/j.gloenvcha.2021.102353

Hornig, S. (1992). Framing Risk: Audience and Reader Factors. *The Journalism Quarterly, 69*(3), 679–690. doi:10.1177/107769909206900316

Hulme, M. (2009). *Why We Disagree About Climate Change: Understanding Controversy, Inaction and Opportunity*. Cambridge University Press. doi:10.1017/CBO9780511841200

ICRW. (2022). *SRHR is a Climate Issue*. ICRW. Retrieved from https://reliefweb.int/sites/reliefweb.int/files/resources/ICRW_SRHR-is-a-Climate-Issue_04.22.pdf

Khatri, D. B., Bhusal, R. P., Paudel, N. S., & Gurung, N. (2012). *REDD+ politics in the media: A case study from Nepal*. CIFOR.

Koirala, M., & Bhatta, R. (2010). *Communities Challenging Climate Change*. CSRC and NEFEJ.

Lück, J., Wessler, H., Wozniak, A., & Lycarião, D. (2016). Counterbalancing global media frames with nationally colored narratives: A comparative study of news narratives and news framing in the climate change coverage of five countries. *Journalism, 19*(12). https://journals.sagepub.com/doi/10.1177/1464884916680372

Maeseele, P., & Pepermans, Y. (2017). Climate Change Communication, Media, and Ideology. *Climate Change Communication*, 1-22.

Methmann, C., & Rothe, D. (2012). Politics for the day after tomorrow: The logic of apocalypse in global climate politics. *Security Dialogue*, 323-344. doi:https://journals.sagepub.com/doi/10.1177/0967010612450746

Ministry of Defence. (2016). *National Security Policy, 2016*. Law Books Management Board.

Ministry of Home Affairs. (2018). *Nepal Disaster Report, 2017: The Road to Sendai*. Kathmandu: Ministry of Home Affairs. Retrieved from http://drrportal.gov.np/uploads/document/1321.pdf

Mittal, R. (2012). Climate Change Coverage in Indian Print Media: A Discourse Analysis. *The International Journal of Climate Change: Impacts and Responses*, *3*(2), 219–232. doi:10.18848/1835-7156/CGP/v03i02/37105

MoFE. (2019). *Climate Change Scenarios for Nepal: National Adaptation Plan (NAP)*. Kathmandu: Ministry of Forests and Environment. Retrieved from https://lib.icimod.org/record/34554

MoFE. (2021). *Vulnerability and Risk Assessment and Identifying Adaptation Options*. Ministry of Forests and Environment.

NDI. (2022). Climate Change. Kathmandu: National Democratic Institute for International Affairs (NDI).

NEFEJ. (2011). *Nepal: Strengthening Capacity for Managing Climate Change and the Environment*. NEFEJ.

Ng, Y.-L. (2022). Uses and Gratifications of and Exposure to Nature 2.0 and Associated Interdependence With Nature and Pro-Environmental Behavior. *Social Science Computer Review*, *40*(1), 61–76. doi:10.1177/0894439320901490

Nisbet, M. C. (2009). Communicating Climate Change: Why Frames Matter for Public Engagement. *Environment: Science and Policy for Sustainable Development*, *51*(2), 12-23. https://www.tandfonline.com/doi/abs/10.3200/ENVT.51.2.12-23

Nisbet, M. C., & Mooney, C. (2007). Science and society. Framing Science. *Science*, *316*(5821), 56. Advance online publication. doi:10.1126cience.1142030 PMID:17412941

Onta, P. (2006). *Mass media in post-1990 Nepal*. Martin Chautari.

Painter, J. (2013). *Climate change in the media – reporting risk and uncertainty*. London: I.B. Tauris & Co. Ltd.

Pepermans, Y., & Maeseele, P. (2014). Democratic Debate and Mediated Discourses on Climate Change: From Consensus to De/politicization. *Environmental Communication*, *8*(2), 216–232. doi:10.1080/17524032.2014.906482

Pepermans, Y., & Maeseele, P. (2018). Manufacturing Consent: Rereading News on Four Climate Summits (2000-2012). *Science Communication*, 621-649. https://journals.sagepub.com/doi/10.1177/1075547018798119

Pepermans, Y., & Maeselle, P. (2017). Climate Change Journalism: From Agony to Agonistic Debate. *Desenvolvimento e Meio Ambiente*, *40*, 125–140. doi:10.5380/dma.v40i0.49257

Pidgeon, N., & Fischhoff, B. (2011). The role of social and decision sciences in communicating uncertain climate risks. *Nature Climate Change*, *1*(1), 35–41. doi:10.1038/nclimate1080

Schafer, M. S. (2015). Climate Change and the Media. International Encyclopedia of the Social & Behavioral Sciences, 3, 853-859.

Schmid-Petri, H., Adam, S., Schmucki, I., & Häussler, T. (2017). A changing climate of skepticism: The factors shaping climate change coverage in the US press. *Public Understanding of Science (Bristol, England)*, *26*(4), 498–513. doi:10.1177/0963662515612276 PMID:26553047

Shoemaker, P. J., & Resse, S. D. (1991). *Mediating the message: theories of influences on mass media content*. Longman.

Shove, E. (2010). Social Theory and Climate Change. *Theory, Culture & Society*, *27*(2-3), 277–288. doi:10.1177/0263276410361498

Stecula, D. A., & Merkley, E. (2019). Framing Climate Change: Economics, Ideology, and Uncertainty in American News Media Content From 1988 to 2014. *Frontiers in Communication*, *4*, 6. Advance online publication. doi:10.3389/fcomm.2019.00006

Swain, K. A. (2021). Media Framing of Climate Change Mitigation and Adaptation. In M. Lackner & B. Sajjadi (Eds.), *W.-Y. Chen, Handbook of Climate Change Mitigation and Adaptation*. Springer. doi:10.1007/978-1-4614-6431-0_6-3

Swyngedouw, E. (2010). Apocalypse Forever? *Theory, Culture & Society*, *27*(2-3), 213–232. doi:10.1177/0263276409358728

UN Climate Change Conference. (2021). *COP 26 Goals*. Retrieved from COP 26: https://ukcop26.org/cop26-goals/

UNFCCC. (2002). *United Nations Framework Convention on Climate Change.* United Nations. Retrieved from https://unfccc.int/files/essential_background/background_publications_htmlpdf/application/pdf/conveng.pdf

Valkenburg, P. M., Semetko, H. A., & Vreese, C. H. (1999). The effects of news frames on readers' thoughts and recall. *Communication Research*, *26*(5), 550–569. doi:10.1177/009365099026005002

Vu, H. T. (2013). The online audience as gatekeeper: The influence of reader metrics on news editorial selection. *Journalism.* Advance online publication. doi:10.1177/1464884913504259

Weimann, G. (2015). Chapter. In J. D. Wright (Ed.), International Encyclopedia of the Social & Behavioral Sciences (2nd ed., pp. 291-293). Elsevier. doi:10.1016/B978-0-08-097086-8.95051-7

Women, U. N. (n.d.). *Women, Gender Equality and Climate Change.* UN Women. Retrieved from https://www.un.org/womenwatch/feature/climate_change/downloads/Women_and_Climate_Change_Factsheet.pdf

Y. K. Hanusch, F., Ramaprasad, J., Skjerdal, T., Hasim, M. S., Muchta, N., . . . Less, S. B. (2017). Journalists' Development Journalism Role Perceptions: Select countries in Southeast Asia, South Asia, and sub-Saharan Africa. *Journalism Studies*, 576-594. https://www.tandfonline.com/doi/full/10.1080/1461670X.2016.1254060

KEY TERMS AND DEFINITIONS

Climate Change: Long-term shifts in temperatures and weather patterns. Such shifts can be natural, due to changes in the sun's activity or large volcanic eruptions.

Discourse: Interchange, discussion, and conversation over an idea or topic.

Media: A communication channel through which we disseminate various forms of information such as news, music, movies, education, among others.

Chapter 7
Climate Change and Its Implications on School Education in Kenya

George Mokua Ogendi
https://orcid.org/0000-0003-2279-8423
Kisii University, Kenya

ABSTRACT

Climate change is the greatest challenge of the 21st century. It is threatening all aspects of societal life. The right to education of children is negatively impacted by these climate-induced challenges. This chapter presents findings of a study on the effects of climate change on the education of the younger generation. The study used a mixed research design method and was done within Kisii Municipality, Kenya. The findings were that climate change and its effects have brought about increased waterborne diseases, reduced available income for families, and compromised living standards, thereby negatively impacting on children's education. To address the effects of climate change, however, provision of appropriate education can go a long way in mitigating these climatic challenges. Secondly, topics on climate change should be included in the school curriculum.

INTRODUCTION

Climate change is the greatest challenge of the 21st century which is threatening all aspects of societal life (Campbell-Lendrum et al., 2018) and globally, every creature is at a high health risk (Watts et al., 2018). Warming of the globe and subsequent climate change has brought about various effects on the environment whereby

DOI: 10.4018/978-1-6684-9099-0.ch007

majority of them are severe (IPCC, 2014). These impacts include but not limited to deaths, high rise of sea levels, temperature increase, severe droughts, floods, rampant wildfires, emigrations and immigrations, among others, that often cause havoc. Humanity and other organisms worldwide are experiencing extraordinary effects on their ecosystems. These climate hazards are undoubtedly affecting human mobility and migrations as they go about in search of water, food for themselves and their animals (Christian Aid, 2007; Biermann and Boas, 2010). These migrations have caused cross-border displacements and it is projected that come 2050, it will affect a human population of between 25 million and 1 billion (IOM, 2012). What we are certain of is that climate change is with us and causing far-reaching effects in many regions, including large numbers of people and animals being displaced. For example, in 2019, about 24 million people were displaced by weather related disasters across 140 countries worldwide (IDMC, 2020).

The right to education of our children is negatively impacted with these climate induced migrations (UNICEF, 2021). In fact, children are at risk of not only losing in education but also their homes and shelter. Therefore, accessing education which is key in opening opportunities for their future is compromised (OECD, 2022). The inclusion of climate change education is key because currently the status of its inclusivity is marginal (UNICEF, 2019). The pledge "to leave no one child behind" which is well articulated in the 2030 Global Development Agenda, and particularly the Sustainable Development Goal 4, are devoted to an all-inclusive education which addresses disasters, pandemics and conflicts, resultant cross-border displacement and uneducated populace (UNESCO, 2015). Because of these, all nations are called upon to have it as their responsibility in developing inclusive, responsive and resilient education systems which will meet the needs of children, youth and adults in climate-crisis-affected contexts.

Among others, classrooms are becoming hotter, thereby affecting meaningful and progressive learning (Lala and Hagishima, 2023), while teachers and learners are missing the essential commodity (water) of life at a time when it is needed most. Waterborne diseases such as typhoid, salmonella, schistosomiases, and vector-transmitted diseases like malaria, leishmaniases and trypanosomiases are on the increase and so is displacement of several communities from their habitable locations (UNDP, 2017).

Indeed, all these effects are associated with climate change, and the world is greatly disturbed. So, what next? Will provision of relevant education help? Yes, education will play a key role. But, the pertinent question remains "is education provision under threat from climate change?" This question can best be answered if two critical issues are taken into account. It needs to be borne in mind that while education is key to ending climate change, it remains one of the biggest threats to education as well. The threats are many, for example, climate change and poverty exist in a vicious cycle, especially for the most vulnerable people. The effects of climate change are leading to food and water scarcity in schools and many communities, lost livelihoods, lower education levels, among others (Urama & Ozor, 2010). This challenge of lower education levels is worrisome, as education is one of the greatest tool people have to use to escape the cycle of climate change effects. Reports from UNICEF indicate that half of the world's 2.2 billion children are at "extremely high risk" of climate change, including its impact on education. Clearly, education is part of the answer to these challenges. Increasing environmental awareness and education on climate change can contribute to effective adaptation and mitigation. The UN Framework Convention on Climate Change (UNFCCC) Article 6 on education, training and public awareness support that member countries need to develop and implement educational and public awareness programmes on climate change and its effects in educational institutions. The Paris Climate Agreement (UNFCCC COP 21, 2015) Article 12 further supports the role education can play in enhancing climate actions.

Amidst all these commitments from key world bodies (UN, UNICEF, UNESCO, WHO, etc) and conferences, not all countries have considered and fully embraced these guidelines, including developing a relevant curricular on climate change education. Inclusion of climate change aspects is invisible in most academic curricular. Therefore, the voice of climate change education is hardly heard. Countries are not including educational issues in national priorities for climate mitigation and adaptation actions, which are a core element of the Paris Agreement. Out of the 196 Parties to the UNFCCC, only 32 per cent of them mention climate change and related issues in their education systems (UNICEF, 2019). In other words, most countries' participation in climate change education and related discussion is usually marginal. Hence, urgent actions are needed to enhance climate change and related issues in the education sector.

The climate crisis affects the social and environmental determinants of health like clean air, safe drinking water, sufficient nutritious food and secure shelter. Humanity and more specifically the children and adolescents are poised to bear the additional burden of disease related to climate change, compromising their physical and physiological development. The availability of water is another challenge in a changing climate. The scarce water supply comprises a risk of contracting food-borne and water-borne diseases which can generate long-term health problems and increase absenteeism from schools. Food security and nutrition is also threatened as children and adolescents may not be able to obtain sufficient and nutritious food either at home or at school, which has proven to have compounding effects on their academic wellbeing.

Indirectly, climate change negatively impacts on the wellbeing, income and security of children and their parents and/or guardians. For example, across the tropics, livelihoods are predominantly dependent on rain-fed agriculture and livestock rearing, making these practices particularly susceptible to the climate crisis (Morton, 2007). With bad weather seasons, the situation reduces household income and purchasing power to the lowest ebb making parents/guardians not able to afford school costs among them fees, transport, and school stationery materials and even to a larger extent managing their children, especially adolescents. At times, the children are made to miss school and support livelihood or household activities. In more extreme cases, households may decide to migrate. Evidence suggests that relocation and attending classes in a different school generally translates to dropouts or lower academic performance.

For example, in 2018, more that 17 million people were internally displaced because of natural disasters including climate change (Internal Displacement Monitoring Centre, 2021) and in the year 2020, the number jumped to 30.7 million due to natural extreme weather events. The catastrophic effects of climate change are no longer isolated emergencies but have become the new global norm. People displaced by the effects of climate change face significant vulnerabilities with regard to accessing education, saturated school capacity, destroyed infrastructure, language barriers, exacerbated poverty, school dropouts and many more.

Because of all these effects, there is need to incorporate climate change issues into education planning, education system priorities, climate change policy and financing (UNDP- Zimbabwe Human Development Report, 2017). Climate finance requirements for the education sector need to be met so as to take care of scenarios where livelihoods are affected through the provision of food and support to such families. This will limit school absenteeism and promote continuous learning in the climate affected areas.

At present, only a fraction of contingency funds is allocated to education ministries (and education authorities report that most funds go to agriculture and infrastructure). The climate investment needs for the education sector should also include mitigation needs, such as the provision of reliable, affordable and sustainable water and food for schools so as to curb absenteeism. Secondly, energy efficiency needs to be taken care of and provided to schools more especially during the dry spell when waters in hydropower stations are in their lowest levels hence affecting power production. To cushion this scarcity, concerned governments need to provide generators to schools which will produce energy for use during periods of need.

Climate change and poverty exist in a vicious cycle, especially for the poor and vulnerable people in society. The effects of climate change are leading to food and water scarcity in schools and most communities. This has further led to lost livelihoods, lower education levels, among others. Apparently, education is under threat from climate change. Yet, education is key to tackling the climate change crisis. Against this background, the researcher was prompted to undertake water quality analysis, including collecting secondary data from Kisii Teaching and Referral Hospital (KTRH) in Kenya to determine how climate change has affected learning in schools as well as inhabitants within Kisii municipality of Kisii County in Kenya and beyond.

METHODOLOGY

The study used a mixed research method which combined the qualitative and quantitative data collection techniques. The water quality research used the purposive transect method in sampling Nyakomisaro-Riana River water for quality testing. Questionnaires were completed by street children, while hospital records were used to obtain data concerning patients who visited the facilities to seek medication as well as treatment for conditions or diseases of interest to the researcher. Finally, the direct observations and secondary data were used to enrich the research. The study area was Kisii Municipality in Kisii County, while the questionnaires sought to gather information on why the street children opted to leave studies and hospital records provided data on waterborne and related diseases.

FINDINGS AND DISCUSSIONS

The research found out that climate change has contributed to the increase of people in Kisii town (urbanization). This increase is putting pressure on Kisii town and other cities to provide education, water and other services to the growing numbers. This finding is in agreement with the prediction that climate change effects in rural areas could amplify migration to urban centers (IPCC, 2014). Sometimes, this type of movement can have positive impacts on education, depending on the characteristics of the receiving areas (World Bank, 2018), but the negative ones are many more especially for the developing countries. Therefore, it is important that all-inclusive education need to be offered to learners and leaders in these countries to put into place measures to curb this trend of rural to urban migration lest there be conflicts to do with urban resources and services.

One of the negative social challenges climate change has brought into Kisii town is the rise of street children (urchins) (Georgeta, 2020). In the context of unattended children, it was found that children who are pushed out of schools and with homes which are challenged foodwise and lack parental guidance, tended to shift to Kisii and other towns. These "urchins" not only become a nuisance to the cleanliness of the town but also jeopardize the town's security (Case 1). This finding corroborated well with what was found out in one of the rural areas of China (UNESCO, 2018).

Any disturbance to the academic progression of children by anything may bring trauma and create a feeling that these children are abandoned by their parents (UN Human Rights Council, 2010). It was further realized that the majority of Kisii town street "urchins" were boys.

Figure 1. Spiraling gang violence threatens to bring back Kisii's feared Sungu Sungu days
Source: Standard newspaper of June 10th, 2023 [not copyrighted]

Case 2
Learning fails to take off as flood victims camp in classes

A submerged house in the flooded Lake Baringo. Photo/PD/Raphael Munge
Source: People daily: Wednesday, May 10th, 2023

Learning has been paralysed in several schools within Nyanza region due to floods which have displaced thousands of people. Reports also show that six people could have died in the past one week due to the heavy rains pounding the country. Authorities reported the deaths in Uasin Gishu and Narok counties, with the victims including school children.
In Nyando, Kisumu County, many learners failed to report for the second term on Monday as a result of floods caused by heavy rains. Many families in the lower Nyando river basin are camping in schools on higher grounds.Among the affected schools are Ombaka, Ogenya, Kandaria, Ugwe, Osep, Kibarwa and Odienya, Nyamasao and Magina primary schools. Others are Kanyagwal, Kandaria and Odienya secondary schools…

For the case of girls, upon reaching Kisii town, they were easily swayed into the prostitution business, while some were taken as house helpers within or without the town. Efforts of integrating majority of these street children with their families realized little success. Among others, such children sneaked out of their homes or schools and came back to the streets where they undertake begging and mugging. This finding was similar with that of the Global Education Monitoring Report /

UNESCO (2019a). A small fraction of these street children who remained with their reunited families or continued with schooling in respective schools where they were taken struggled to complete their studies because of lack of fees or money to buy such items as uniform and stationery. Indeed, Global Education Monitoring Report / UNESCO (2019b) reported a similar scenario as well.

To ensure that kids remain in school despite a changing climate, there is need for more research to understand how climatic conditions impact education in different places and among different age groups. For example, climate change challenges may bring in agricultural knowledge that could bring in the adoption of drought- or heat-tolerant crop varieties or short-term varieties like the “Katumani” maize hybrid, cassava among other crops which may help the farmers to have sufficient income and food, hence not negatively impacting on the learning of their children (FAO, 2017). Further, climate change knowledge can provide sources of income to farmers during periods of adverse weather to prevent income losses. Knowledge about climate change may be a solution that would help improve the nutrition of young children and pregnant women, thereby lessening the need for agricultural families to pull their kids out of school. There are many other potential pathways to be explored to promote education provision in the wake of climate change.

Learners who have knowledge about climate change and its effects may be able to understand periods in which they are in and try to be vigilant about the kind of water they drink and from which sources. Learners will be aware that during heavy rains, most of the streams and brooks are contaminated with water-borne diseases causing organisms. Therefore, they will avoid drinking water directly from instreams but rather boil or treat it before drinking. Further, instream bathing will be avoided because learners will be knowledgeable about the varied toxins released by several species of algae which flourish well in nutrient rich water bodies because or leaching caused by heavy rains due to climate change. When water was tested, the following results were recorded (Table 1).

Table 1. Percentage abundance of algae in the Nyakomisaro – Riana River during the period May 2018 to October 2019

Species	Class	Counts	Percentage Abundance
Microcystis aeruginosa	Cyanophyceae	31139	33.89
Anabaena flos-aquae	Cyanophyceae	8114	8.83
Aphanocapsa rivuralis	Cyanophyceae	5436	5.92
Anabaena cirnalis	Cyanophyceae	4143	4.51
Coelomoron vestitus	Chlorophyceae	3379	3.68
Coelomoron mierostoldes	Chlorophyceae	3239	3.53
Coelomoron reguraris	Chlorophyceae	3065	3.34
Microcystis flos-aqua	Cyanophyceae	3015	3.28
Chroococcus dispersus	Cyanophyceae	2716	2.96
Chroococus turgidus	Cyanophyceae	2077	2.26
Merismopedia tennuissimma	Cyanophyceae	1854	2.02
Anabaenopsis tanganyikae	Cyanophyceae	1392	1.51
Coelastum microporum	Chlorophyceae	1120	1.22
Romeria elegans	Chlorophyceae	993	1.08
Aulacoseira nyansenssis	Diatom	992	1.08
Others			< 1.00

Source: Researcher's own work; not copyrighted

Excessive algal growth due to nutrient overload in River Nyakomisaro-Riana can have undesirable and detrimental effects on water end users. These effects concern taste and odour, increased levels of cyanotoxins such as microcystins, and increased turbidity levels in source water. Certain algal species known to produce the water odours (for example, musty, earthy, fishy, or grassy) are often the culprits of taste and odour problems in eutrophic waters. Nutrient enrichment to waterbodies due to heavy rains associated with climate change can cause undesirable algal species such as cyanobacteria to dominate in a waterbody.

Cyanotoxins produced by cyanobacteria or blue green algae are potentially lethal to animals and humans even in small quantities. These cyanotoxins are of two types; hepatotoxins such as microcystins that can cause serious damage to the liver; and neurotoxins which are toxic to the nervous system. For example, health effects from massive algal blooms triggered a shutdown of drinking water intakes in Kansas, USA (Kansas Department of Health and Environment, 2008). Therefore, prevention from the use of waters enriched by cyanobacteria is necessary for preventing ailments of both school children, their parents and the community as a whole.

For malaria, it was clear from the findings that the malarial complaints addressed in Kisii Teaching and Referral Hospital followed the sporadic rainfall trends associated with climate change (Figure 1). This corroborates well with the findings of Biseko and Jae (2022) who found out that the spread of malaria is related to climate change because temperature and rainfall are key parameters of climate change. Fluctuations in temperature affect the spread of malaria by lowering or speeding up its rate of transmission while the amount of rainfall also affects the transmission of malaria by offering a lot of sites suitable for mosquitoes to breed in. The high malarial complaints indicate that the learners or their parents who were infected directly or indirectly affected learning. Therefore, with the climate change knowledge in the curriculum, the learners and the community will be well prepared and know how to undertake precautionary measures to avoid such ailments or reduce them.

Figure 2. Rainfall amounts and malarial related complaints during the May 2018 - October 2019 sampling period within Nyakomisaro-Riana River
Source: Researcher's own work; not copyrighted

Case 1

Spiralling gang violence threatens to bring back Kisii's feared Sungu Sungu days

NEWS By: Erick Abuga | Jun 10th 2023
Source: Standard Newspaper

From murder, and robbery to attacks from machete-welding gangs, Kisii town is slowly creeping back to crime. Youths seem to have taken over the town, terrorising residents in nearly all the estates at night. Community policing members have not been spared either. Last Friday, at least four of the members were attacked around the Daraja Mbili area by machete-welding street gangs. They were treated at the Kisii Teaching and Referral Hospital (KTRH) with deep panga cuts.

The same day in Nyanchwa area near Kitutu Central Police Station, a businessman who was heading home after closing his M-Pesa shop was attacked by two people armed with a pistol. According to a police report, the businessman was attacked as he was entering his compound. He run back to the road as they chased after him up to a nearby shop where he sought safety. Locals responded, forcing the attackers to open fire with a bullet injuring a 28-year-old man. The victim is being treated at the KTRH.

A police officer privy to the case said they have retrieved CCTV footage from a nearby shop that will help them identify the attackers. Two weeks ago, a journalist and a medical officer attached to KTRH were attacked in two separate incidents. They both lost their phones, money and other valuables. On Wednesday last week, a county employee was discharged from the hospital after admission at KTRH. He was treated for deep panga for cuts that he sustained from his attackers along the Nyanchwa-National Housing Road. Locals headed home as early as 10 pm are attacked by machete-wielding gangs.

The gangs, walking in a group of two to five have been terrorising locals in the town in the last five months. They steal money and any other valuables. Charles Kenganda, a member of the County Assembly and who stays in Nyanchwa says the number of young men roaming the area at night is alarming. "People are now forced to go home latest 8 pm. This seems like an organised crime that will soon kill the economy of this town. Security apparatus must come out to tame this kind of attacks and robberies."

In the recent past, several street children have also relocated to the town's CBD after their habitat; the town's dumping site was destroyed. A few months ago, the Kisii County government began the process to reclaim the area. More than 150 street children who have been camping in the area for more than a decade have since moved to the town posing a security risk to the residents. Peterson Ondimu, a manager at one of the restaurants in town says the street urchins have been attacking female revellers. "They break into private cars and steal music systems as well other properties. It is becoming impossible to walk with a female in town, they will attack you and demand some money."

Kisii County Commander Charles Kases said they are dealing with the criminals and some of the politicians funding some of the criminal activities. "We are handling these cases one by one. Some murders are related to family conflicts and others are assault cases. We have deployed a 24-hour security patrol system to weed out all the muggings. This will be a thing of the past."

Destruction of Schools and Other Infrastructure

Change in climatic conditions can affect schools in different ways. In the most direct sense, extreme weather events such as heavy rains may destroy or damage school buildings, or schools may be used to shelter people who have been displaced from their homes (Case 2). This leaves children temporarily unable to attend school, with some kids never returning to continue with their studies.

Figure 3. Learning fails to take off as flood victims camp in classes
Source: People daily: Wednesday, May 10th, 2023 [not copyrighted]

Case 3

Floods Displace Thousands of Households in Mandera and Kisumu

Floods in Mandera, Kenya, April 2023. Photo: Kenya Red Cross Society
Source: Daily Nation

Floods continue to cause hardship for communities in Kenya, where authorities report flooding has impacted around 5,000 families in the west and northeast of the country over the last week. Kenya Red Cross reported that data collected on 27 March showed 36,432 people affected across the country. In recent days flooding has impacted areas of Mandera County in the northeast of the country, close to the border with Somalia. Kenya Red Cross reported over 2,000 families were impacted after flooding began around 21 April. At the beginning of April, flooding cut roads in areas around Kutulo. Heavy rain since 23 April has caused several rivers to overflow in Kisumu County in the west of Kenya. On 24 April heavy rain caused the Maenya River to break its banks, flooding areas of Kolwa Central and displacing over 600 families. Local government and media later reported other areas were affected in Kisumu West and Kisumu East, with a total of 725 families displaced.
The Nyando River broke its banks on 30 April, flooding areas of Nyando Sub-County, in particular around Ahero. Areas of Nyakach were also flooded. In a statement of 30 April, Kisumu governor Professor Anyang Nyong'o asked all residents in flood-prone areas to move to safe ground immediately. The Kisumu government reported 1,500 households displaced in Nyando and a total of 2,700 across the county, as of 30 April. Many of the displaced families have lost material possessions, crops and livestock. The governor said the flood victims were in dire need of water, medicine, blankets, mattresses, mosquito nets and food. Heavy rains were expected to continue and more families were likely to be affected, the governor added.

School buildings and other infrastructure like roads are critical to education provision. When floods due to heavy rains hit the schools and connecting infrastructure (murram roads serving schools within the Kisii Municipality), learning was affected because learners struggled to reach school and some never turned up to learn. For example, within the same period, the climate change induced rains led to schools' destruction (Case 3). Similarly, some schools served as temporary shelters for families who had lost their homes and linkage roads. This is among the most visible and direct effects that climate change has brought on education, which heavily affects school children.

Learners and teachers struggle wading through the dilapidated muddy roads, while some are completely unable to go to school, hence missing the foundational teachings because of climate change related disruptions. Further, teachers are affected as they struggle to meet the syllabus requirements and recovery from so much time lost. Indeed, it is not a mean task to get kids back into the classroom. Some children simply never go back to school! With the number of climate events increasing in frequency and scope each year, these interruptions to education are expected to increase as well. This corroborates well with the findings of Friedrich (2009) whereby he found out that over 1.8 million children were out of school irrespective of the Kenyan government putting aside 30% of its public expenditure for education. Therefore, floods may affect teaching and learning as most of the learners are displaced and drop out of school. The floods affect most schools and bring a lot of damages to the schools' infrastructure such as classrooms, toilets/ latrines, books and playground which are essential amenities for the provision of quality education.

Figure 4. Floods displace thousands of households in Mandera and Kisumu
Source: Daily Nation: April 2023 [not copyrighted]

Case 4

Two schools in Budalang'i face closure over flooding

DISASTER: *Flood victims in Budalangi, Busia*
Image source: The Star Newspaper, 5th March 2020

Musoma primary and secondary schools in Bunyala South, Budalang'i constituency, are likely to be closed by the public health department due to flooding. The floodwater is from the rains currently falling in most parts of the country. Pit latrines in the two learning institutions are flooded, putting the health of the children at risk. Pupils said they were forced to wade through the floodwater to reach their respective schools.

Early childhood education pupils are most affected as they are unable to wade through the floodwater and need their parents and guardians to take them to school. The situation has hampered parents from taking their children to Form One at Musoma Secondary School. Teachers are said to be seeking transfers to other schools.

Parents urged the public health department to move with speed and close the two learning institutions to save their children from contracting waterborne diseases such as cholera and bilharzia. Busia deputy governor Moses Mulomi who toured the two schools told the principals the national and county governments were working on a solution to the problem. Over 400 people from Khajula location have so far been forced out of their homes by floodwater.

Impact on Sources of Income and Education

Findings from the questionnaires indicated that 68% of the street children did attribute their exit from school because of lack of stable income which was to support them finish their education. Quoted in verbatim, one street child did say "when the dry spell came about, the tea bushes were no longer yielding, the plucking we used to undertake twice a month was no more, therefore our parents got no earnings at the end of the month from the "Chai Sacco" which is contracted by Kenya Tea Development Authority (KTDA) to pay farmers for their weighted tea leaves". In fact, Sub-Saharan Africa has experienced several years of failed rains that have complicated the lives and livelihoods of millions of farmers and pastoralists. For example, across Kenya, the Maasai, Samburu, Turkana, Rendile, Somali communities have gone through difficult times of their life perennially.

Agricultural households suffering losses to income and food security caused by droughts or heat waves may not have enough money to pay school fees or may pull their children out of school to help earn additional income. During severe droughts, girls may miss school because they must travel long distances to collect water or may be married off at an early age, which often coincides with dropping out of school. In addition, entire families may migrate in search of food, water, and employment, pulling their kids out of school.

When families lose their reliable sources of income, parents/guardians are often unable to pay school fees, buy uniforms, and other needed academic stationery materials. Sometimes they are pushed to a level of pulling their children out of school so that they can either help out at home or earn money to help the family. The worst-case scenario is where some parents/guardians withdraw their children from schools. For the case of boys, they are utilized in taking care of animals and the worst case is where girls are pushed into early marriages with hopes of giving them a chance for a better life with someone who can support them and lower their own expenses. None of these are easy decisions to make, but parents/guardians find themselves pushed by the realities of the climate crisis, especially among the Maasai, Samburu and other ethnic communities in Kenya.

Water Quality and Teachers' Ability to Provide Quality Education

Water quality of River Nyakomisaro-Riana was highly compromised. The water quality testing results did indicate that climate change brought about flooding which increased the microbial load and nutrients into the Kisii Municipality rivers (Ogendi, 2022) and other rivers in general. Also, during the dry spell water temperatures were increased. With the increase of microbial load during the heavy floods, it was accompanied with a lot of school absenteeism of leaners because of children suffering from stomach pains associated with waterborne diseases like typhoid, salmonellosis among others (Case 4).

Figure 5. Two schools in Budalangi face closure over flooding
Source: The Star Newspaper: 5th March 2020 [not copyrighted]

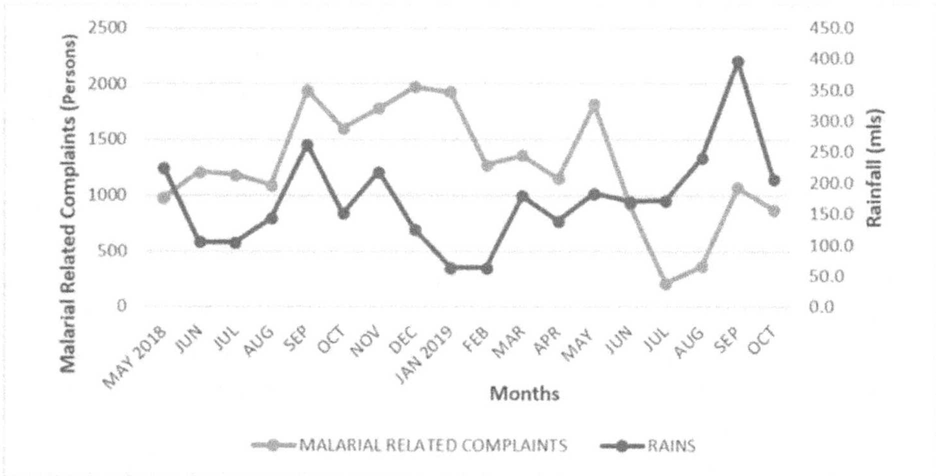

The teacher-student ratio is a thorn in the flesh with regard to provision of quality education. Therefore, absenteeism or coming late of a teacher from/to class means a lot to curriculum implementation and delivery. From observations and questionnaire findings, street children indicated that they left school during the long periods of rains on reasons that teachers came late to school and the attendance register was not strictly marked. So, street children capitalized on this and remained absent from school as they joined their fellow street children in the town streets. Secondly, with climate change and its attendant disasters, among them, upsurge of waterborne ailments and malaria, teachers as well as students are equally affected. In schools where there are inadequate instructors, a climate-related disease erupting in a school may bring learning to a standstill. For example, water sampling done along the Nyakomisaro river which passes through four schools in Kisii municipality did indicate high levels of coliforms (Figure 2). The rise of coliforms in this river followed the sporadic rainfall which was due to climate change.

Figure 6. Fecal coliforms and average monthly rainfall in the Nyakomisaro-Riana River during the May 2018 – October 2019 sampling period
Source: Researcher's own work; not copyrighted

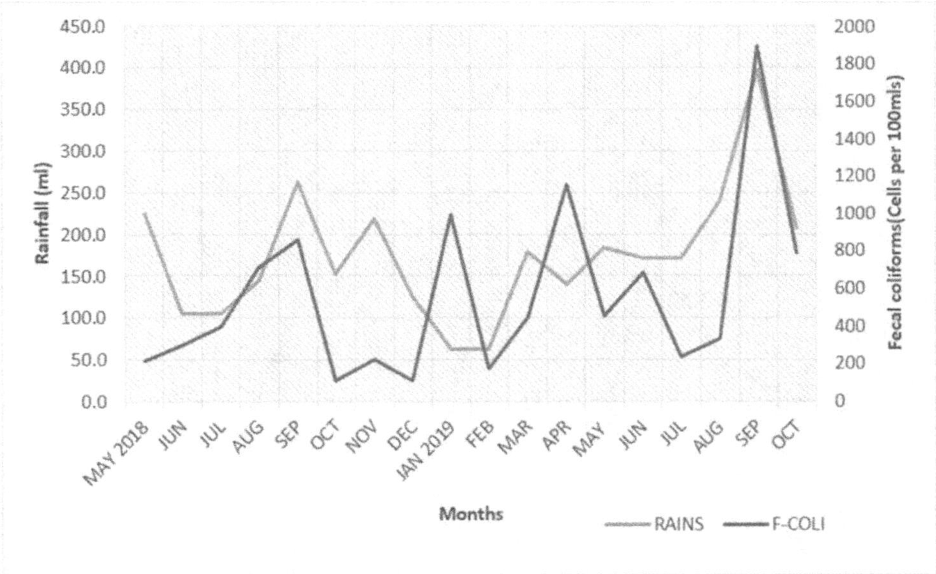

Whenever there was a fall, the coliforms also increased in the river. The high fecal coliform counts at the river can be attributed to the sporadic heavy rains which led to sewage leakages and the transportation of the wastes into nearby waterbodies. Further, dumped animal waste which had fecal coliforms from nearby bushes along the river, the roads and paths were transported by the raging floods (Ogendi et al. 2015). This rise of fecal coliforms in River Nyakomisaro waters translates to sickness among learners or parents/guardians upon consuming water drawn directly from the river during the floods. These coliforms bring about varied stomach upsets and pains, typhoid, salmonella, among others. For example, data collected from Kisii Teaching and Referral Hospital (KTRH) in the period May 2018 through October 2019 did indicate that typhoid related complaints followed the trend of fecal coliforms in this river (Figure 3). These ailments cause learners to miss school and stay in their homes or visit hospitals where their health will be attended to.

Figure 7. Fecal coliforms in the Nyakomisaro-Riana River and typhoid-related complaints in KTRH during the May 2018 - October 2019 sampling period
Source: Researcher's own work; not copyrighted

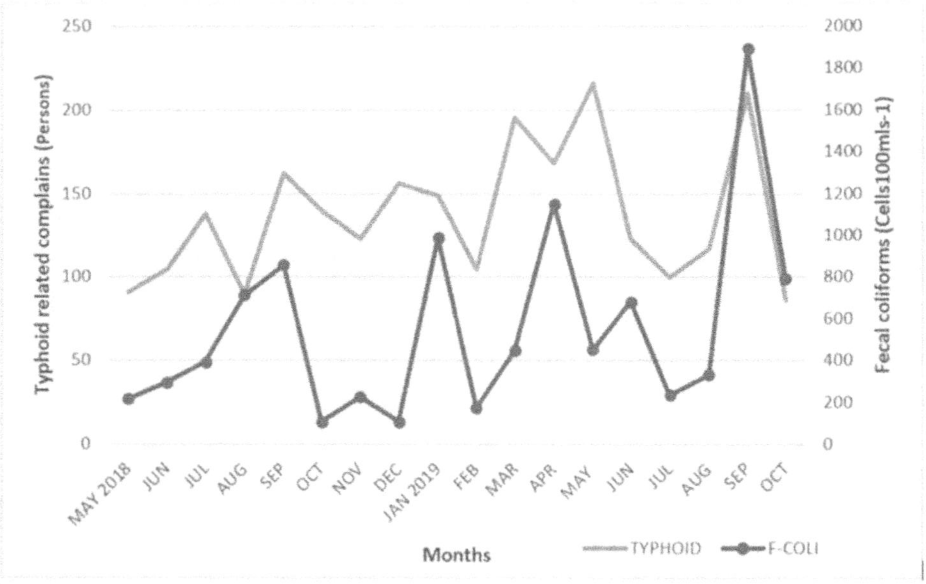

Climate-related school closures threaten quality education provision. Even the most qualified teachers struggle to help students who have missed school accordingly, especially if it's a prolonged period. This therefore, requires such schools to reset their teaching timetables to include remedial sessions in the morning and night. At times, teachers and leaners are forced to utilize the weekends in covering the lost time in as much it affects the freedom of worship of children and teachers. Sometimes, situations may worsen making it hard to honor such arrangements for both students and teachers; for example, when sources of their water like brooks or nearby rivers dry up. This drying up of water sources leads to lack of water, thus forcing schools to close or students and teachers emigrate to other places until such a time when water will be available, hence affecting quality education delivery. Extreme water stress already has affected over a quarter of the world's population (Hofste et al., 2019) and is expected to increase with global climate change which not only aggravates famine but also increases forced migration and war (Hoffmann et al., 2020). For example, an estimated population of 22.5 million people have been displaced by weather-related disasters (Internal Displacement Monitoring Centre, 2021), and so are increases in mental health vulnerability (Siriwardhana & Stewart, 2013).

Psychological Torture on Children

As per data collected through questionnaires administered to the children, nearly 80% of the respondents indicated their fear of climate change and its related consequences. This corroborated with the findings of Vergunst and Berry (2022) who found out that there is a heightened stress and anxiety about the consequences of living on a warming planet experienced by young learners especially of the ages between 6 and 12 years. When these students are shown clips on how climate change affects people and their animals, the emaciated animals and people disturb them. Therefore, the introduction of climate change teachings to these youngsters at an early age will prepare them for such happenings and make them prepare and appreciate in case of such climate change effects face them. Climate change education will help them have a know-how on dealing with impacts of climate change for they are the most likely to experience higher psychological torture than the elderly.

Climate change is worsening the situation as it is increasing the frequency, intensity and duration of weather-related stressors such as drought, heat waves, cyclones and floods (USGCRP, 2018). Exposure to these events is linked to increased anxiety and depression, which compromise long-term mental health resilience of younger people as well as the elderly. This can be intensified by other physical effects of climate change, such as disrupted sleep due to a heat wave, and thus affecting the health and education of learners hence compromising their long-term social and economic prospects.

In the tropics, especially the Arid and Semi-Arid Lands (ASALs), too much exposure to extreme temperature or rains is affecting the unprepared young learners by either making them lose good class hours of learning because of going to school late, learning under trees or forced out of classes because of extreme temperatures. As climate change is leading to increasingly severe heat waves, floods, droughts, and hurricanes, it is important for policy makers to understand how this extreme weather impacts students' education and develop appropriate policies and solutions which will keep children in school in a globe of increasing climate variability.

FUTURE RESEARCH DIRECTIONS

There is need to undertake research on how the global climate change policies have been tailored to climate change education and their implementation on the education and training programs in the rest of the countries in the world. Further research is also necessary on the strategies to be employed to enhance environmental awareness in schools since the effects of increasing global warming and climate change are becoming more apparent and complex. Also, a study needs to be conducted on how the immediate, rapid, and large-scale enactment of national and county governments' climate-related laws and policies have guided them in the mainstreaming of climate change in the school curricular. Finally, research needs to be conducted to establish the effects of school feeding on reducing school dropouts related to lack of food in learners' homes.

Way Forward

Education is one of the best tools to mitigate climate change challenges. If all students and the relevant stakeholders are to have sound knowledge about climate change and its effects, then, it will be easier to cope with climatic stresses than it would be with an uninformed citizenry. Because of the failure of most of our institutions to incorporate climate change aspects in their curricula, the understanding of climate change and its consequences is minimal among many people. Therefore, there is urgent need for schools to integrate climate change in the curricula, and offer it at all levels of education. Incorporating climate change and its effects into the education curricula will enable communities to not only adapt to, but address the resultant challenges, including the related disasters. In the wake of climate change challenges, an empowered and engaged society in the same stands to respond most appropriately, thereby fostering sustainable development.

ACKNOWLEDGMENT

The author registers a vote of thanks to the Kisii University Research Department for providing some funds which made this research, and the Kisii Teaching and Referral Hospital (KTRH) laboratory technical manager for the provision of the disease incidences data.

REFERENCES

Biermann, F., & Boas, I. (2010). Preparing for a warmer world: Towards a global governance system to protect climate refugees. *Global Environmental Politics*, *10*(1), 60–88. doi:10.1162/glep.2010.10.1.60

Biseko, J. M., & Jae, W. L. (2022). Relationships between transmission of malaria in Africa and climate factors. *Scientific Reports*, *12*(1), 14392. Advance online publication. doi:10.103841598-022-18782-9 PMID:35999450

Campbell-Lendrum, D., Wheeler, N., Maiero, M., Villalobos Prats, E., & Nevelle, T. (2018). *World Health Organization COP24 special report on health and climate change*. WHO. https://www.who.int/publications/i/item/cop24-special-report-health-climate-change

Christian Aid. (2007). *Human tide: The real migration crisis.* Christian Aid Report.

FAO. (2017). *The Future of Food and Agriculture – Trends and Challenges.* https://www.fao.org/3/a-i6583e.pdf

Friedrich, H. (2009). *International Education Statistics.* http://www.childinfo.org.statistics

Georgeta, S. M. (2020). The Social Communication of the Street Children. *Scientia Moralitas Conference Proceedings.* 10.5281/zenodo.4385036

Hoffmann, R., Dimitrova, A., Muttarak, R., Crespo Cuaresma, J., & Peisker, J. (2020). A meta-analysis of country-level studies on environmental change and migration. *Nature Climate Change, 10*(10), 904–912. doi:10.103841558-020-0898-6

Hofste, W., Reig, R., & Schleifer, L. (2019). *17 countries, home to one-quarter of the world's population, face extremely high-water stress.* World Resources Institute. https://www.wri.org/blog/2019/08/17-countries-home-one-quarter-world-population-face-extremely-high-water-stress

IDMC. (2020). Global report on internal displacement 2020. IDMC.

Internal Displacement Monitoring Centre. (2021). *Internal displacement in a changing climate*. Norwegian Refugee Council. Retrieved March 13, 2023 from https://www.internal-displacement.org/sites/default/files/publications/documents/grid2021_idmc.pdf

IOM. (2012). *International dialogue on migration 18: climate change, environmental degradation and migration.* IOM.

IPCC. (2014). Climate change 2014: synthesis report. In R. K. Pachauri & L. A. Meyer (Eds.), Contribution of working groups I, II and III to the fifth assessment report of the intergovernmental panel on climate change (pp. 1–168). Academic Press.

Kansas Department of Health and Environment Bureau of Water. (2011). *Water Quality Standards White Paper Chlorophyll-a Criteria for Public Water Supply Lakes or Reservoirs.* Retrieved April 14, 2023 from https://ehs.ku.edu/

Lala, B. & Hagishima, H. (2023). Impact of Escalating Heat Waves on Students' Well-Being and Overall Health: A Survey of Primary School Teachers. *Open Access*, *11*(6), 126. doi:10.3390/cli11060126

Morton, J. F (2007). The impact of climate change on smallholder and subsistence agriculture. *Scientific Research.* . doi:10.1073/pnas.0701855104

OECD. (2022). *Building the future of education.* https://www.oecd.org/education/future-of-education-brochure.pdf

Ogendi, G. M. (2022). Effects of nutrients on the water quality and the algal community of River Nyakomisaro-Riana, Kisii, Kenya. *International Journal of Fisheries and Aquatic Studies*, *10*(5), 111–118. doi:10.22271/fish.2022.v10.i5b.2734

Ogendi, G. M., Getabu, A. M., Onchieku, J. M., & Babu, J. M. (2015). Assessment of the Microbial Load of Nyanchwa-Riana and Nyakomisaro-Riana Rivers, Kisii, Kenya. *International Journal of Fisheries and Aquatic Studies*, 182–192. www.fisheriesjournal.com

Siriwardhana, C., & Stewart, R. (2013). Forced migration and mental health: Prolonged internal displacement, return migration and resilience. *International Health*, *5*(1), 19–23. doi:10.1093/inthealth/ihs014 PMID:24029841

UN Human Rights Council. (2010). *Study of the Office of the United Nations High Commissioner for Human Rights on challenges and best practices in the implementation of the international framework for the protection of the rights of the child in the context of migration.* Author.

UNDP. (2017). *Climate change, migration and displacement: The need for a risk-informed and coherent approach.* https://cdn.odi.org/media/documents/11874.pdf

UNDP-Zimbabwe Human Development Report. (2017). *Climate Change and Human Development: Towards Building a Climate Resilient Nation.* https://www.undp.org/sites/g/files/zskgke326/files/migration/zw/UNDP_ZW_2017ZHDR_Briefs---Climate-Change-and-Education.pdf

UNESCO. (2015). *SDG4 - Education 2030: framework for action.* UNESCO.

UNESCO. (2018). *Ensuring the right to equitable and inclusive education: results of the ninth consultation of Member States on the implementation of the UNESCO Convention and Recommendation against Discrimination in Education.* UNESCO.

UNESCO. (2019a). *Enforcing the right to education for refugees.* UNESCO.

UNESCO. (2019b). *Indigenous peoples' right to education: overview of the measures supporting the right to education for indigenous peoples reported by Member States in the context of the Ninth Consultation on the 1960 Convention and Recommendation against Discrimination in Education*. UNESCO.

UNFCCC COP 21. (2015). *The twenty-first session of the Conference of the Parties (COP) and the eleventh session of the Conference of the Parties serving as the meeting of the Parties to the Kyoto Protocol*. Author.

UNICEF. (2019). *It is getting hot: call for education systems to respond to the climate crisis, perspectives from East Asia and the Pacific*. UNICEF East Asia and Pacific Regional Office.

UNICEF. (2021). *The climate crisis is a child rights crisis Introducing the Children's Climate Risk Index*. UNICEF Division of Communication. www.unicef.org

Urama, K. C., & Ozor, N. (2010). *Impacts of climate change on water resources in Africa: the role of adaptation*. https://www.researchgate.net/publication/267218899_Impacts_of_climate_change_on_water_resources_in_Africa_the_role_of_adaptation#fullTextFileContent

USGCRP. (2018). *Fourth national climate assessment, Vol. 2: Impacts, risks, and adaptation in the United States*. Rep, US Glob. Change Res. Progr. https://nca2018.globalchange.gov/downloads/NCA4_2018_FullReport.pdf

Vergunst, F., & Berry, H. L. (2022). Climate Change and Children's Mental Health: A Developmental Perspective. *Clinical Psychological Science. Volume*, *10*(4), 767–785. doi:10.1177/21677026211040787 PMID:35846172

Watts, N., Amann, M., Ayeb-Karlsson, S., Belesova, K., Bouley, T., Boykoff, M., Byass, P., Cai, W., Campbell-Lendrum, D., Chambers, J., Cox, P. M., Daly, M., Dasandi, N., Davies, M., Depledge, M., Depoux, A., Dominguez-Salas, P., Drummond, P., Ekins, P., ... Costello, A. (2018). The Lancet Countdown on health and climate change: From 25 years of inaction to a global transformation for public health. *Lancet*, *391*(10120), 581–630. doi:10.1016/S0140-6736(17)32464-9 PMID:29096948

World Bank. (2018). *Groundswell: Preparing for internal climate migration*. Author.

ADDITIONAL READING

Bedford, D. (2016). Does Climate Literacy Matter? A Case Study of U.S. Students' Level of Concern about Anthropogenic Global Warming. *The Journal of Geography*, *115*(5), 187–197. doi:10.1080/00221341.2015.1105851

Christensen, R., & Knezek, G. (2015). The Climate Change Attitude Survey: Measuring Middle School Student Beliefs and Intentions to Enact Positive Environmental Change. *International Journal of Environmental and Science Education*, *2015*(10), 773–788.

Deisenrieder, V., Kubisch, S., Keller, L., & Stötter, J. (2020). Bridging the Action Gap by Democratizing Climate Change Education-The Case of k. i.d.Z.21 in the Context of Fridays for Future. *Sustainability*, *12*(5), 1748. doi:10.3390u12051748

Frick, J., Kaiser, F. G., & Wilson, M. (2004). Environmental Knowledge and Conservation Behavior: Exploring Prevalence and Structure in a Representative Sample. *Personality and Individual Differences*, *37*(8), 1597–1613. doi:10.1016/j.paid.2004.02.015

Hung, C. C. (2014). *Climate Change Education: Knowing, Doing and Being*. Routledge. doi:10.4324/9781315774923

IPCC Climate Change. (2021). *The Physical Science Basis. Contribution of Working Group I to the Sixth Assessment Report of the Intergovernmental Panel on Climate Change*. Cambridge University Press.

IPCC Climate Change. (2022). *Impacts, Adaptation, and Vulnerability. Contribution of Working Group II to the Sixth Assessment Report of the Intergovernmental Panel on Climate Change*. Cambridge University Press.

Kuthe, A., Keller, L., Körfgen, A., Stötter, H., Oberrauch, A., & Höferl, K. M. (2019). How Many Young Generations Are There? –A Typology of Teenagers' Climate Change Awareness in Germany and Austria. *The Journal of Environmental Education*, *50*(3), 172–182. doi:10.1080/00958964.2019.1598927

Sipos, Y., Battisti, B., & Grimm, K. (2008). Achieving Transformative Sustainability Learning: Engaging Head, Hands and Heart. *International Journal of Sustainability in Higher Education*, *9*(1), 68–86. doi:10.1108/14676370810842193

KEY TERMS AND DEFINITIONS

Climate Change: Long-term shifts in temperatures and weather patterns.

Disaster: Sudden calamitous event that disrupts the functioning of a community or society and that exceeds its ability to cope using its own resources.

Impact: Something having a strong effect on someone or something.

School Education: Knowledge and skills gained through a study of a curriculum.

Street Children or Urchins: Homeless children living in the town streets.

Waterborne Diseases: Illnesses caused by microscopic organisms, like viruses and bacteria, that are ingested through contaminated water.

Chapter 8
Building Bridges to Sustainability:
Overcoming Barriers and Promoting Integration in Architectural Education

Seyda Emekci
https://orcid.org/0000-0002-5470-6485
Ankara Yildirim Beyazit University, Turkey

ABSTRACT

The consideration of environmental factors plays a key part in the design of sustainable buildings within their respective contexts. The global community is currently grappling with the adverse impacts of climate change, which can be attributed to the release of greenhouse gases into the atmosphere. Architecture is a unique area of study that enables the development of spatial solutions to address human requirements, while also carrying the essential duty of ensuring the creation of a sustainable built environment. Although the inclusion of sustainability in architectural education is widely acknowledged, the majority of curricula lack a comprehensive integration of this topic. In order to conceptualize a sustainable built environment, it is imperative for the architectural curriculum to incorporate a foundational comprehension of the interconnection between sustainability, technology, and design. This chapter analyzes the barriers to sustainable environmental design in architectural education. It also identifies worldwide integration promotion efforts.

DOI: 10.4018/978-1-6684-9099-0.ch008

INTRODUCTION

Climate change has a significant impact on various regions across the globe. The impact of increasing sea levels and the occurrence of extreme weather events, such as intense precipitation, flooding, severe storms, and unanticipated heat waves, significantly influence the overall well-being of human populations. Based on the findings presented in the United Nations World Population Prospects, it is projected that approximately three-quarters of the world's population will reside in urban areas by the mid-21st century (United Nations, 2022). The report indicates that climate change will lead to a decline in the provision of water services and overall living conditions in urban regions. The increasing prevalence of impermeable surfaces leads to the occurrence of flooding events and the contamination of surface waters. Urban heat islands have been found to have detrimental effects on energy consumption, greenhouse gas emissions, and human health and comfort (US EPA, 2022). The reduction in the groundwater table has detrimental effects on vegetation growth and can lead to the intrusion of saline water into coastal aquifers (Şen, 2015). The availability of freshwater is diminishing. The European Commission (EC) highlights the significant ramifications for Europe and identifies the various challenges associated with a climate crisis (European Commission, 2022).

Under such circumstances, it is imperative to reconsider urban areas as experimental spaces for pioneering climate adaptation and mitigation strategies. The existence of such aspirations is substantiated by a multitude of documents that have been signed by state representatives and municipalities. As a result, numerous research and development initiatives are currently directed toward addressing the challenges posed by the climate crisis. The United Nations 2030 Agenda for Sustainable Development, regarded as a pivotal framework for addressing poverty and environmental conservation, encompasses a comprehensive set of 17 goals (UN, 2022).

Goal 11: make cities and human settlements inclusive, safe, resilient, and sustainable;

Goal 13: take urgent action to combat climate change and its impacts. (UN, 2022)

According to current statistics, approximately half of the world's total energy consumption is accounted for by buildings, making them a significant contributor to the acceleration of global warming and the transformation of natural ecosystems (IPCC, 2007). Within the framework of the present climate crisis, it is crucial to acknowledge the influence of buildings on the environment (IPCC, 2017) and the increasing emphasis on ecological consciousness in relation to regulations pertaining to the construction industry (e.g. the European Directive on Energy Performance of Buildings, 2003). Consequently, the role of higher education in effectively educating future architects on the principles and methodologies of architectural education is gaining considerable importance. However, this endeavor encounters various obstacles of both an educational and professional nature. The necessity to commence a transformation in the education and training of building professionals to facilitate the effective integration of environmental considerations within the field of architecture, encompassing aspects such as climatic design, material selection, construction methods, passive and hybrid strategies, resource efficiency, and impact reduction, is primarily driven by three key factors (Altomonte, 2009):

- The current practices in the construction industry have exhibited a lack of consistent responsiveness to the imperative of improving sustainable environmental design within the context of innovative architectural discourse.
- The accreditation and qualification criteria established by professional organizations have not effectively contributed to the systematic advancement and widespread adoption of environmentally sustainable principles in building design.
- University curricula have demonstrated limited efficacy in systematically incorporating the principles of sustainable environmental design into the education of architecture students.

Architects and urban planners are actively engaged in the ongoing process. Nevertheless, one must consider the following inquiry: to what extent does the architectural studies curriculum effectively tackle concerns related to climate change, and do graduates possess the necessary skills to actively engage in the development of strategies that are sensitive to climate change? Can architectural education serve as an efficacious mechanism for addressing climate change for the purpose of achieving a sustainable future? The objective of this book section is to conduct a critical analysis of the obstacles that impede the effective incorporation of sustainable environmental design in higher education and architectural practice. Additionally, it aims to identify the necessary efforts and initiatives needed to promote this integration on a global scale.

LITERATURE REVIEW

Existing literature has demonstrated a significant emphasis on evaluating and overseeing the integration of climate change within various higher education establishments globally (Cowan & Van Der Ryn, 1994; Dumitru, 2017; Hegarty et al., 2011; Vincent & Dutton, 2014). The research community has initiated investigations into the integration of sustainability education within architecture programs, in response to the significant influence of sustainability and sustainable development on the built environment (Alcázar & Chavez, 2014; Hyland & Grierson, 2012). Numerous research studies have been conducted to explore the integration of sustainability education and architectural programs in different countries. Mavromatidis et al. (Mavromatidis, 2018) emphasized the pressing necessity of establishing a multidisciplinary decision-making framework encompassing building and urban design. This framework aims to enhance comprehension of the influence of city development on climate change by employing Bejan's "constructal law" as a means to describe the inherent inclination of flow systems in an interdisciplinary fashion. In 2014, a pioneering postgraduate study on this subject was initiated at Cracow University of Technology in partnership with the Polish Green Building Council (Celadyn, 2018). The researchers effectively emphasized the significance of integrating sustainability-focused coursework into postgraduate architecture programs.

To address the challenges posed by rapidly growing building and construction markets in emerging economies like Indonesia, a proposed method aims to facilitate the institutional development of built environment professions. This approach brings the discussion closer to the professional realm. The findings of this study indicate that relying solely on curriculum development is insufficient for the successful implementation of low-carbon futures. It is imperative for educational institutions to collaborate with various other entities, such as government bodies, practitioners, and industry stakeholders, in order to effectively drive, complement, and facilitate the necessary changes (Iyer-Raniga & Dalton, 2017). Various methodologies for incorporating sustainability in architectural curricula has been adopted on a global scale. Several effective strategies include incorporating sustainability elements into existing courses, developing dedicated courses focused on sustainability, and implementing activities to enhance students' awareness of sustainability. Extensive published research has suggested new approaches and methods for incorporating sustainability into studio teaching, as design studios are considered the fundamental component of architectural programs (Davis, 2010) To facilitate the integration of sustainable design principles into studio instruction, Naresuan University in Thailand devised strategies within the design studio that promote self-awareness, self-evaluation, and self-critique. The aforementioned activities were structured as games, exercises, and group discussions (Hengrasmee & Chansomsak, 2016). In their study, Mavromatidis (2018) devised a methodology for incorporating sustainability principles into the field of architecture education. This involved the amalgamation of three distinct educational frameworks pertaining to architectural synthesis. The seminar-workshop employed an innovative and pedagogical approach, which yielded highly promising outcomes while preserving the students' creative abilities throughout the entirety of the process.

Meeting The Challenge

The attainment of this vision necessitates a persistent and enduring endeavor to revolutionize education across all tiers. Despite the concerted endeavors of numerous individuals and collectives operating within the confines of the formal educational system, the pursuit of education aimed at fostering a just and sustainable world does not occupy a prominent position in the hierarchy of priorities. According to McIntosh et al.(2001), undoubtedly, individuals emerging from prestigious educational institutions are playing a significant role in steering society toward its present state, characterized by detrimental consequences, social disparities, and an unsustainable trajectory. According to Glyphis (2001), sustainable design has been incorporated into the curriculum and professional activities of only a limited number of architecture schools. This statement holds true in the realm of education across a wide range of intellectual disciplines and professions. The compelling evidence supporting the necessity for educational transformation is derived from the current global conditions, as well as the extensive endeavors undertaken by numerous non-governmental organizations (NGOs) and educational institutions focused on environmental and sustainability education. These efforts are aimed at rectifying the shortcomings inherent in the conventional educational system.

In relation to the question of why this particular occurrence occurs, the problem is influenced by various structural elements within the existing system. The complex and interdependent issues that society must address include the interactions between population, human activities, and the environment, as well as the strategies, technologies, and policies required for a secure, just, and environmentally sustainable future. These issues transcend disciplinary boundaries. In the realm of higher education, there exists a prevailing structure wherein knowledge is compartmentalized into distinct and highly specialized domains, as well as traditional fields of study.

The establishment of a sustainable trajectory for the future of humanity necessitates a fundamental change in mindset, shifting towards a systemic approach that places significant emphasis on collaboration and cooperation. A significant portion of higher education places emphasis on individualized learning and competition, leading to professionals who may lack adequate preparation for collaborative endeavors. The process of acquiring knowledge is often characterized by fragmentation, wherein faculty members encounter obstacles that discourage them from engaging in interdisciplinary collaboration or expanding their research beyond their own disciplines. These obstacles are primarily rooted in long-standing incentives such as tenure and research, as well as prevailing professional practices.

Higher education institutions have a significant ethical obligation to enhance the consciousness, understanding, competencies, and principles necessary for the establishment of an equitable and environmentally responsible future. Higher education plays a pivotal yet frequently disregarded role in actualizing this vision. This program caters to most individuals who are involved in the development, leadership, management, teaching, employment, and impact of societal institutions.

ARCHITECTURAL CURRICULA

In recent years, there has been a reevaluation of the correlation between architectural design and climate change within the realm of architectural education. One crucial aspect of this re-evaluation is the recognition that design and construction are no longer regarded as separate and disconnected domains that will inevitably and passively converge in the awareness of architectural students. Furthermore, it is acknowledged that not only architectural students should have a voice in design, but also in all stages of construction. The prevailing consensus in contemporary discourse is that the incorporation of construction, and consequently its influence on climate, is an indispensable and intrinsic component of the design process (McIntosh et al., 2001). As a result, there has been a recent examination of the architecture curriculum in relation to European Higher Education policies (Bologna Declaration 1999), with a particular emphasis on recognizing the significance of design and construction in an architect's education. Current discussions on architectural education not only address the theoretical aspects that underlie the intricate connection between design and construction (Herr, 2014) but also highlight the need for teaching methods that foster students' understanding of sustainability issues in a creative manner (Shareef & Farivarsadri, 2020).

The establishment of a symbiotic relationship between construction and design necessitates forging connections between these two domains. To date, the focus of construction in architecture schools has primarily revolved around the presentation of preconceived ideas in the form of technical solutions (Gelernter, 1988). The approach of viewing design as a predicament that can be resolved solely through sequential modifications, rather than as a cohesive process that can effectively bridge this divide, merely sustains its existence. Furthermore, the delineation between construction, design, and sustainability can be ascribed to the lack of direct involvement of teachers in the implementation of ideas, leading to a potential underestimation of the significance of their interconnection(Grover et al., 2020).

Architectural curricula in Europe are designed with significant reliance on the guidelines provided by esteemed organizations such as the Royal Institute of British Architects (RIBA), UNESCO, the Architects' Council of Europe (ACE), various accreditation boards, local chambers of architects, and numerous other institutions, committees, and stakeholders (Borucka & Macikowski, 2017). Furthermore, there exists a multitude of debates and disputes surrounding the conceptualizations of the architectural disciplines, which exert an impact on pedagogical approaches and subject matter (Celadyn, 2019). The European Directive 2005/36/EC on the recognition of professional qualifications is the primary legislative document that outlines the essential areas of knowledge and skills that architectural education should impart to its graduates. Regrettably, notwithstanding the disconcerting nature of the reports from the United Nations and the European Commission, this influential regulation fails to address the necessary competencies required for graduates to effectively contribute to the advancement of urban resilience and sustainability amidst a climate crisis (Nyka, 2019).

Within the European Directive, it can be observed that only a limited number of the 11 designated areas of knowledge and skills possess a somewhat ambiguous connection to the aptitudes required for a postgraduate student to cultivate innovative ideas centered around the concepts of sustainability and urban resilience.

- *An ability to create architectural designs that satisfy both aesthetic and technical requirements;*
- *An adequate knowledge of urban design, planning, and the skills involved in the planning process;*
- *An understanding of the structural design, construction, and engineering problems associated with building design;*
- *An understanding of the relationship between people and buildings, and between buildings and their environment; and of the need to relate buildings and the spaces between them to human needs and scale;*
- *An understanding of the profession of architecture and the role of the architect in society, in particular in preparing briefs that take account of social factors. (Directive 2005/36/EC of the European Parliament and of the Council of 7 September 2005 on the Recognition of Professional Qualifications (Text with EEA Relevance), 2005)*

The absence of explicit mention of sustainability in the European Directive should not be interpreted as a lack of consideration for this topic in the education of architects. On the contrary, sustainability occupies a prominent position in the discourse surrounding architectural education, albeit in a more comprehensive manner(Nyka, 2019). Certain issues explored by students can be perceived as being potentially influenced by the climate crisis. Students and teachers place significant emphasis on various aspects, such as the innovative repurposing of materials and structures (Celadyn, 2019), the preservation and modification of architectural heritage (Szczepański, 2019), and the incorporation of participatory methods in urban design (Smatanova & Urban, 2018). Nevertheless, a prominent inadequacy within architectural education is the limited integration with climate change. The development of urban adaptation and mitigation strategies should be integrated between various competencies and disciplines. However, the current architectural studies curricula, which adhere to the European Directive 2005/36/EC, tend to isolate future graduates from environmental concerns due to their highly intensive nature.

Architect's Role

Architects have a distinct need to possess a comprehensive understanding of space production, as their professional endeavors primarily involve the intricate coordination of various factors that contribute to the creation of novel spatial environments, setting them apart from other disciplines. Architects necessitate a foundational understanding of scientific principles, encompassing construction methodologies, effective communication, and keen observational aptitude, in order to proficiently conceive architectural designs. Additionally, they must have a discerning intuition to select and convey visual representations through the manipulation of forms and colors. Architects, in isolation, can be perceived as individuals who assume the role of managers in overseeing the process of creating a spatial environment on a limited scale. The development of a building's design necessitates the integration of three primary qualities: ecological design, user responsiveness, and meaningful imagery. The three qualities can be attributed to the conceptualized, perceived, and experiential triad. Architects require a comprehensive knowledge base to make informed decisions regarding appropriate construction techniques. Additionally, they must diligently observe human behavior and engage in effective communication with potential users to successfully incorporate their needs. Furthermore, architects must have the intuition necessary to skillfully amalgamate various design elements into a coherent and meaningful ensemble (Wiedmann & Salama, 2012).

Glyphis (2001) reflected the academic, commercial, and public understandings of three separate conceptions of the architect as three fractal triangles. The academic,

- The term "commodity," which is typically understood to indicate "fitness for use," is expanded to mean "effectiveness," in terms of the environment, the economy, and programs.
- Firmness combines long-term environmental sustainability, comfort, and longevity in addition to structural reliability.
- Delight incorporates deeper meaning in addition to aesthetic pleasure.

The second triangle represented the architect as a professional working with the customer to achieve the following three goals:

- The economy of means (using architectural means to ensure the lowest lifetime cost) and best effectiveness of outcomes (including environmental quality).
- Identity (increase knowledge of place and sense of self).
- Public acclaim (fame, esteem, and position).

The third triangle addressed the issues surrounding the architect's role as a protector of the public welfare (basic legal issues):

- Health is generally referred to as environmental health (clear air and water, sanitation, environmental quality, and stress reduction).
- Neighborhood security and crime prevention must now be considered in addition to fire and structural standards for safety.
- Welfare has been broadened to include issues of social fairness, harassment-free living conditions, and private property rights, privileges, and restrictions (Glyphis, 2001).

When an architect effectively incorporates these three triangles, they inherently contribute to the sustainable world. Therefore, through the implementation of appropriate construction methodologies for environmentally philosophical, and intellectual issues that are crucial to architecture education were reoriented using this representation.

The first fractal triangle emphasized the basic ideas put forward by Vitruviusfriendly structures, architects play a pivotal role in augmenting urban efficiency. When an architect engages in the design of a building that fulfills the needs of both individuals and communities, their contribution to urban diversity becomes evident. The role of an architect in shaping urban identity can be substantial when their design aligns with the subjective preferences of a given society. Architects assume a crucial role in the creation of built environments that embody ecological principles, exhibit diversity, and possess significant meaning.

RECOMMENDATIONS

Bringing Sustainable Design to Architecture and Design Schools

The concept of sustainability is widely discussed and debated across various disciplines. However, the impact of these debates on fields like architecture is often biased and lacks comprehensive consideration of its multifaceted nature. The phenomenon arises from either the expansive and debatable nature of sustainability or the insufficient interdisciplinary understanding of its various aspects. To bolster this argument, it is imperative to direct attention toward the perception of sustainability within the realm of architectural practice. The Brundtland Report (1987) provides a comprehensive explanation of the concept of sustainable development, encapsulated in a frequently cited definition: "Development that effectively addresses the requirements of the current generation while safeguarding the capacity of future generations to fulfill their own needs." This definition has effectively become the widely recognized phrase for research, debates, and reports that seek to incorporate sustainability into the discourse of a particular field of study. Therefore, the utilization of the term 'symbolized word' provided a clear and effective means of communicating the concept to the public, although it did deviate somewhat from the true scope of sustainability. In 1987, this particular definition undoubtedly offered valuable insight into the appropriate manner in which humans ought to engage with the environment and its resources (Murray & Cotgrave, 2007).

The incorporation of sustainability into education has emerged as a growing area of interest in recent decades, prompted by the Brundtland Report's critique of educational systems for their approach to teaching sustainability concepts (Figueiró & Raufflet, 2015). The prioritization of the architecture curriculum holds significant importance in fostering sustainability within the built environment. Nevertheless, there are still obstacles hindering the incorporation of sustainability into curricula. These barriers stem from various factors, including a lack of interest, awareness, compatibility in knowledge, and the ongoing challenge of establishing a definitive approach to teaching sustainability (Guy & Farmer, 2001). The fact that almost all architectural schools have classes on environmentally conscious design in their curricula should be emphasized (Frederick, 2012). Despite this, most schools work to teach sustainable architectural techniques because of fear that the field may become overly dominated by engineering features and that graduates from the education system may wind up working as "whole building engineers" (Hartman, 2012).

Considering the inherent nature of architectural design as a meticulous amalgamation of creativity, scientific expertise, technological advancements, and considerations of social and environmental factors, it is incumbent upon architects to possess a comprehensive understanding of the subject matter. This knowledge equips them with the necessary tools to effectively navigate the intricate web of interdisciplinary design choices, particularly those pertaining to sustainability. Alternatively, the absence of knowledge leads to a disparity, resulting in a lack of integration of environmental design decisions within the realm of architectural education, thereby impeding the comprehensive understanding of architectural history.

The Adaptive Architectural Education Program

The field of architectural education is characterized by a constant evolution of creative methods and innovative techniques. Expanding the perceptual range of students within architectural design studios affords them the opportunity to engage in critical thinking, assessment, and growth. Designers commonly draw inspiration from various sources in order to effectively tackle complex design challenges. The role of architectural education in the attainment of sustainable development is of the highest priority, as it equips prospective professionals with the necessary knowledge and skills to actively contribute to sustainability in their field of practice. Sustainable architecture is commonly regarded as the overarching concept that encompasses energy-related concerns in the constructed environment.

The primary obstacle to the incorporation of the concept into architectural education is the prevailing misperception. The universal application of assessment methods (i.e. LEED, BREEAM, DGNB) seems to be focused on the elaboration of sustainability. 'Green' or 'sustainable' buildings are frequently alienated from important elements like design considerations and theoretical backgrounds by these methodologies since they have a tendency to treat the building as a subject for scientific investigation (Farmer & Guy, 2010). Therefore, the dominance of sustainable technologies over ideas like locality, minimal resource use, and environmental design approaches may lead to stereotype sustainability measures in the built environment that are only technological fixes and don't take into account design aesthetics (Keitsch, 2012).

All courses should incorporate environmental issues through a discussion-based methodology. To facilitate a comprehensive understanding of the discussion-based approach, it is imperative to commence by introducing the concepts as a cognitive framework (Gucyeter, 2016). In order to enhance interest and awareness surrounding the principles of sustainability, it would be more efficacious to establish a connection between these concepts and real-life applications. To illustrate, when examining the environmental sustainability of building materials, it may be beneficial to consider the recycling practices of individuals in their everyday lives. This approach can effectively underscore the significance of utilizing building materials that are capable of being recycled. The significance of this approach lies in its direct correlation with individuals' lifestyle choices and their perception of the environment. As students assimilate this viewpoint and begin to modify their daily habits, it would be more advantageous for them to cultivate a heightened sense of awareness as designers who have agency in shaping the built environment.

In numerous instances, prioritizing the cultivation of an evolving awareness holds greater significance within educational lectures than the mere recitation of stereotypical statistical data.

The primary objective of an adaptable architecture curriculum should be to foster a continuous process of learning and application of sustainability principles throughout an individual's lifetime. To attain this objective, the initial measure entails prioritizing the adoption of sustainable architecture and acknowledging that sustainable design constitutes a fundamental concern within the realm of architectural education.

Gaining Knowledge From the Natural Environment

Architects find nature and the environment to be consistently inspiring sources of knowledge, which holds particular significance. The field of architecture recognizes the significant importance of the natural environment, emphasizing the need for the built environment to align harmoniously with its surroundings in order to achieve optimal architectural outcomes. Therefore, a significant objective in the field of education is to attain proficiency in the natural environment. It is imperative for individuals who have completed their studies in architecture to have a comprehensive understanding of environmental challenges (Mahmoodi, 2001). The insufficient attention given to the functioning of nature within architectural education diminishes the inclination and consciousness of aspiring architects toward the natural environment. According to Bashir et al. (2013), the promotion of environmental awareness serves to validate and legitimize our professional field.

The act of examining the natural world and deriving solutions from it holds significant value for designers. By carefully selecting the most suitable materials for design purposes and implementing recycling strategies that are tailored to the specific local conditions. In the realm of natural systems, optimization takes precedence over maximization, as it entails the utilization of minimal resources and energy to achieve optimal performance. Nature employs minimal resources and precisely allocates them in appropriate locations. The exploration of nature-inspired solutions to design challenges is an innovative approach that merits support within the realm of design education (Amer, 2019). One of the primary objectives of architectural education is to produce graduates who possess a heightened awareness and commitment toward environmental and sustainable concerns.

It is imperative for students to recognize and embrace the architects' duty in upholding the ecological well-being of the planet. This can be achieved by altering the prevailing notion of architectural aesthetics, emphasizing the integration of buildings with the natural environment, rather than supplanting nature with concrete structures. Architects are required to enhance the sustainability and adaptability of buildings in response to their surrounding environment.

CONCLUSION

Addressing climate change acknowledges the shared obligation of individuals and communities. The construction industry, according to the International Energy Agency (2020), accounts for 50% of global energy consumption. Consequently, the involvement of key stakeholders such as architects, engineers, planners, and contractors become increasingly crucial in delivering sustainable and appropriate design for the built environment. The efficacy of the process necessitates effective communication between engineers and architects, a facet that is frequently acknowledged by professional organizations as an area requiring further enhancement.

Architectural education bears a significant responsibility to foster a transformative shift in the mindset of architectural engineering students. This shift should encompass not only the enhancement of identity, functionality, and aesthetic values in design but also the interpretation and incorporation of sustainable strategies and methodologies that align with the principles of ecological integration.

The subject matter of climate change in buildings has been incorporated into the curriculum of numerous schools of architecture worldwide for an extensive period. Nevertheless, most of these programs have operated concurrently with, and frequently in rivalry against, more conventional alternatives. Over an extended period, the field of architectural education has exhibited a sluggish response in adapting to a fresh set of demands, often operating under the assumption that the engineering profession is primarily responsible for addressing the environmental aspects of buildings. The recognition of climate change issues in architectural education has undergone a notable transition, moving from a specialized technical focus to a more prominent position on the agenda. Nevertheless, although this shift in perspective is commendable, there remains a lack of consistent integration of sustainable environmental design principles into the fundamental structure of architectural education.

To address the challenges posed by the contemporary climate crisis and ensure the long-term sustainability of human settlements, there is a growing demand in the marketplace for graduates with expertise in architectural disciplines. These individuals must possess a diverse skill set that encompasses not only creative design abilities but also a deep understanding of environmental issues. To ensure that the promotion of sustainable environmental design is integrated into the architectural curriculum, a thorough process of revising higher education and professional development is necessary.

REFERENCES

Alcázar, M., & Chavez, J. R. (2014). Educational Program for Promoting the Application of Bioclimatic and Sustainable Architecture in Elementary Schools. *Energy Procedia*, *57*, 999–1004. Advance online publication. doi:10.1016/j.egypro.2014.10.083

Altomonte, S. (2009). Climate Change and Architecture: Mitigation and Adaptation Strategies for a Sustainable Development. *Journal of Sustainable Development*, *1*(1), 1. Advance online publication. doi:10.5539/jsd.v1n1p97

Amer, N. (2019). Biomimetic Approach in Architectural Education: Case study of 'Biomimicry in Architecture' Course. *Ain Shams Engineering Journal*, *10*(3), 3. Advance online publication. doi:10.1016/j.asej.2018.11.005

Bashir, F. M., Ahmad, M. H., & Hamid, M. (2013). Design Studio as Problem Based Learning in Architectural Education in Universiti Teknologi Malaysia. *4th International Research Symposium on Problem-Based Learning (IRSPBL) 2013*, 373–379.

Borucka, J., & Macikowski, B. (2017). Teaching Architecture – Contemporary Challenges and Threats in the Complexity of Built Environment. *IOP Conference Series. Materials Science and Engineering*, *245*(8), 082058. doi:10.1088/1757-899X/245/8/082058

Celadyn, W. (2018). Postgraduate studies and sustainable architecture. *Global Journal of Engineering Education, 20*(1), 54–58. https://www.scopus.com/inward/record.uri?eid=2-s2.0-85041920632&partnerID=40&md5=173be32b0608a8b3a41039d80aef3daf

Celadyn, W. (2019). Controversy over the visions of the architectural profession. *World Transactions on Engineering and Technology Education*, *17*, 71–75.

Cowan, S., & Van Der Ryn, S. (1994). *Ecological Design*. Island Pres.

Davis, D. (2010). Integrating sustainability into studio design curriculum. *ASEE Annual Conference and Exposition, Conference Proceedings.*

Directive 2005/36/EC of the European Parliament and of the Council of 7 September 2005 on the recognition of professional qualifications (Text with EEA relevance), CONSIL, EP, 255 OJ L (2005). https://data.europa.eu/eli/dir/2005/36/oj/eng

Dumitru, D. (2017). Reorienting higher education pedagogical and professional development curricula toward sustainability– a Romanian perspective. *International Journal of Sustainability in Higher Education*, *18*(6), 894–907. doi:10.1108/IJSHE-03-2016-0046

European Commission. (2022). *Consequences of climate change*. https://climate.ec.europa.eu/climate-change/consequences-climate-change_en

Farmer, G., & Guy, S. (2010). Making morality: Sustainable architecture and the pragmatic imagination. *Building Research and Information, 38*, 368–378. doi:10.1080/09613218.2010.482236

Figueiró, P. S., & Raufflet, E. (2015). Sustainability in higher education: A systematic review with focus on management education. *Bridges for a More Sustainable Future: Joining Environmental Management for Sustainable Universities (EMSU) and the European Roundtable for Sustainable Consumption and Production (ERSCP) Conferences, 106*, 22–33. 10.1016/j.jclepro.2015.04.118

Frederick, C. P. (2012). *Curriculum Improvement in Education for Sustainable Development: Measuring Learning Outcomes in an Introductory Urban Planning Course* [Master's Thesis]. Arizona State University. doi:10.1080/10464883.1988.10758475

Gelernter, M. (1988). Reconciling Lectures and Studios. *Journal of Architectural Education, 41*(2), 46–52. doi:10.1080/10464883.1988.10758475

Glyphis, J. (2001). How can the architect contribute to. A sustainable world. *Proceedings of the Wingspread. Conference,* 24–26.

Grover, R., Emmitt, S., & Copping, A. (2020). Critical learning for sustainable architecture: Opportunities for design studio pedagogy. *Sustainable Cities and Society, 53*, 101876. doi:10.1016/j.scs.2019.101876

Gucyeter, B. (2016). The Place of Sustainability in Architectural Education: Discussion and Suggestions. *American Journal of Archaeology*, *2*, 237–256.

Guy, S., & Farmer, G. (2001). Reinterpreting Sustainable Architecture: The Place of Technology. *Journal of Architectural Education*, *54*(3), 140–148. doi:10.1162/10464880152632451

Hartman, H. (2012). Is Sustainability Just Another “Ism”? *Architectural Design,* 136–140. https://doi.acm.org/10.1002/ad.1444

Hegarty, K., Thomas, I., Kriewaldt, C., Holdsworth, S., & Bekessy, S. (2011). Insights into the value of a ‘stand-alone’ course for sustainability education. *Environmental Education Research, 17*(4), 451–469. doi:10.1080/13504622.2010.547931

Hengrasmee, S., & Chansomsak, S. (2016). A novel approach to architectural education for sustainability: A quest for reformation and transformation. *Global Journal of Engineering Education, 18*(3), 160–166. https://www.scopus.com/inward/record.uri?eid=2-s2.0-84994635754&partnerID=40&md5=64352ee45c35192231936be993344a4f

Herr, C. (2014). *Radical constructivist structural design education for large cohorts of Chinese learners*. Academic Press.

Hyland, C., & Grierson, D. (2012). Learning Lessons from the Scottish School Building Programme. *The International Journal of Environmental Sustainability, 8*(1), 63–76. doi:10.18848/2325-1077/CGP/v08i01/55025

IEA. (2020). *Buildings – Topics*. IEA. https://www.iea.org/topics/buildings

IPCC. (2007). *Climate Change 2007: Mitigation. Contribution of Working Group III to the Fourth Assessment Report of the Intergovernmental Panel on Climate Change*. Cambridge University Press.

IPCC. (2017). *Climate Change 2007: Synthesis Report.* Intergovernmental Panel on Climate Change, Climate Change. https://www.nrc.gov/docs/ML0932/ML093220680.pdf

Iyer-Raniga, U., & Dalton, T. (2017). Challenges in Aligning the Architecture Profession in Indonesia for Climate Change and Sustainability. *International High-Performance Built Environment Conference – A Sustainable Built Environment Conference 2016 Series (SBE16), IHBE 2016, 180*, 1733–1743. 10.1016/j.proeng.2017.04.336

Keitsch, M. (2012). Sustainable Design: A Brief Appraisal of its Main Concepts. *Sustainable Development, 20*(3), 180–188. Advance online publication. doi:10.1002d.1534

Mahmoodi, A. S. (2001). *The design process in architecture: A pedagogic approach using interactive thinking* [PhD thesis]. University of Leeds.

Mavromatidis, L. (2018). Coupling architectural synthesis to applied thermal engineering, constructal thermodynamics and fractal analysis: An original pedagogic method to incorporate "sustainability" into architectural education during the initial conceptual stages. *Sustainable Cities and Society*, *39*, 689–707. doi:10.1016/j.scs.2018.01.015

McIntosh, M., Cacciola, K., Clermont, S., & Keniry, J. (2001). *State of the Campus Environment—EcoLeaders*. National Wildlife Federation. https://www.nwf.org/Home/EcoLeaders/Campus-Ecology-Resource-Center/Reports/State-of-the-Campus-Environment

Murray, P., & Cotgrave, A. (2007). Sustainability literacy: The future paradigm for construction education? *Structural Survey*, *25*(1), 7–23. doi:10.1108/02630800710740949

Nyka, L. (2019). Bridging the gap between architectural and environmental engineering education in the context of climate change. *World Transactions on Engineering and Technology Education*, *17*, 204–209.

Şen, Z. (2015). Chapter One—Introduction. In Z. Şen (Ed.), *Applied Drought Modeling, Prediction, and Mitigation* (pp. 1–41). Elsevier. doi:10.1016/B978-0-12-802176-7.00001-8

Shareef, S. S., & Farivarsadri, G. (2020). An Innovative Framework for Teaching/Learning Technical Courses in Architectural Education. *Sustainability*, *12*(22), 9514. Advance online publication. doi:10.3390u12229514

Smatanova, K., & Urban, J. (2018). Outcomes of the experimental assignment in Studio Project I - Urban Design. *World Transactions on Engineering and Technology Education*, *16*, 115–120.

Stupar, A., Mihajlov, V., & Simic, I. (2017). Towards the Conceptual Changes in Architectural Education: Adjusting to Climate Change. *Sustainability*, *9*(8), 1355. Advance online publication. doi:10.3390u9081355

Szczepański, J. (2019). Sustainable monument preservation in architectural education. *World Transactions on Engineering and Technology Education*, *17*, 42–47.

UN. (2022). *UN Agenda 2030—Directorate of Programme Co-ordination*. Directorate of Programme Co-Ordination. https://www.coe.int/en/web/programmes/un-2030-agenda

United Nations. (2022). *World Population Prospects 2022*. Department of Economic and Social Affairs Population Division. https://population.un.org/wpp/

US EPA. (2022). *Heat Island Impacts* [Overviews and Factsheets]. https://www.epa.gov/heatislands/heat-island-impacts

Vincent, S., & Dutton, K. (2014). Integration of interdisciplinary environmental and sustainability education and research and urban sustainability. *Journal of Environmental Studies and Sciences*, *7*(1), 112–120. Advance online publication. doi:10.100713412-014-0202-z

WCED. (1987). *Report of the World Commission on Environment and Development: Our Common Future*. http://www.un-documents.net/A372BC4E-993E-4C10-B368-9E75816E59DB/FinalDownload/DownloadId-1990161B7AFC5CB358F7280210D7A48F/A372BC4E-993E-4C10-B368-9E75816E59DB/our-common-future.pdf

Wiedmann, F., & Salama, A. (2012). *The Role of Architecture in Producing Urban Qualities for Sustainability: Implications for the Future of Architectural Education*. Academic Press.

ADDITIONAL READING

Crysler, C. G. (1995). Critical pedagogy and architectural education. *Journal of Architectural Education*, *48*(4), 208–217. doi:10.1080/10464883.1995.10734644

Dutton, T. A. (1987). Design and studio pedagogy. *Journal of Architectural Education*, *41*(1), 16–25. doi:10.1080/10464883.1987.10758461

Türkkan, S. (2023). From Disruptions in Architectural Pedagogy to Disruptive Pedagogies for Architecture. In *Disruptive Technologies: The Convergence of New Paradigms in Architecture* (pp. 131–146). Springer International Publishing. doi:10.1007/978-3-031-14160-7_10

KEY TERMS AND DEFINITIONS

Accreditation: A formal recognition process conducted by authorized bodies to evaluate the quality and standards of educational programs, ensuring they meet specific criteria and deliver effective learning outcomes.

Architectural Curriculum: A structured plan of study encompassing various courses and educational experiences aimed at equipping students with the knowledge and skills required for a career in architecture.

Higher Education: The level of education beyond secondary school, typically offered by universities and colleges, aimed at providing advanced knowledge and skills in specific disciplines.

Innovative Architectural Discourse: Progressive and forward-thinking discussions and dialogues within the field of architecture that challenge conventional practices and explore novel approaches to design, construction, and environmental sustainability.

Interdisciplinary Collaboration: Collaboration among individuals from different academic disciplines to address complex issues, fostering the integration of diverse perspectives and expertise.

Symbiotic Relationship: A mutually beneficial relationship between two entities, in this context, referring to the interconnectedness of architectural design and construction, emphasizing their integration for effective environmental and climatic considerations.

Chapter 9
Promoting Environmental Sustainability Among Students and Citizens for a Sustainable Future

Seema Yadav
https://orcid.org/0000-0002-8477-5895
Bhopal School of Social Sciences, India

ABSTRACT

Climate change is a dangerous problem that endangers the continuation of life. The sixth major extinction catastrophe in recorded history is being brought on by climate change. There is an existential threat to life on Earth from the climate changes brought on by the use of fossil fuels. Education increases young people's knowledge while simultaneously fostering awareness, enabling informed decision-making, and fostering changes in attitudes and behaviors. Education is essential to the global response to climate change. Climate change education (CCE) is one of the most significant approaches to increasing awareness of mitigating climate change. The majority of students must receive the knowledge, critical thinking skills, scientific understanding, and ethical frameworks necessary to prevent, adapt, and reverse climate change as a result of climate change education. Climate change education initiatives can play a significant role in developing communities of the kind of pragmatic visionaries needed to address new issues as they emerge and build on the successes of others.

DOI: 10.4018/978-1-6684-9099-0.ch009

INTRODUCTION

Industrialization fueled by fossil fuels has made possible the greatest human boom in recorded history. However, it also contributes to extraordinarily severe, long-lasting pollution that is upsetting the planet's ecosystems. The pollution is cumulative, dating back to the start of the industrial period, 200 years ago, and takes between 12 to more than 50,000 years for greenhouse gas emissions to be reabsorbed from the atmosphere (Irwin, 2020).

By equipping students with the skills necessary for green employment, which are specialised occupations, education plays a crucial role in empowering learners with competences for sustainability (Okada & Gray, 2023). Education plays a crucial role in the worldwide response to climate change since it increases young people's knowledge while also promoting awareness, enabling informed decision-making, and promoting changes in attitudes and behaviors toward adopting sustainable lifestyles (Ouariachi & Wim, 2020). The greatest value of education is found in the transformative potential of adaptation learning support curriculum, pedagogy, and technological tools that assist people in learning while making complicated adaptive decisions (Feinstein & Mach, 2020). One of the most important ways to raise awareness about mitigating climate change is through climate change education (CCE) (Park, Choe, & Kim, 2020). Researchers have recently begun to stress the significance of including emotional components in education for sustainable development (ESD). Two justifications are frequently presented: The first worry is that students may experience negative feelings like worry and guilt due to the complexity and importance of sustainability challenges like climate change (Ojala, 2021). Climate Change and Sustainability Education (CCSE) is necessary for societies all over the world to support responsible development, including environmental protection, welfare, and the green economy (Okada & Gray, 2023). Developing green skills to ensure healthy living, new occupations, and just societies is seen to require putting the appropriate educational frameworks in place (Okada & Gray, 2023).

Even though addressing climate change (CC) through education remains a top priority on the international agenda, educationalists, climate scientists, and those who shape or make policy related to climate or education still have insufficient knowledge of what this entails (Mochizuki & Bryan, 2015). The sustainability of ecosystems and the socioeconomic system that supports human health are at risk due to climate change, which is a major problem. To prepare for climate change, increase public health and eco-health literacy, deepen learning, and strengthen graduate qualities, they have provided evidence for the need for a climate change curriculum in education (Maxwell & Blashk, 2016). The case for incorporating climate change education into medical education is compelling. Internationally, scientists, governments, and health authorities all acknowledge the existence of anthropogenic climate change.

Climate change has serious ramifications for human health, making it important for aspiring doctors (Maxwell & Blashk, 2016). Future doctors have the chance to gain the knowledge and skills necessary for clinical practice and a public health role in a world with climate change by integrating climate change into medical school (Maxwell & Blashk, 2016). Discovering the degree of curriculum integration is necessary to rectify any gaps and ensure that climate change education is implemented properly in schools (Boakye, 2015). An important part of sustainability education is teaching students to recognise, understand, and resolve issues that have an impact on the planet's life now and in the future (Okada & Gray, 2023).

ENVIRONMENTAL PHENOMENON AND ITS IMPACT ON HUMANKIND

The continuation of life is threatened by the wicked challenge of climate change. The world is currently experiencing its sixth major extinction event due to climate change. For animals, insects, plants, and reptiles, the climate changes brought on by the burning of fossil fuels pose an existential threat. This failure in the climate raises several questions about the cause of the issue, including those about chemical compounds, market factors, and social standards (Irwin, 2020). The greatest possible risk to modern society is climate change, one of the most important socio-scientific challenges (Park et al., 2020). All attempts are being undertaken to stop climate change and the associated global warming, which is an environmental problem that is greatly feared and threatens human survival (Boakye, 2015). Climate experts agree that human activity is most likely to be accountable for the current state of global warming, which is happening and already having effects. This consensus is not always shared or understood, according to polls of the general public, high school instructors, and college students (Aubrecht, 2018). The effects of climate change, which are now manifesting quickly and will only worsen over time, as well as the success of our efforts to avert a global disaster by reducing emissions, all have an impact on the well-being of young people (Karsgaard & Davidson, 2023). The health of people and the environment are significantly impacted by climate change (Boyko, Desak, Fleming, Diplock, & Pons, 2020).

The injustices connected to the causes, consequences, and politics of climate change have inspired youth all around the world to organize in reaction to these realities (Karsgaard & Davidson, 2023). Every component of the ecosystems that civilizations are a part of is under threat from global warming, and emissions are still rising. Impacts are already being seen, and they include rising sea levels, higher average temperatures, and most visibly a rise in the frequency and severity of extreme events like flooding, drought, and forest fires (Karsgaard & Davidson, 2023). The

Anthropocene Age, as it is now commonly referred to by geologists, is one that has been brought about by high population expansion and more materialistic lifestyles. Human activities are the principal thing preventing civilization from surviving in the Anthropocene. Many of the advancements in development and health over the past century are in danger due to climate change. It might also catalyze the essential societal change toward a more enduring and healthy future (Maxwell & Blashk, 2016). The cognitive process behind the technology is challenged by climate change. The majority of us are therefore unprepared to critically consider it or implement positive reforms because it is exposing the severely flawed metaphysical assumptions of modernity (Irwin, 2020).

CLIMATE CHANGE EDUCATION: KNOWLEDGE, ATTITUDE, AND SKILL ENHANCEMENT

Over the past few years, there has been a significant rise in the general public's knowledge of how humans affect the environment on a worldwide scale (Leichenko & O'Brien, 2020). Through a focus on physical processes, the documenting of increasing emissions, and empirical evidence of a changing climate, climate change courses and curricular interventions largely emphasize scientific literacy. In-depth discussions of climate change remedies in the classroom are frequently confined to "business as usual" legislative alternatives, fresh technologies, and behavioral interventions to cut emissions or encourage adaptation (Leichenko & O'Brien, 2020).

Universities offer an unparalleled opportunity to inform students about climate change challenges and actively involve them in climate affairs, both as citizens and as potential future workforce influencers (Filho et al., 2023). The implementation of climate change education has drawn more attention recently since it has emerged as one of the most significant global challenges (Ouariachi & Wim, 2020). When properly conceived, education may be a potent tool for facilitating successful adaptation to climate change (Feinstein & Mach, 2020). Education is frequently implicitly used to mean either school or learning in discussions about climate policy. The second is too nebulous for practical policy action, and the first view excludes other significant forms of education (Feinstein & Mach, 2020). In many parts of the world, moralizing and politicizing discourses about climate change have also permeated children's and youth's daily lives (Rousell & Cutter-Mackenzie-Knowles, 2020).

Climate change Education must catch up with more extensive social movements and climate change-responsive discourses (Rousell & Cutter-Mackenzie-Knowles, 2020). Instead of avoiding the Earth's impending runaway climate change, the teaching opportunity can be used to encourage students to reflect on what is truly and profoundly important. Students can then experiment with innovative replacements for scientific educational methods based on their attitudes and beliefs (Rousell & Cutter-Mackenzie-Knowles, 2020). Recently, academic interest in climate change education (CCE) and communication has increased in tandem with the rise in interest in climate change education for practitioners (Armstrong & Krasny, 2020). By fusing knowledge/skill development with contemplation on the necessity of actively altering present social systems and individual behavior, climate change education (CCE) programs should promote citizen responsiveness to climate change (Park et al., 2020). Programs focused on action-emphasized climate change education (AECCE) should build positive values/attitudes, encourage involvement and action at all grade levels, balance theoretical and practical elements, and address both individual and societal levels of action (Park et al., 2020).

The perceived injustices related to the causes, effects, and politics of climate change have recently sparked youth mobilization throughout the world (Karsgaard & Davidson, 2023). This project offers a platform for exploring climate change education from the perspectives of different adolescents, who inform possibilities for climate change education that addresses justice, individualization, and emotionality. The project is grounded in a critical global civic education paradigm (Karsgaard & Davidson, 2023). In addition to the scientific sciences, several fields of the social sciences and humanities are also used in discussions about the impacts of climate change, the steps needed to limit the danger of further climate change, and the steps needed to adapt to climate change (Aubrecht, 2018).

It is crucial to understand how effective the pre-tertiary curriculum is in teaching about climate change so that, if necessary, improvements can be implemented. So that they can teach the general public, and students can be better prepared (Boakye, 2015). Understanding the effects of climate change on people depends heavily on education. Because a white, middle-class hegemony dominates school curricula and CCE, students from other social groups may not be appropriately served. Being a new area of education, CCE lacks a substantial body of implementation models that have been well-tested (Ross, Rudd, Skains, & Horry, 2021). In several nations around the world, including the United States, the United Kingdom, and Germany, CCE has been examined and put into practice as a significant topic (Park et al., 2020). The Sustainable Development Goals (SDGs) and global citizenship education, for instance, have lately been incorporated into environmental education programs in the United Kingdom. To encourage students to actively seek answers to environmental problems, such as climate change, the United Kingdom also suggested using an integrated curriculum, community, and campus-3C strategy (Park et al., 2020).

A classroom is a crucial place for the growth of youth knowledge, creativity, efficacy, and collective action skills in the face of climate change, even as young people take to the streets (Karsgaard & Davidson, 2023). The public frequently envisions or anticipates that children and young people will be the leaders of the future generation who will end the effects of past environmental and climate inaction (Cutter-Mackenzie & Rousell, 2019). When it includes and is directed by young people, education about climate change can open up a completely new area of inquiry and educational experience (Cutter-Mackenzie & Rousell, 2019). The science classroom is where climate change is most frequently covered as a subject in K–12 schools. It is a human problem, though, and needs social as well as technological and scientific solutions (Siegner & Stapert, 2020). To stop climate change, humans must critically examine the world they are creating and think about the underlying causes, such as the interdependence of humans and nature, individualism, and consumerism. Another crucial aspect of combating climate change is public participation. People should be able to react to climate change not only on an individual level but also on a communal one (Park et al., 2020). The way institutes educate about climate change in higher education needs to be revised to reflect the Anthropocene, a newly suggested epoch. Understanding the Anthropocene requires a fundamental rethinking of how we perceive the relationships between people and nature (Leichenko & O'Brien, 2020).

Effective climate change education is a possibility due to rising interest in the topic and growing awareness of the difficulties involved in resolving this issue (Monroe, Plate, Oxarart, Bowers, & Chaves, 2019). Countries are becoming more conscious of the significance of climate change education since it has grown to be one of the greatest environmental concerns facing the planet. Just as climate change is a wicked problem with no easy answers, climate change education can also be categorized as a wicked problem (Cantell, Tolppanen, Aarnio-Linnanvuori, & Lehtonen, 2019).

Climate change education presents a growing field of practice and study that is inherently responsive and accountable to the fast-changing environmental conditions of daily living, instead of merely slipping into one of these disciplines or sub-disciplines (Cutter-Mackenzie & Rousell, 2019). Climate change education is to advance, it must transcend the confines of the fields of environmental education, education for sustainable development, education for sustainability, and other forms of sustainability-focused education (Cutter-Mackenzie & Rousell, 2019). The future curricula for aspiring public health professionals may be influenced by understanding the consequences of climate change within this profession and the necessity for climate change education, resulting in professionals entering the sector ready to face the problems of the future (Boyko et al., 2020).

Various issues should be balanced while developing and implementing climate change programs. An all-encompassing educational strategy, the climate change education approach emphasizes the significance of the following elements: knowledge, critical-thinking abilities, values, identity, and worldview; action; motivation; participation; future orientation; hope and other emotions; and operational hurdles (Ouariachi & Wim, 2020). Students exhibit outstanding behavior and are eager learners. Because instruction is not sufficiently tailored to match each student's needs and skills, some students progress is slower than anticipated. Students have a global perspective, but they still have limited opportunities to participate in decision-making in their schools and communities (Ross et al., 2021). The CCE program is already a distinctive and compelling approach to CCE, as evidenced by the enthusiastic reactions it has received from educators and students, as well as the numerous institutions that have sought to include it in their curricula. However, with the developments described, it can become an even more powerful and positive model for involving young people in the fight against catastrophic climate change (Ross et al., 2021).

Over the last century, the field of climate change has gradually grown and established itself. Finally and long overdue, climate change needs to be included in the curriculum as a core, highly regarded discipline (Eilam, 2022). As public health education is established, climate change needs to be included in the curriculum (Boyko et al., 2020). The need for a workforce that should be able to recognize the effects of climate change on the community and comprehend the role of public health in reacting to these effects (Boyko et al., 2020). To support aspiring public health professionals as they enter the sector, education should address climate change, which is relevant to the work done in the area (Boyko et al., 2020). Three separate but interconnected pathways education infrastructure, general education, and adaptation learning support climate change adaptation. Each of these pathways offers specific policy alternatives (Feinstein & Mach, 2020). Education can be a potent tool for facilitating effective adaptation to climate change, but its policy worth has been constrained by ambiguous definitions of education and by inadequate cross-field communication that stops crucial ideas from one subject or policy domain from reaching another (Feinstein & Mach, 2020).

Educating students about the causes, effects, and solutions to climate change is a national necessity for public and private education due to the reality of climate change, both as it is currently being experienced and as it is predicted to be in the future. To support the movement for youth climate education, numerous organizations and research institutions are already dedicated to this goal and are developing curricula, materials, papers, and evaluations (Siegner & Stapert, 2020). Climate Change and Environmental Education (CCEE) are essential for instilling in people and communities a sense of connection to nature and for indirectly raising awareness of ways to mitigate and adapt to the environment's degradation brought on by climate change. Imparting the values, abilities, and behavior needed to adapt to and reduce environmental deterioration as well as climate change is insufficient. Teachers constantly use innovation to teach CCEE (Karim et al., 2022).

(Kunkle & Monroe, 2019) make suggestions for possible directions the field of environmental education could take to create climate change resources that could lessen unintentional biases among science educators and more successfully enlist their support in educating about its causes and potential remedies regardless of worldview differences. A unique opportunity exists for environmental education (EE), with its holistic approach to interventions based on the most recent findings in educational psychology, critical skill development, and behavior change theories, to contribute to the creation of curricula and professional development that more successfully engage students in climate change education (Kunkle & Monroe, 2019). The worldview tends to be dismissive of environmental risks like climate change since addressing such a problem would necessitate government regulations and restrictions on business and industry changes that run counter to their worldview's perspectives on how society should function by fostering competition among individuals and maximizing free enterprise unhindered by governmental control (Kunkle & Monroe, 2019).

There is a serious issue as a result of the teachers' disparate levels of SD content knowledge and the local SD management. The students may receive diverse topic knowledge that may not even be congruent with the widely accepted concept of SD, depending on the teachers' definition and choice of subject matter. One cannot claim that students in Swedish schools have an equivalent education in SD (Gustafsson, Engström, & Svenson, 2015). Even though it's vital, education about climate change should go beyond giving people the knowledge to understand it. It must enable students to make decisions, comprehend trade-offs, and come up with creative solutions that can help humans incorporate ecologically friendly choices into a bigger framework for how they live. In addition to educating students about climate change, educators must also assist them in imagining and enacting alternate futures (Karsgaard & Davidson, 2023). Education professionals now urgently need to understand climate change. Education's normative principles and the contemporary, globalized world have long been strongly related (Irwin, 2020). Future professionals may be better able to put their understanding of climate change into practice if their education incorporates climate change activism (Boyko et al., 2020). Engaging policymakers, understanding the value of teamwork, and using the media to increase support for climate change are all skills that are essential for climate change advocacy (Boyko et al., 2020).

Although difficult and divisive, secondary science education should cover the vital subject of climate change, according to politicians and educators. Uncertain opinions exist among teachers on the teaching of climate change. Teachers' decisions on how to teach can be influenced by the prevalence of disputes, which can lead to misunderstandings about the science behind climate change. As a result, when analyzing a teacher's instruction, the function of their views must be taken into account (Nation & Feldman, 2021). From the viewpoints of the instructors, they looked into the breadth of climate change and environmental education (CCEE) in the current Malaysian curriculum. The research produced five themes: improving instructors' capacities, empowering and mobilizing youth, stepping up local-level initiatives, and promoting policy (Karim et al., 2022). The majority of instructors thought that while the Secondary School Standards-based Curriculum had improved, the emphasis should still be on students rather than an exam-focused approach. Climate change and environmental protection were not adequately covered in the science textbook (Karim et al., 2022).

Research in other education subfields suggests that teachers' beliefs and perspectives about emotions in the learning process of climate change are important to investigate because they may have a significant impact on how teachers approach emotions in the learning process and result in the development of a meta-emotion philosophy (Ojala, 2021). It's also noteworthy that some teachers' meta-emotion beliefs include the idea that feelings are largely irrational and have no role in understanding climate change. This understanding of emotions appears to be somewhat out of date, as evidenced by basic research on the subject of emotions in general, which demonstrates that feelings can occasionally serve as a necessary predicate for making sensible decisions (Ojala, 2021).

INNOVATIVE WAYS TO DEAL WITH CLIMATE CHANGE WITH EDUCATION

Climate change education must give the majority of people the knowledge, critical thinking abilities, scientific comprehension, and ethical frameworks they need to mitigate, adapt, and reverse climate change. Under the umbrella of "Climate Change Education," there is a wide range of ideas and techniques that coexist. A cursory comprehension of the phrase would have pupils learn about a current, scientifically proven phenomenon that is the outcome of how people interact with their surroundings. Programs for educating people about climate change can play a crucial part in creating communities of such pragmatic visionaries that are required to handle upcoming issues as they develop and build on the success of others (Monroe et al., 2019).

Although interdisciplinary classes might be the best setting for students to learn about climate change right now, there are currently more students enrolled in general chemistry or introductory biology classes (Aubrecht, 2018). The human connection to climate change is highlighted and student participation, climate literacy, and action are encouraged when climate education is approached through social studies and language arts (Siegner & Stapert, 2020). Social studies and language arts-based approaches to climate education encourage student engagement, climate literacy, and action while emphasizing the link between human activity and climate change

(Siegner & Stapert, 2020). Climate experts agree that human activity is most likely to be to blame for the current state of global warming, its effects, and its very likely cause. Surveys of the general population, high school educators, and college students show that this agreement is not always shared or understood (Aubrecht, 2018). Even though climate science and climate change are intrinsically interdisciplinary, chemistry is nevertheless necessary to grasp many of the principles that are necessary to comprehend how the Earth's atmosphere affects its climate and the effects of increased greenhouse gas emissions (Aubrecht, 2018).

Factual data and fresh materials are frequently given to science educators so they can introduce students to novel ideas or get them involved in solving real-world problems (Kunkle & Monroe, 2019). Both more relevant and engaging chemistry lessons and more climate-literate citizens can result from making linkages between chemistry ideas and the principles of climate literacy (Aubrecht, 2018). Critical Global Climate Education-GCE encourages students to go beyond individual responses to collective action from their differential positions while addressing the interrelationships, structures, and processes underlying global climate injustices while acknowledging the power relations inherent to the global system (Karsgaard & Davidson, 2023).

Integrative, phenomena-based, and collaborative learning constructs a systemic understanding of the world at its best. They provide the student with useful building blocks to create a comprehensive understanding of the multifaceted problems the globe faces and how they are connected. They provide a student with the building blocks to create her worldview and teach her what makes for a healthy existence and communal well-being. Systematic understanding necessitates content area comprehension, but even more so necessitates innovative learning methodologies, making it a crucial tool for coping with a changing world.

Phenomena-based learning is intensely collaborative since it is an active learning process that fosters thinking through dialogue and teamwork. To create an egalitarian learning environment, all students contribute their knowledge, ideas, and experiences to a group process. Every concept is important since it advances our understanding of the problem as a whole. Ideas are shared and developed through a relational process. There is more than one correct response; there are no bad questions. The implied conclusion is that local management, including the principal, must assume more accountability for SD teaching. This includes a standard planning process used by the entire faculty at the school, but maybe to varying degrees. At this level, an evaluation is also necessary to ensure that the students receive the intended education (Gustafsson et al., 2015). The model appears to be an accurate representation of climate change teaching and is possibly applicable in many situations. In teacher education, the model may be extremely useful. The model can assist pre- and in-service instructors in reflecting on their teaching strategies and evaluating how holistic their instruction is while creating lessons and learning modules on climate change (Cantell et al., 2019). Teachers can consider whether their methods in the classroom encourage adopting preventative measures and whether they are equipped to handle students' emotions, such as nervousness. The methodology can also be applied to the modification and creation of curricula. It can help assess programs in the field of sustainability education (Cantell et al., 2019).

Education has a special and revolutionary contribution in the area of adaptation learning support, which includes curricula, pedagogical tools, and technology resources that enable people to learn while making complicated adaptive decisions (Feinstein & Mach, 2020). The human ability to methodically support the learning that underpins adaptation may be the limiting factor as human societies seek to strike a balance between the traditional social processes that preserve stability and modern ones that allow change (Feinstein & Mach, 2020). (Rousell & Cutter-Mackenzie-Knowles, 2020) call for the creation of new types of climate change education that directly involve children in responding to the complex scientific, social, ethical, and political issues surrounding climate change. They identify the need for participatory, interdisciplinary, creative, and affect-driven approaches to climate change education that, to date, have been largely absent from the literature.

Social Studies can be used to teach and learn about climate change issues. Education should go beyond merely imparting information and focus on the growth of problem-solving abilities that will help solve the human species' survival problems in an environment that is constantly changing (Anyanwu & Njoku, 2023). (McGregor, Pind, & Karn, 2021) argue that the difficulty in teaching history is not just recognizing how climate has changed over time and the effects it has, but also that the Anthropocene is a multifaceted phenomenon that requires adaptation in our ways of being and perceiving ourselves. (McGregor et al., 2021) proposed several novels and even radical directions to examine as a community of history educators, drawing from insights from Indigenous studies, environmental history, and climate change education. In doing so, it is hoped that a strategy for navigating ambiguity will have been put out. With a collection of common questions, presumptions, and goals as a guide, it might be possible to discover approaches to teaching and learning that more effectively address the precarity of the present. To recognize and respond to climate change in all of its complexity, education can and should play a role. It outlines the arguments in favor of discussing CCE within the framework of education for sustainable development (ESD) (Mochizuki & Bryan, 2015). Enhancing learners' comprehension of the causes and effects of climate change as well as their willingness to take action to address it are necessary for advancing CCE within the context of ESD, also known as Climate Change Education for Sustainable Development (CCESD) (Mochizuki & Bryan, 2015).

The science classroom is where climate change is most frequently covered as a subject in K–12 schools. It is a human problem, though, and needs social as well as technological and scientific solutions. In a middle school classroom, a curriculum on climate change was introduced using an integrated social studies and language arts framework (Siegner & Stapert, 2020). Administrators provide options for scaling and implementing the program in different schools and contexts (including public schools), while teachers describe its triumphs and obstacles. Researchers, educators, and curriculum makers working to include a multidisciplinary, socioscientific approach to climate change education in their work should consider the findings from this study (Siegner & Stapert, 2020). Teachers of science are not exempt from presumptions and bias. studies examining the impact of the creation-evolution controversy. Even in many classrooms, there is evidence of a systematic undermining of science (Kunkle & Monroe, 2019). It will be easier for teachers if a website dedicated to online learning is launched and made available to both students and teachers. There is a pressing need to mandate climate education at all levels. By incorporating a realistic strategy, like the 3R practice in daily life at all levels, policymakers need to demonstrate that they are serious about their work (Karim et al., 2022). The relevant authorities should set up enough resources to assist 3R at the neighborhood level. For the creation of useful innovation, industry and academic professionals must work together. In the future, climate policy must be more scientific (Karim et al., 2022).

Student self-reports and teacher interview data show encouraging trends and aspirations to action, albeit it may take several years to fully analyze the entire impact of the curriculum on enabling student action (Siegner & Stapert, 2020). The action and solutions focus can be strengthened by better integrating genuine and significant student climate action projects in the local community. Students receiving the humanities-focused climate curriculum exhibit academic gains in reading comprehension and enthusiasm for the curriculum content (Siegner & Stapert, 2020). Possible initiatives include field trips, practical schoolyard greening projects, contacting local elected officials, and creating community-school relationships that enable service learning projects to have significant effects (Siegner & Stapert, 2020). Future research should concentrate on the education into action pathway and the student action components of CCE in particular, as the 'action' component is still elusive and challenging to assess in many CCE curriculum studies and climate literacy evaluations (Siegner & Stapert, 2020). Giving educators and academics the chance to discuss climate education using humanities frameworks at professional gatherings and conferences is one of the recommendations this study's other authors made. At conferences, educators researching this socio-scientific topic are already debating methods for helping their pupils cultivate "constructive hope" as opposed to hope based on denial or ignorance. The development of more climate education-related conference themes or tracks will advance CCE opportunities outside of science classrooms (Siegner & Stapert, 2020). By interacting between these potential and real-world elements, CCE programs may promote the education of citizen activists who may take voluntary, sustainable action in response to climate change (Park et al., 2020).

The development of critical abilities (such as data inference and systems thinking), raising student achievement, and meeting state and federal criteria could all be the emphasis of teacher training (Kunkle & Monroe, 2019). Templates and lesson plans for student-led research and projects, critical thinking activities, enrichment activities with a systems perspective, and other resources that teachers can use to promote depth of understanding and student-centered learning could be included in curricula (Kunkle & Monroe, 2019). (Ross et al., 2021) analyze interactive digital narratives (IDNs) produced during the You and CO2 Climate Change Education Programme and present a new engagement model for CCE as a result, and they demonstrate how engagement was correlated with children's perceptions of their capacity to affect change on an individual, local, and governmental level. A model for modifying CCE courses to take into account the diverse cultures and viewpoints of young people is provided by the h-ACE.

CHALLENGES IN THE IMPLEMENTATION OF CLIMATE CHANGE EDUCATION

Due to the dominance of the scientific sciences in climate change education, significant social, political, cultural, and economic aspects of climate change have been left out, despite their importance in understanding the causes, effects, and possible responses to global warming (Karsgaard & Davidson, 2023). Globally, climate change (CC) is rarely covered in school curricula, and the majority of graduates are uninformed about CC (Eilam, 2022). Youths' ability to absorb their growing knowledge of and personal experiences with climate change, as well as their ability to develop important civic skills, are compromised by this exclusion (Karsgaard & Davidson, 2023).

The framing of climate change as a scientific and technical issue is not limited to the classroom; it also appears in Western institutions of research and policy, which exclude the knowledge, experience, and worldviews of "non-experts" such as Indigenous peoples, people of color, residents of the Global South, and young people (Karsgaard & Davidson, 2023). Scientific exclusion can be seen as a kind of colonialism, wherein colonized peoples have been used as research and policy targets to their cultural and material harm, in addition to the exclusion of valuable knowledge. Decolonization can therefore be achieved in part by ensuring that various groups, particularly youth, actively contribute to the creation of climate change education (Karsgaard & Davidson, 2023). The education sector is still underutilized as a strategic resource to mitigate and adapt to climate change, even though the significance of education in tackling climate change concerns is becoming more widely acknowledged. A logical framework for climate change education (CCE) has not yet been created by stakeholders in the education sector in many countries (Mochizuki & Bryan, 2015).

Climate change education increasingly links social, political, and economic factors with the environment through holistic, cross-curricular, and transformational learning to lessen climate change and address related vulnerabilities and injustices, even though much of it still concentrates on science-related disciplines (Karsgaard & Davidson, 2023). Climate change education when viewed independently from more established disciplines like environmental education, science education, and education for sustainable development, education remains a relatively new and under-theorized topic of study (Cutter-Mackenzie & Rousell, 2019). Administrators present prospects for scaling up and implementing the curriculum in various schools and situations, while teachers discuss the accomplishments and difficulties of teaching the program (Siegner & Stapert, 2020). Eleven upper secondary school teachers from a variety of topics were questioned to map out their perspectives on SD, and how they teach SD to explore how this political purpose is demonstrated in knowledge and activities among the instructors. All of the teachers agreed that it is very important to teach SD, and they all said they did it according to their definitions (Gustafsson et al., 2015). Additionally, they all noted that despite the steering documents' emphasis on SD, the problem's local management is inadequate. A major issue is created by the divergent perspectives of the teachers on SD and the ineffective local management. The students may receive a variety of topic information, possibly knowledge that isn't even consistent with the accepted concept of sustainable development, depending on the teachers' definition and choice of subject matter. Therefore, it cannot be said that the schools provide all students with an equal education in the field of SD (Gustafsson et al., 2015). Education will not be able to fulfill its responsibility of information transfer unless it keeps up with climate change challenges (Irwin, 2020).

In the transition to a post-carbon society, education can be crucial. To make the required adjustments to the school curriculum, educationalists must have a clear understanding of the issues and the course of action for corrective action (Irwin, 2020). Science is being discussed globally, and this discussion is beginning to spark the kind of thinking that will be required to transform our culture and our educational systems into ones that are more focused on the environment and one another. Externalities related to the environment and pollution have historically been minor concerns. But as they are today, they pose serious problems for the function of education (Irwin, 2020). There are many misconceptions concerning the causes of climate change, despite educators' best efforts to dispel them through school programs, Extension programs, and informal settings (Monroe et al., 2019). In addition to the difficulties brought on by the complexity of climate change, educators who engage with a range of audiences, from children in classrooms to people in communities, describe facing other obstacles to effectively educate their audiences about the issue (Monroe et al., 2019). Programs for educating people about climate change can be crucial in creating communities of the kind of pragmatic visionaries needed to tackle upcoming challenges as they build on the achievements of others (Monroe et al., 2019).

Models for teaching sustainability and the environment do not, meet the requirements for illustrating climate change education since they do not sufficiently capture the broad nature of climate change (Cantell et al., 2019). Escape room applications in educational contexts are quickly gaining popularity as a way to involve students in their learning environment in response to the need for engaging and interactive approaches to climate change communication and instruction (Ouariachi & Wim, 2020). These applications were found to be one of the two most successful methods in teaching climate change. They can work on soft as well as hard skills, broaden knowledge, boost motivation, and promote attitude adjustment (Ouariachi & Wim, 2020). The addition of a timer helps students comprehend the urgency of taking action in the real world and that, with prompt action; it is still possible to limit the rate of climate change. In actuality, it is advised that after the challenge is finished, there be time set apart in the classroom for debriefing and reflection (Ouariachi & Wim, 2020). Due to ambiguous terminology and inadequate inter-disciplinary communication, the value of education for climate change adaptation policy has been constrained (Feinstein & Mach, 2020). Although not simultaneously, research on climate change education (CCE) has advanced alongside that of climate change science. For many years, the Intergovernmental Panel on Climate Change (IPCC) has brought together the most recent scientific findings from different agencies, countries, and academic fields (Busch, Henderson, & Stevenson, 2019).

For educating children and adolescents about climate change, participative and innovative methods are still mostly untapped resources (Rousell & Cutter-Mackenzie-Knowles, 2020). (Rousell & Cutter-Mackenzie-Knowles, 2020) investigation suggests innovative strategies for engaging students in climate change through participatory and artistic styles of participation. Although it acknowledges that climate and environmental change have an impact on humans and that they are adjusting to it, human activities are nevertheless presented as "external" stressors that threaten "natural" systems. In other words, the social and human components of environmental change have not been completely incorporated into many climate change courses (Leichenko & O'Brien, 2020).

Teachers value climate change education and hold steadfast opinions regarding the anthropogenic sources, effects, and implications of climate change. However, due to the contentious nature of the subject, the current political environment, and stakeholder opposition, teachers were unable to promote these viewpoints in their lessons (Nation & Feldman, 2021). Due to space limitations in the current secondary science curricula, science teachers are unable to easily incorporate the issue of climate change into their lessons, which is a barrier to climate change teaching. Many secondary science teachers lack the necessary training to adopt the pedagogical strategies required to teach climate change science in an efficient manner (Nation & Feldman, 2021). Their views on the contentious nature of climate change science may restrict the information they teach. Teachers are not using the pedagogical techniques they perceive to be efficient ways of teaching science even though climate change is a contentious issue or they want to be objective in their teachings (Nation & Feldman, 2021). It can be challenging to include climate change in traditional science classes and environmental education programs because the topic is still divisive and highly politicized (Kunkle & Monroe, 2019).

EFFECTIVE IMPLEMENTATION OF CLIMATE CHANGE EDUCATION

The opportunity to reflect on what truly and profoundly matters, envision a better future together, and then take on the role of practical visionaries in bringing that future into reality can be taken advantage of at this time (Monroe et al., 2019). Scholars have highlighted several research areas related to climate change education that go beyond educational strategies and the precise (and frequently scientific) environmental knowledge that students should have (McGregor et al., 2021). Education about climate change can challenge the human-centered perspective that has led to issues like excessive consumption, pervasive pollution, and habitat destruction. It can also help bring about the necessary shift in thinking and learning for the Anthropocene (McGregor et al., 2021).

To answer "why" questions about climate change, education must go beyond the traditional science classroom's focus on "what" and "how" inquiries. This can be done by addressing climate change as a social issue in which students themselves are involved (Karsgaard & Davidson, 2023). It is crucial to investigate educational approaches that improve understanding of the diverse yet interconnected contributions to, effects of, and responsibilities for responding to climate change in light of the global nature of the climate crisis, including issues of intra and intergenerational justice (Karsgaard & Davidson, 2023). Despite the future-focused nature of climate action, climate change education may engage in the critical historical study; for instance, it may trace the origins, growth, difficulties, and effects of collective action, such as the recent youth climate strikes and their anti-racist and decolonizing elements, rather than praising government leadership or highlighting inspirational figures like Greta Thunberg (Karsgaard & Davidson, 2023).

The overt and covert curricula have been supported by the industrial model. But the post-carbon world is starting to take on a new eco-centric orientation in terms of economics, technology, and social organization (Irwin, 2020). When properly conceived, education can be a potent tool for effective climate change adaptation (Feinstein & Mach, 2020). Improved general education, as measured by literacy, school attendance, and overall academic attainment, can enhance adaptive capacity, and research-based adaptation learning support can accelerate social and policy change by maximizing learning before and during adaptive decision-making. These actions can reduce vulnerability and build resilience (Feinstein & Mach, 2020). (Monroe et al., 2019) reported a variety of educational program results, including an increased understanding of the science of climate change and its possible local and global repercussions. Some of the initiatives sought to address issues beyond knowledge and mindsets on climate change. These courses were created to give students the knowledge and tools they need to evaluate the validity of scientific findings and to encourage them to take initiatives to slow down or adapt to climate change. (Cutter-Mackenzie & Rousell, 2019) argue that when children and young people's voices, political agency, and creativity are included, climate change education can create a completely new area of educational experience and study. As a tool to include students in their learning environment and promote both hard skills and soft skills like teamwork, communication, and leadership, the use of recreational escape rooms in educational contexts has grown in popularity (Ouariachi & Wim, 2020). A foundation for achieving professional alignment between educators with divergent worldviews could be provided through a student-centered approach to climate change education (Kunkle & Monroe, 2019).

By addressing the underlying causes, education can only assist calm a dangerous state. Multidisciplinary and interdisciplinary frameworks must be used for climate change education. A social and all-encompassing learning process must be incorporated into climate change education. The ramifications of the invisible and uncertain nature of climate change require educators to urgently and fundamentally reassess their approaches (Reid, 2019). The most significant effects of our approach relate to learning assistance for adaptation. Together, education and adaptation researchers must create, test, and implement learning tools that fulfill the aforementioned purposes (Feinstein & Mach, 2020). How could we redesign adaptive decision-making scenarios to encourage learning among both inexperienced and experienced participants? What types of educational materials, technological aids, and pedagogical techniques support learning while assisting individuals in making adaptive decisions in various contexts and scales? In light of the ongoing climate change, finding the answers to these issues could fundamentally alter how communities can adapt (Feinstein & Mach, 2020). (Rousell & Cutter-Mackenzie-Knowles, 2020) exhort educational experts to have the guts to go beyond pointless studies probing kids' and teenagers' understanding of climate science. Through the developing transdisciplinary area of climate change education, such a turn may very well see educational experts working directly with kids and young people themselves in genuinely collaborative, imaginative, and creative ways.

Students can better understand their skepticism about climate change by looking at various discourses and frames related to it (Leichenko & O'Brien, 2020). Understanding the dismissive rhetoric enables students to see why actors with substantial stakes in the fossil fuel or industrial agriculture sectors could not take climate change seriously and, in some circumstances, actively oppose climate and energy policy (Leichenko & O'Brien, 2020). As more students are willing to engage with transformative solutions, teaching about transformations is essential for climate change education. Critical thought, the ability to take viewpoints, and education for transformation are necessary (Leichenko & O'Brien, 2020). When entrenched interests fight strategies and interventions that challenge social and cultural norms, laws, regulations, and institutions, it frequently sparks social movements, political activism, and support for political candidates that support climate action (Leichenko & O'Brien, 2020).

(Rousell & Cutter-Mackenzie-Knowles, 2020) propose that rather than emphasizing what people don't know or comprehend about climate change, education about it should become adaptable to the pre-existing ideas, attitudes, and situational settings of individual audiences. It is possible to make emotional connections between a variety of experiences and information about climate change, including scientific data, time-lapse photography, digital simulations, maps, fictional stories, and other affect-driven educational interactions (Rousell & Cutter-Mackenzie-Knowles, 2020). ESD must consider emotions to prevent hopelessness and resolve disagreements healthily. Some ESD scholars appear to subscribe to the notion that emotions and cognition are intertwined, and that observing emotional responses might occasionally reveal novel, crucial insights on how to address sustainability-related problems (Ojala, 2021). Personal experiences, as well as more generally cultural, societal, and emotional norms, can have an impact on teachers' meta-emotion philosophies and ideas about emotions in education. Today's neoliberal society controls through fostering new mentalities regarding, for instance, what is crucial to learn in addition to regulating through policy texts (Ojala, 2021). Integrative approaches can assist students in identifying responses that appeal to many understandings of climate change and its solutions by bringing attention to worldviews and values, the influence of framings and discourses, and opportunities for transformative action (Leichenko & O'Brien, 2020).

To improve both the quality and quantity of CCEC globally, collaboration and communication between researchers and policymakers are essential, whether at the UNFCCC level, at regional or organizational levels, or both (McKenzie, 2021). Anthropocene perspectives urge for an integrated approach to climate change education by highlighting connections between environmental and social aspects of the issue. Such a strategy connects climate change to core worldviews and assumptions about how people see themselves concerning one another, other species, the environment, and to the future (Leichenko & O'Brien, 2020). Students can examine how possibilities, a sense of purpose, and constructive hope can drive action by learning about the emotional aspects of climate change. This will enable them to relate to a variety of emotions (Leichenko & O'Brien, 2020). The perception is that the best way to handle the environmental problem of climate change is through the implementation of managerial, behavioral, and technical solutions through international accords and climate regulations (Leichenko & O'Brien, 2020).

Future generations must carefully examine the current status of secondary science classroom climate change education, both the curricula being taught and the people delivering it (Nation & Feldman, 2021). Education on climate change is a new societal and scientific concern, and science curriculum is beginning to take it into account. For the first time, teaching about climate change is a challenge for many science instructors (Nation & Feldman, 2021). The instructor, as well as other members of the community, such as the students, can have an impact on how the students develop their scientific ideas and practice (Nation & Feldman, 2021).

Teacher educators need to focus more on assisting teachers in addressing aspects of climate change that are outside of their areas of expertise, such as the requirement for teachers to be able to question conventional subject-matter curricula and assessment techniques (Nation & Feldman, 2021). It is not sufficient to just include climate change in the existing science curriculum; rather, teachers must have a deeper understanding of good pedagogical methods to successfully teach material that is climate-centered. Teachers need to be aware that presenting climate change information in a way they believe to be objective results in prejudice that reinforces the unfounded ideas of climate change deniers (Nation & Feldman, 2021). Our science teachers need additional professional development to be better equipped with successful teaching techniques. This can involve pre-service teacher preparation in university education programs, but it's more significant because of the vast number of secondary science teachers with other certifications. To better equip them to teach climate change science, in-service training for teachers is required. This can be accomplished through professional development programs and workshops (Nation & Feldman, 2021).

Students' proactive climate action can be sparked by a comprehensive curriculum, effective community participation, and CCEE in a supportive environment that empowers teachers and students. More research is needed to evaluate the effectiveness of pedagogy and CCEE in terms of time spent implementing them and the skills that kids in both primary and kindergarten levels are developing (Karim et al., 2022). Materials and training could more effectively engage educators across worldview differences by emphasizing the opportunities they have to hone their skills in communicating the nature of science and empowering their students to be active participants in their education rather than framing climate change education as an urgent necessity for raising awareness of environmental risks and reducing carbon dioxide emissions, a strategy that is likely to raise cultural red flags and disengagement (Kunkle & Monroe, 2019). Through a shared commitment to professional values as opposed to divisive commitments to a worldview-driven goal, (Kunkle & Monroe, 2019) emphasizes a shared purpose for supporting climate change education.

Materials that explore these tactics as potential answers may be more engaging to educators than those that only concentrate on changing people's behavior and enforcing emission laws (Kunkle & Monroe, 2019). With any of these strategies, the objective of reshaping climate change education would not be to persuade teachers to accept it as necessary or to completely omit critical content, but rather to create a space for more objective, unbiased considerations of top-notch materials and professional development (Kunkle & Monroe, 2019). A practical and efficient way to boost interest in climate change education across the worldview spectrum may be to enroll educators in professional development programs that provide instruction and resources for enhancing these professional skills and assisting their students to do the same (Kunkle & Monroe, 2019). Focusing on shared professional obligations like helping teachers navigate their own biases and educating students about the nature of science should be the main goal of professional development. To lessen hostility surrounding disagreements and increase engagement from educators who feel marginalized because of their ideas, it will be crucial to foster trust and discussion among educators, curriculum developers, and trainers. In other words, we need to do a better job of applying what we know about great science education to its facilitators (Kunkle & Monroe, 2019).

CONCLUSION

The continuation of life is endangered by the dangerous issue of climate change. Climate change will likely result in the sixth significant extinction catastrophe in recorded history. The climate changes brought on by the use of fossil fuels pose an existential threat to life as we know it on Earth. Education plays a crucial role in the worldwide response to climate change since it increases young people's knowledge while also developing awareness, enabling informed decision-making, and encouraging changes in attitudes and behavior. Subject curricula are beginning to take climate change education into account as a significant societal and scientific concern. One of the most important methods for raising knowledge of climate change mitigation is climate change education (CCE). As a result of climate change education, the majority of students must acquire the knowledge, critical thinking abilities, scientific knowledge, and ethical frameworks required to avoid, adapt to, and reverse climate change. Initiatives to educate people about climate change can have a big impact on creating communities of the pragmatist visionaries needed to tackle new problems as they arise and build on previous achievements. Climate change education is a growing social and scientific concern, and science curricula are starting to take this into account.

REFERENCES

Anyanwu, J., & Njoku, C. (2023). Integrating Climate Change Education into Social Studies Curriculum at The Basic Education Level. *International Journal of Education, Learning and Development*, *11*(2), 33–40. Advance online publication. doi:10.37745/ijeld.2013/vol11n23340

Armstrong, A. K., & Krasny, M. E. (2020). Tracing Paths from Research to Practice in Climate Change Education. *Sustainability*, *12*(4779), 1–21. doi:10.3390u12114779

Aubrecht, K. B. (2018). Teaching relevant climate change topics in undergraduate chemistry courses: Motivations, student misconceptions, and resources. *Current Opinion in Green and Sustainable Chemistry*, *13*, 44–49. doi:10.1016/j.cogsc.2018.03.008

Boakye, C. (2015). Climate Change Education: The Role of Pre-Tertiary Science Curricula in Ghana. *SAGE Open*, *5*(4), 1–10. doi:10.1177/2158244015614611

Boyko, M., Desak, T., Fleming, C., Diplock, K., & Pons, W. (2020). Climate change education: The need for comprehensive climate change education in environmental public health curriculum. *Environmental Health Review*, *63*(4), 114–120. doi:10.5864/d2020-029

Busch, K. C., Henderson, J. A., & Stevenson, K. T. (2019). Broadening epistemologies and methodologies in climate change education research. *Environmental Education Research*, *25*(6), 955–971. doi:10.1080/13504622.2018.1514588

Cantell, H., Tolppanen, S., Aarnio-Linnanvuori, E., & Lehtonen, A. (2019). Bicycle model on climate change education: Presenting and evaluating a model. *Environmental Education Research*, *25*(5), 717–731. doi:10.1080/13504622.2019.1570487

Cutter-Mackenzie, A., & Rousell, D. (2019). Education for what? Shaping the field of climate change education with children and young people as co-researchers. *Children's Geographies*, *17*(1), 90–104. Advance online publication. doi:10.1080/14733285.2018.1467556

Eilam, E. (2022). Climate change education: The problem with walking away from disciplines. *Studies in Science Education*, *58*(2), 231–264. doi:10.1080/03057267.2021.2011589

Feinstein, N. W., & Mach, K. J. (2020). Three roles for education in climate change adaptation. *Climate Policy*, *20*(3), 317–322. doi:10.1080/14693062.2019.1701975

Filho, W. L., Yayeh Ayal, D., Wall, T., Shiel, C., Paco, A., Pace, P., Mifsud, M., Lange Salvia, A., Skouloudis, A., Moggi, S., LeVasseur, T., Vinuesa Antonio, G., Azeiteiro, U. M., Ioannis, N., & Kovaleva, M. (2023). An assessment of attitudes and perceptions of international university students on climate change. *Climate Risk Management*, *39*(100486), 1–16. doi:10.1016/j.crm.2023.100486

Gustafsson, P., Engström, S., & Svenson, A. (2015). Teachers' View of Sustainable Development in Swedish Upper Secondary School. *Procedia: Social and Behavioral Sciences*, *167*, 7–14. doi:10.1016/j.sbspro.2014.12.635

Irwin, R. (2020). Climate change and education. *Educational Philosophy and Theory*, *52*(5), 492–507. doi:10.1080/00131857.2019.1642196

Karim, N., Othman, H., Zaini, Z., Rosli, Y., Wahab, M. I. A., Al Kanta, A. M., Omar, S., & Sahani, M. (2022). Climate Change and Environmental Education: Stance from Science Teachers. *Sustainability (Switzerland)*, *14*(24), 1–18. doi:10.3390u142416618

Karsgaard, C., & Davidson, D. (2023). Must we wait for youth to speak out before we listen? International youth perspectives and climate change education. *Educational Review*, *75*(1), 74–92. doi:10.1080/00131911.2021.1905611

Kunkle, K. A., & Monroe, M. C. (2019). Cultural cognition and climate change education in the U.S.: Why consensus is not enough. *Environmental Education Research*, *25*(5), 633–655. doi:10.1080/13504622.2018.1465893

Leichenko, R., & O'Brien, K. (2020). Teaching climate change in the Anthropocene: An integrative approach. *Anthropocene*, *30*, 100241. Advance online publication. doi:10.1016/j.ancene.2020.100241

Maxwell, J., & Blashk, G. (2016). Teaching about climate change in medical education: An opportunity. *Journal of Public Health Research*, *5*(673), 37–40. doi:10.4081/jphr.2016.673 PMID:27190980

McGregor, H. E., Pind, J., & Karn, S. (2021). A 'wicked problem': Rethinking history education in the Anthropocene. *Rethinking History*, *25*(4), 483–507. doi:10.1080/13642529.2021.1992159

McKenzie, M. (2021). Climate change education and communication in global review: Tracking progress through national submissions to the UNFCCC Secretariat. *Environmental Education Research*, *27*(5), 631–651. doi:10.1080/13504622.2021.1903838

Mochizuki, Y., & Bryan, A. (2015). Climate Change Education in the Context of Education for Sustainable Development: Rationale and Principles. *Journal of Education for Sustainable Development*, *9*(1), 4–26. doi:10.1177/0973408215569109

Monroe, M. C., Plate, R. R., Oxarart, A., Bowers, A., & Chaves, W. A. (2019). Identifying effective climate change education strategies: A systematic review of the research. *Environmental Education Research*, *25*(6), 791–812. doi:10.1080/13504622.2017.1360842

Nation, M. T., & Feldman, A. (2021). Environmental Education in the Secondary Science Classroom: How Teachers' Beliefs Influence Their Instruction of Climate Change. *Journal of Science Teacher Education*, *32*(5), 481–499. doi:10.1080/1046560X.2020.1854968

Ojala, M. (2021). Safe spaces or a pedagogy of discomfort? Senior high-school teachers' meta-emotion philosophies and climate change education. *The Journal of Environmental Education*, *52*(1), 40–52. doi:10.1080/00958964.2020.1845589

Okada, A., & Gray, P. (2023). A Climate Change and Sustainability Education Movement: Networks, Open Schooling, and the 'CARE-KNOW-DO' Framework. *Sustainability (Switzerland)*, *15*(3), 2356. Advance online publication. doi:10.3390u15032356

Ouariachi, T., & Wim, E. J. L. (2020). Escape rooms as tools for climate change education: An exploration of initiatives. *Environmental Education Research*, *26*(8), 1193–1206. doi:10.1080/13504622.2020.1753659

Park, N.-E., Choe, S.-U., & Kim, C.-J. (2020). Analysis of Climate Change Education (CCE) Programs: Focusing on Cultivating Citizen Activists to Respond to Climate Change. *Asia-Pacific Science Education*, *6*(1), 15–40. doi:10.1163/23641177-BJA00004

Reid, A. (2019). Climate change education and research: Possibilities and potentials versus problems and perils? *Environmental Education Research, 25*(6), 767–790. doi:10.1080/13504622.2019.1664075

Ross, H., Rudd, J. A., Skains, R. L., & Horry, R. (2021). How Big Is My Carbon Footprint? Understanding Young People's Engagement with Climate Change Education. *Sustainability, 13*(1961), 1–19. doi:10.3390/su13041961

Rousell, D., & Cutter-Mackenzie-Knowles, A. (2020). A systematic review of climate change education: Giving children and young people a 'voice' and a 'hand' in redressing climate change. *Children's Geographies, 18*(2), 191–208. Advance online publication. doi:10.1080/14733285.2019.1614532

Siegner, A., & Stapert, N. (2020). Climate change education in the humanities classroom: A case study of the Lowell school curriculum pilot. *Environmental Education Research, 26*(4), 511–531. doi:10.1080/13504622.2019.1607258

KEY TERMS AND DEFINITIONS

Anthropocene Age: It is an acronym for a new geologic era, one that is characterized by our own profound influence on the earth.

Anthropogenic Climate Change: Emissions of greenhouse gases are caused by human-induced activities, such as the use of energy, industrial operations, the use of solvents and other products, agriculture, land use change and forestry, and garbage.

Climate Change: Long-term modifications to weather and temperature patterns are referred to as climate change. Such changes may occur naturally as a result of variations in the sun's activity or significant volcanic eruptions.

Climate Change Education: Giving people the knowledge, abilities, values, and attitudes necessary to be change agents, it enables them to comprehend and address the effects of the climate catastrophe.

Education for Sustainable Development (ESD): Every person has the opportunity to learn the knowledge, skills, attitudes, and values required to create a sustainable future through education for sustainable development.

Green House Gas Emission: Human-related greenhouse gas emissions increase the greenhouse effect and fuel climate change. Coal, oil, and natural gas combustion produce the majority of carbon dioxide.

Industrialization: Industrialization is the process of moving a country or region's economy away from reliance on agriculture and towards one that is driven by industry.

Sustainable Development Goals (SDGs): Our world is supposed to change thanks to the Sustainable Development Goals (SDGs). They are a call to action to eradicate poverty and injustice, safeguard the environment, and guarantee the prosperity, justice, and health of all people. It is essential that nobody is forgotten.

Sustainable Lifestyle: Understanding how the world around us is impacted by our lifestyle decisions is essential to sustainable living, as is figuring out how to make everyone's life better and lighter.

Chapter 10
Pedagogical Strategies to Enhance Climate Change Education Outcomes

Billiah Nyamoita Gisore
https://orcid.org/0000-0002-5116-7615
Kisi University, Kenya

Evelyn Njurai
Kisii Univrsity

ABSTRACT

Global temperatures have soared since the Industrial Revolution some 140 years ago, with the ecosystems having a severe impact globally. Environmental issues are changing not only the normative lifestyles, but also the ethics and understanding of the world. Educating people on ways to relate to the habitat calls for adopting sustainable practices that reduce the impact on climate. Climate change education (CCE) is a new field of pedagogy that addresses the issues of climate change, but has no established model for implementation. The need for transformative learning with pedagogical approaches that incorporate diverse climate change learning experiences is required. The chapter therefore addresses the need for various aspects of climate change to be addressed as part of educational pedagogy and the need for a curriculum with climate-friendly, age-appropriate content. The possible strategies and the barriers that seem to hinder the achievement of educational outcomes are also examined.

DOI: 10.4018/978-1-6684-9099-0.ch010

INTRODUCTION

Climate change is an alteration of climate that is directly or indirectly caused by human activity which changes the composition of the earth's atmosphere (UNFCCC, 1992). Some of the effects of climate change include increased flooding, severe storms, changes to the environment and wildlife populations, higher energy costs, as well as an increase in the concentration of greenhouse gases (GHGs) in the atmosphere, which has an impact on a number of climate parameters (The National Research Council, 2011). This is in line with Carleton and Hsiang (2016) who indicate that as a result of climate change, temperatures have risen, which will eventually result in a variety of negative effects, including an increase in mortality, financial harm, extreme changes in temperature and precipitation, a decrease in agricultural production, an increase in electricity use, and a decrease in labor productivity.

Numerous myths exist about climate change, with a general lack of public knowledge of the problems it poses as well as a lack of understanding of how to adapt to it. There also seems to be a lack of preparation or reluctance to act despite many aspects of global climate change having been discussed for a long time (Leiserowitz & Smith, 2010). The public must be informed about the activities necessary at all levels in order to deal with the complexity of climate change and its ramifications (Mochizuki & Bryan, 2015) as the final patterns and magnitudes of the impacts at the local, regional, and global levels depend on the direction, amount, and intensity of the climate changes (Chang & Pascua, 2016).

To enable prudent decisions to be taken at the home as well as at the general society level, education on the human and behavioral elements of climate change is necessary (Gardner & Stern, 2008). Young people need to safeguard their future and they must be motivated to deal with climate change and be sufficiently informed about the science behind it (Yli-Panula, Jeronen, Koskinen & Mäki, 2022). Initiatives that can help people make informed decisions and change their behavior should be put in place so that people can adapt their beliefs and conduct in response to the new environment (Gardner & Stern, 2008). The prevalent non-ecological worldview that dominates human behavior and ideas needs to be tested in light of how bad the situation is (Trott, 2022) and to develop environmental ethos, specific environmental challenges should be addressed at a young age. Education, especially which is gained in a systematic manner, is one of the essential avenues for dealing with and averting such global challenges (Anderson, 2012).

Schools are the settings where moral development of children can be encouraged through the teaching of values (Biesta & Tedder, 2007). It is a place where young people receive help as they interact with local community structures and values as they gain knowledge of and interaction with larger global structures. It has a big impact on how well people understand how climate change affects them, helps people alter their habits over time, and makes it easier for people to comprehend how to deal with the repercussions of climate change (UNESCO, 2017). To modify behavior and encourage climate action, education is essential.

Education that places more of an emphasis on sustainable actions will help to improve the environmental condition, according to Goal 4 (Quality Education) which is aligned with Goal 13 (Climate Action) of the Sustainable Development Goals (SDGs) (UNESCO, 2017). Climate Change Education (CCE) is addressed in the context of education for sustainable development in line with Goal 13 (Climate Action). This highlights the value of CCE, which, as a newly developed area of pedagogy, depends on educational institutions for both implementation and success. In light of the current climate emergency, it is essential that teaching and learning are refocused (UNESCO & UNFCCC, 2016).

EXPECTED CLIMATE CHANGE EDUCATION OUTCOMES

Climate Change Education (CCE) is a relatively new branch of education which differs from other disciplines as it focuses on the problem of the present (Fahey, 2012). The objectives of CCE include; expanding and organizing information, fostering critical thinking, and promoting action. It emphasizes getting students to think about what they can each do in the face of risk, ambiguity, and rapid change (Stevenson, Nicholls, & Whitehouse, 2017). According to Tolppanen et al., (2017), CCE necessitates a transformative education based on an ecological vision, with holistic, systemic, multi- and interdisciplinary design and implementation. These abilities are a part of the critical competencies for the twenty-first century, which are connected to the 2030 Agenda's target 4 on high-quality education. CCE will aid in preparing learners who understand the social and economic repercussions, complexities of change, and adaptation to such change. CCE also has the potential to be one of the most effective mitigation measures of climate change with long-term effects (McKeown & Hopkins, 2010).

The purpose of CCE, according to Kagawa and Selby (2010), is to deeply consider what is significant and to jointly envisage future views. The CCE goals can be seen as a continuum, with social transformation that will minimize societal impacts on the climate at one end and climate literacy, or knowing how and why climate change is occurring, at the other (McKeown & Hopkins, 2010). The expected goals of CCE are the growth of knowledge by comparing the numerous sources of information on climate change concerns, the development of skills by taking values into consideration, inquiry-based learning, and taking environmental action (Chang & Pascua, 2017). A learner who possesses the four competences will be successful, self-assured, a responsible citizen, and an effective contributor (Priestley

& Biesta, 2013). To be able to handle uncertainty, develop and analyze alternative futures, as well as examine their own values and behavioral patterns, CCE will assist learners in thinking critically, systematically, and appropriately (Cutter-Mackenzie & Rousell, 2019).

As a link between the present and next generations, learners are essential to socioeconomic development, especially tackling climate change. They must become aware of, respect, and understand the interconnectedness of their surroundings. One of the most important and successful methods to develop skills for tackling the climate problem is by integrating climate change education into formal education systems (Stevenson, Nicholls, & Whitehouse, 2017), particularly in regard to adaptation and mitigation. This is due to the fact that education has multiplier effects, meaning that when people collaborate with others outside of a school or other educational institution and share what they have learned, families and communities benefit as well (Bangay & Blum, 2010).

The current education tends to emphasize scientific information or the "climate" while ignoring the "change" (Kagawa & Selby, 2010), reflecting how society has responded to the phenomenon. Since CC is a contentious, complex, and contextual term that may not be simple to understand, there is a lot of material and inconsistencies surrounding it that can be confusing and frustrating for learners (Chang & Pascua, 2017). The material that is available has not provided recommendations on what should be taught and what should make up CCE (Tolppanen et al., 2017). As a result, the effects of climate change and the need for actions to combat it receive minimal attention (Stevenson et al., 2017).

Kimberly (2022) claims that despite the increased urgency, education of the young continues to be carried out in the same manner as it was more than ten years before. For learners to engage with information, inquire, understand, pose probing questions, and take whatever actions are determined to be appropriate to respond to climate change, there is a need for a type of learning that is critical, creative thinking, and capacity building (Keller, Stötter, Oberrauch, Kuthe, Körfgen & Hüfner, 2019). CCE aims to develop knowledge of and skills for addressing climate change, climate injustice, and the impacts of global warming on biodiversity (UNESCO & UNEP, 2011). It should be seen as learning in the face of risk, uncertainty, and rapid change (Stevenson et al., 2017). Public education about the gravity of the problem is necessary, as is the formulation of effective response plans.

STRATEGIES TO ACHIEVE THE CLIMATE CHANGE EDUCATION OUTCOMES

Climate change is not just a scientific phenomenon but a complex socio-scientific issue that demands more than the teaching of content (Stevenson, et al., 2017). It needs relevant content to enable one to cope with climate change, and that will impart skills such as future thinking, critical analysis, and deepened understanding of the impacts of climate change (Sterling, 2011). The complex nature of climate change means that mitigation and adaptation must encompass a continuous stream of activities, actions, decisions, and attitudes that inform decisions about all aspects of life and that reflect existing social norms and processes (Nelson, Adger & Brown, 2007). To realize the objectives of CCE, one must think beyond the question of "how do we change individuals" and take into account generational differences in behavior as well as available options.

In order to enable transformation, everyone needs to understand climate change, be aware of how decisions are made, and seek to assist those making decisions more (Kagawa & Selby, 2010). To deal with the effects and actions of global climate change, society needs to adapt its attitudes and behaviors rather than taking a "business as usual" approach with traditional content distribution (Sterling, 2011). There is a need for climate change education strategies that strike a balance between creating critical thinkers who can engage fresh information about the phenomenon and compassionate people who are determined to take action to improve their living environment (Chang, 2015). The way that knowledge is currently structured (CCE) does not appear to encourage activities that address climate change.

With curriculum and pedagogies that enable learners to analyze the source of the problem, consider appropriate next steps, and take constructive action, classrooms need to be geared toward "out-of-the-box" thinking (Glasser, 2007). Chang and Pascua (2017) claim that learners' background knowledge of climate change appears lacking and is made up of false information that is ingrained in logical and structurally sound mental models. Therefore, the primary goal of the teaching approach should be to shift the mental models that learners have regarding climate change. The pedagogical interventions ought to include exploratory, immersive, and participatory methods that let students actively participate in the process of change (Sharplin, 2014).

The teacher's role in communicating the science of climate change is crucial for changing the mental models that learners have developed about it (Ratinen, 2016). A number of studies indicate that there are deficiencies in socio-environmental issues or competences in sustainability in preservice teachers' training (Desjean-Perrotta, Moseley & Cantu, 2008; Valderrama-Hernández, Rubio, Sánchez-Carracedo, Franco, González, Gil-Doménech, Vidal-Raméntol, Miñano, 2019; Morote, Hernández &

Olcina, 2021), as well as in that of in-service teachers (Teixeira, Silva & Neves, 2018). In order for teachers to feel comfortable implementing environmental education programs in the classroom that primarily focus on personally relevant and meaningful information using a variety of engaging, active, and learner-centered teaching methods, there is a need to intervene in initial and in-service teacher training (Pogacnik, nidarcic, & Strgar, 2014).

Inadequate teacher training has been linked to low environmental literacy and unsuccessful environmental education initiatives (UNESCO, 1997). This necessitates spending money on instructional materials development and pedagogy training for effective teaching techniques, as an increase in instructors' CCE knowledge and teaching skills directly impacts students' CC knowledge, skills, values, and attitudes (Lotz-Sisitka, 2010). How teachers structure their lessons depends on the features of the disciplines they value (Brooks, 2010). Values are fundamental beliefs that guide behavior and act as yardsticks for evaluating particular issues (Halstead & Taylor, 1996).

To encourage teachers to teach CC, the subject should be connected to their professionals (Hannula, 2022). This development of one's professional identity should be a reflection of a teacher's competence, which includes subject-matter knowledge and comprehension, social performance of relevant disciplinary practices, such as talking and acting in certain ways, and acceptance as a teacher of climate change by oneself, colleagues, and others (Carlone & Johnson, 2007). It follows that the teachers should be recognized as CC education experts since once a community of specialists is developed, other forms of help will inevitably emerge through institutional initiatives (Young, 2014). Additionally, this will deal with the issue of teachers' fragmented and inadequate grasp of CC, which could result in a misinterpretation of CCE (Hannula, 2022).

If teachers do not obtain training, they will not be confident in how they are teaching climate change issues (Fahey, 2012). It is crucial to keep in mind that teachers do not necessarily need to have a complete understanding of the subject before introducing climate change into their lectures. In order to feel more comfortable in settings when content and pedagogical practices are scarce, teachers can benefit from regular, brief professional development courses that provide background information (Stevenson et al., 2017).

Educational programs should place a high focus on transdisciplinary thinking in order to prepare future learners to address challenges that are not yet recognized to exist, such as the potential effects of climate change (Fahey, 2012). Each course in the program, according to Biggs and Tang (2009), needs to present and create course-specific intended learning outcomes and processes for gauging their effectiveness. To make climate change knowledge meaningful and relevant to participants, the activities or educational interventions utilized in the programs should be developed with learners in mind.

With an emphasis on misconceptions and "alternative conceptions" regarding climate change, climate change education is connected to a few disciplines like the physical sciences, technology, engineering, and mathematics (Hallar, McCubbin, & Wright, 2011). According to Chang and Pascua (2017), as CCE is multidisciplinary in nature, it is necessary to convert to an interdisciplinary teaching style that would incorporate CCE knowledge into the curricula of all disciplines. According to Wise (2010), the different knowledge silos must cooperate in order to fully comprehend the nature, potential, and hazards of climate change, which is a trans-disciplinary issue.

The authors Laesse, Schnac, Breiting, and Rolls (2009) suggest that although it is related to science subjects, teaching content on climate change should be viewed as a separate field of practice, distinct from other disciplines, and operate outside the confines of the science subjects. Other points of view are different from this one. Young (2014) contends that given how important CCE is, it should be regarded as specialized knowledge with a particular structure and goal. This knowledge is unique in the ways that it is developed, put to use, valued, and assessed, as well as in the ideas and guidelines that relate to the fundamental epistemological tenets. The dispersal of climate change across the curriculum, according to Lehtonen, Salonen, and Cantell (2019), results in the lack of crucial connections between the numerous pieces of information taught in various areas.

The addition of climate change as a taught subject would provide important mechanisms like assessments and accreditation as well as funding which will distinguish curricular courses and support effective teaching (Ross, 2000). By approaching climate change as a discipline, education scholars may be able to begin researching climate change literacy in a more in-depth and pertinent way by rigorously analyzing the set of skills and knowledge associated with such literacy. Teaching through fragmentation effectively may be challenging given the complexity of climate change and the interdependence of its components under the current cross-curricular dispersion scenario.

The introduction of climate change as a disciplinary-subject will allow the learners to develop their epistemic cognition which varies according to one's field (Hofer, 2000). According to Greene and Yu (2016), a full awareness of the normative, or accepted, epistemic practices across a number of fields is required in order to evaluate information sources and build an argument supported by evidence. In the end, disciplines provide learners with benchmarks and boundaries that support conceptual development. Because of these disciplinary restrictions, students will no longer need to piece together the fragmented. (Chang, 2015)

Although many teachers are doing a great job educating students about climate change, formal structures that guarantee the delivery of CCE are lacking (Priestley et al., 2015). When it does exist in school curricula, climate change is generally not acknowledged and is treated as an undefined issue (Cutter-Mackenzie & Rousell, 2019). Beyond the issue of climate change content, curriculum developers should reconsider their work and take into account innovative and avant-garde alternatives that depart from anthropocentric and scientific paradigms of learning, inquiry, and thought (Dawson et.al, 2021). Despite the incorporation of CCE in some curricula, there are no systematic methods incorporated on how to use it as a tool for adaptation and mitigation (Trott, 2022).

Clear, doable solutions that concentrate on the learning process should be available. Focus must be placed on influencing the learners' views and emphasizing the significance of climate change (Faria et al., 2015). The overall achievement of CCE objectives should be supported by pedagogies that can encourage learners' critical and action-oriented involvement (Keller, Stötter, Oberrauch, Kuthe, Körfgen, & Hüfner, 2019). Conducting open dialogues in class that allow learners to share ideas, challenge their own opinions, and improve their capacity to use evidence to back their opinions are just a few possible ways (Halstead, Parsons, Dunhill, & Parsons, 2021). This brings together the knowledge, skills, and abilities required to identify opportunities and formulate a vision on how to use those opportunities to solve a problem. This strategic thinking is crucial in developing strategies to put solutions to environmental problems into practice (Dragoni, Oh, Vankatwyk & Tesluk, 2011).

Additionally, they can participate in discussions in small groups with students who are collaborating, exchanging ideas and making observations, and arriving at fresh conclusions as a consequence. According to Theobald et al. (2015), flipped classroom design can be utilized to direct group conversations and promote active learning, which may improve conceptual knowledge and pro-environmental attitudes. CCE can encourage emotions of empowerment and the development of action skills by fostering community and teamwork and by presenting chances for engagement (Wenger, 2011). Trainers must encourage learners to take action by gauging their interest in participating in specific behaviors that address climate change (Stapleton, 2015). It is crucial that educators reconsider how to mitigate climate change and adapt to it in ways that are socially transformational, employing methods that foster creativity and give learners agency (Flora et al., 2014).

The chance to execute climate-responsible and climate-friendly projects in the greater context of their schools or communities can be offered to learners by educational interventions that may extend beyond the classroom (Robinson, 2011). They should take part in activities that will assist them in developing justifications for why the climate is changing as well as evidence to back up any competing theories about it (Hannula, 2022). The need for inquiry-based, project-based, and action-oriented learning in the classrooms. For instance, a cost effective initiative like planting trees may be used to educate conservation inside the school (Leigh, 2009).

School gardens can be utilized to teach learners about alternative methods for securing and producing food (Dillon et al., 2016). This is in line with aim 12.2 of SDG 12, which emphasizes the optimal use of natural resources and the achievement of sustainable management. The students may also participate in clubs and competitions that support research and activities pertaining to local climate consequences (Carleton & Hsiang, 2016). Given that information is presented in a setting that is relevant to people, it is crucial to relate the local needs, issues, and concerns (Feola, 2015).

Additionally, schools may organize CCE days that concentrate on what students could do as a strategy to respond to climate change, and make sense of the specific real-life events, this will enable experiential learning (Chang & Pascua, 2017). In order to support learners in developing transferable skills to new, uncertain, unidentified, or poorly defined contexts, climate change education necessitates learning that is inquiry-based, reflective, creative, and participatory (Wals, 2011). The learners' enthusiasm and dedication to achieving the CCE goals may be increased by the school's climate-friendly actions.

Another strategy could be to "climate change proof" the educational infrastructure in order to provide climate-safe learning settings (UNESCO & UNEP, 2011). Environmental risks change depending on one's environment. As learners are susceptible to the effects of natural disasters, schools should be aware of local hazards, conduct risk assessments, improve or maintain buildings as necessary, and inform them accordingly (Bangay & Blum, 2010; UNESCO & UNEP, 2011). They should also be involved in risk reduction processes and be aware of what to do in an emergency. The learning process should go beyond simply imparting information of environmental dangers to include exploration of one's immediate surroundings, risk assessment, and discussion of past extreme weather occurrences in the community. In order to help learners to be ready and cope when a disaster comes, it is necessary to incorporate such material into the school curricula (Cameron & Davies, 2010). Since knowledge of climate change is a context-specific goal, the curriculum should include assessment tasks that allow students to show evidence of their comprehension and achievement of the CCE by reporting their personal observations and experiences from their own surroundings (Brown, 2005). By giving students the chance to critically evaluate and take action on climate change, we can influence the way they perceive the world.

To create CCE techniques that encourage learner's positive involvement with climate change and general well-being, teachers must comprehend the psychological experiences of the learners related to climate change (Halstead et al., 2021). This necessitates the development of a practice-based curriculum that describes the elements of climate change that should be taught as well as what the learner should exhibit in terms of CCE knowledge and skills, age-appropriate at various educational levels (Trott, 2022). In order to design performance activities that will allow them to demonstrate the achievement of those outcomes and specify the criteria by which they will be evaluated, it is necessary to first think through the curriculum's content in order to establish learning outcomes (Monroe, Plate, Oxarart, Bowers & Chaves, 2017).

According to Meymeris et al. (2008), the youth are recognized to be leaders in the use of digital technology and may leverage its many applications to create innovative and empowering climate change education programs. For instance, they can create a website or blog for group contributions relating to climate change issues, use social media networks to discuss climate issues and start climate change initiatives or use digital games in school as a way of peer-teaching about climate change and CCE (Cutter-Mackenzie & Rousell, 2019).

Using reflection journals, learners can compile their learning experiences and activities to show how well they have understood the concepts of climate change (Pruneau, Doyon, Langis, et al., 2006), as well as record their learning achievements. They can also write climate-friendly biographies that detail the kinds of activities they personally engage in. The direct participation strategy can inspire learners to act as change agents in their communities (Anderson, 2012).

Additionally, learners could create networks of communities for learning and apply instructional techniques that might result in engagement and action (Pruneau, Doyon, Langis, et al., 2006). Community members should be informed about the fundamentals of climate change, made aware of the issues of greenhouse gas reduction and natural resource adaptation, and persuaded to take personal and policy action to combat the issue (Chang & Liberty, 2017). In a similar vein, Kagawa and Selby (2010) contend that teaching about climate change necessitates a flexible social holistic learning method that integrates learning with action in local community contexts. Communities will be given the freedom to create their own climate change projects and ways to interact with the issue using a bottom-up participatory approach (Figueiredo & Perkins, 2013). Learners should be urged to participate in extracurricular activities and community projects, particularly those focused on reducing the effects of climate change and adapting to them, and to build their capacity to take effective action in response (Stevenson, Nicholls, & Whitehouse, 2017). Transformative learning is most effective in environments that are learner-centered, interactive, action-oriented, and problem-based (UNESCO, 2017).

BARRIERS TO ACHIEVING THE CLIMATE CHANGE EDUCATION OUTCOMES

Barriers to achieving the climate change outcomes may exist at the personal, societal, institutional, or structural levels and may have an impact on how people think, act, and make decisions (Bostrom, Morgan, Fischhoff, and Read, 1994). In many aspects, sustainability concerns are firmly ingrained in a fundamentally non-ecological worldview that guides all thought and behavior rather than being purely scientific problems (Trott, 2022). The majority of people are therefore ill-equipped to adequately consider it or initiate positive changes since climate change challenges the technological scope of cognition and exposes the deeply ingrained flawed metaphysical assumptions of modernity (Irwin, 2008).

CCE is a result of the dynamic interaction that occurs in teaching-learning institutions between the learner, the subject matter being taught, and the larger social aims of education (Lambert & Hopkin, 2014). Governments must develop curricula to include transformational pedagogies that encourage students' positive participation with climate change (Sterling, 2011). Building climate literacy through education for adaptation and mitigation is a critical strategy (UNESCO & UNEP, 2011; Kagawa & Selby, 2012).

One of the issues facing the curriculum is preparing learners for futures that are unknowable or only partially understood (Fahey, 2012). Students are not given enough information in the schools on the climate problem, and they appear unprepared for the world they will soon enter, despite receiving training for careers and societies founded on the system that generated the catastrophe (UK Student Climate Network, 2019). The difficulties with the curriculum are caused by social, political, and educational perceptions. Political influence may have an impact on how the curriculum is written, but it also has the potential to affect how it is interpreted and applied (Olmos-Gómez, Estrada-Vidal, Ruiz-Garzón, López-Cordero, & Mohamed-Mohand, 2019). For instance, UNESCO's 17 SDGs had political influence which has proven difficult to implement for some of them, which may be the same case as CCE.

Teachers are in charge of implementing CCE (Fuertes, Andrés, Corrochano, Delgado, Herrero-Teijón, Ballegeer, Ferrari-Lagos, Fernández & Ruiz, 2020), and their attitudes and knowledge of CCE may have an impact on how a subject is taught (Chang & Pascua, 2017). According to Boon (2016), teachers have difficulty explaining the complex science behind climate change, which contributes to the general lack of understanding. The development of CCE and teachers' CCE competencies to enable them to acquire proactive competences appear to receive relatively little attention in teacher education (UNESCO, 2009; Varela-Losada, Arias-Correa & Vega-Marcote, 2019). The teachers' understanding of climate change may be incomplete, fractured, or lacking entirely if it is not well-grounded (Borg, Gericke, Höglund, & Bergman, 2013).

CCE is seen as a content issue rather than a process or outcome since teachers frequently concentrate more on cognitive teaching, making sure that learners comprehend the material rather than utilizing it as a tool to study climate (Hannula, 2022). Because of this, teachers may see CCE as an add-on that is unrelated to and outside the purview of their academic specialties (Tolppanen et al., 2017). Teachers also have a tendency to prioritize less in the classroom when there is no assessment with accountability or certification (Field, Schwartzberg & Berger, 2019). It might be difficult to find the time to implement CCE, especially with the present emphasis on grades and scores (Nicholls, 2016). According to Fuertes, Andrés, Corrochano, Delgado, Herrero-Teijón, Ballegeer, Ferrari-Lagos, Fernández & Ruiz (2020). For proper implementation, it is crucial that CCE has certification.

Due to the lack of a formal structure, the teachers are free to determine how to carry out the course, even though the students may not have the academic aptitude to move beyond the curriculum. This implies that there may be some variation in the tasks, projects, and content that students complete (Seowa & Ho, 2016). Unified learning with comparable pedagogical strategies that include pertinent climate change learning experiences is required.

To help students understand the need for mitigation and to become change agents, the content in CCE should be matched to the curriculum's learning outcomes (Stevenson, Nicholls, & Whitehouse, 2017). The academic programs' courses are not in line with one another, which results in little to no contribution to the accomplishment of the program-level learning outcomes. For instance, the curriculum in Australia does not mandate that subjects be taught using cross-curriculum priorities (Salter & Maxwell, 2016). This suggests that teaching about climate change cannot be limited to the formal curriculum and traditional educational structures, but must instead make use of emerging informal and hybrid spaces that offer alternative opportunities for learning and action (Stenhouse, 1975).

The community is crucial in bringing about change. There doesn't seem to be enough understanding or information on climate change to get the people involved. Increasing public knowledge of the effects of and responses to climate change in order to empower the public to support climate change responses (Ardoin et al., 2009).

WAY FORWARD FOR CLASSROOM PEDAGOGY

For CCE to be successful, pedagogical practices must be in line with student activities that include a variety of climate change learning experiences and encourage positive learner participation. To help students understand the need for mitigation and to become change agents, the climate change content should be matched to the curriculum's learning outcomes (Stevenson, Nicholls, & Whitehouse, 2017). The accomplishment of program-level learning outcomes is minimally or not at all impacted by courses that are not in line with other courses in the academic programs (Salter & Maxwell, 2016). Future-thinking abilities in teaching, evaluating the growth of students' climate skills, and giving teachers further climate change trainings are all necessary as well. The quantity of inaccurate information that is available makes it difficult to create knowledge based on scientific evidence, hindering the teaching of components of climate change. Students should be assisted to distinguish between accurate and false information about climate change.

CONCLUSION

Teachers must rethink climate change mitigation and adaptation in ways that are not just technical but socially transformative. They should use teaching and learning approaches that harness creativity and empower students to be able to act. The goal is to prepare learners for an uncertain future by helping them gain the capacities (knowledge, skills, dispositions, and values) to deal with future challenges (Lotz-Sisitka 2010). In order for students to connect new information with what they already know, generate interest and meaning, and pay attention to the information, personal relevance is a requirement for excellent education (Kaplan & Kaplan, 1982).

As there will be no luxury of choice in the future, preparedness for climate disruption will, by necessity, overcome any existing stagnation in state and national curriculum (Stevenson, Nicholls & Whitehouse, 2017). We can rely on human inventiveness and imagination, and educators will find change opportunities in these curriculum areas and in hybrids of formal and informal learning. Experiential learning, active participation, and exploration are essential components of educational strategies (Kolb, 1984).

The best educational responses to climate change should involve active social learning that fosters the ability for individual and societal transformation. According to social-constructivist perspectives, knowledge is created through social contact (Dillon, 2003), hence learning takes place through discussions, debates, deliberations, and chances to compare viewpoints so as to mix various viewpoints in novel ways (Powell & Kalina, 2009). There is also a need for professional development in the field of CC which includes practical problem-solving skills, adaptive environmental assessment, and teaching methods with learner-centered learning activities, aligned with the intended learning outcomes.

Creating teaching strategies that will better support various student populations and integrating guiding principles of CCE into teacher education are both crucial. Cooperation between diverse sciences and other disciplines needs to be improved as well. In order for students to grasp the value of their individual actions and collective role in reducing climate change, CC is brought up in shared values of education (Kagawa & Selby, 2012). Building an online platform for lessons and resources on climate change would assist to guarantee that CCE is available to all.

It should be noted that climate change cannot be categorized under any one discipline as the content cuts across almost all disciplines. Additionally, there should be curriculum frameworks that are open to handling climate change as a disciplinary subject on its own.

REFERENCES

Anderson, A. (2012). Climate change education for mitigation and adaptation. *Journal of Education for Sustainable Development*, *6*(2), 191–206. doi:10.1177/0973408212475199

Ardoin, N. M., Falk, J., Heimlich, J. E., & Foutz, S. (2009). Behavior-change theories and free-choice environmental learning. In J. H. Falk (Ed.), *Free-Choice Learning and the Environment* (pp. 57–76). AltaMira Press.

Bangay, C., & Blum, N. (2010). Education responses to climate change and quality: Two parts of the same agenda? *International Journal of Educational Development*, *30*(4), 359–368. doi:10.1016/j.ijedudev.2009.11.011

Biesta, G. J. J., & Tedder, M. (2007). Agency and learning in the lifecourse: Towards an ecological perspective. *Studies in the Education of Adults, 39*(2), 132–149. doi:.2007.11661545 doi:10.1080/02660830

Biggs, J., & Tang, C. (2009). *Teaching for Quality Learning at University* (3rd ed.). Society for Research into Higher Education/Open University Press.

Borg, C., Gericke, N., Höglund, H. O., & Bergman, E. (2013). Subject- and experience-bound differences in teachers' conceptual understanding of sustainable development. *Environmental Education Research, 20*(4), 526–551. doi:10.1080/13504622.2013.833584

Bostrom, A., Morgan, M. G., Fischhoff, B., & Read, D.Mental Models. (1994). What do people know about global climate change? *Mental models. Risk Analysis*, *14*(6), 959–970. doi:10.1111/j.1539-6924.1994.tb00065.x

Brooks, C. (2006). Geographical knowledge and teaching geography. *International Research in Geographical and Environmental Education, 15*(4), 353369. doi:10.2167/irg200.0

Brown, S. (2005). Assessment for learning. *Learning and Teaching in Higher Education*, *1*, 81–89.

Cameron, C., & Norrington-Davies, G. (2010). *Children and disaster risk reduction in Asia and the Pacific: A way forward.* Discussion Paper presented at the High-Level Meeting on Cooperation for Child Rights in the Asia-Pacific Region. http:// www.unicef.org/eapro/Children_and_Disaster_Risk_Reduction.pdf

Carleton, T. A., & Hsiang, S. M. (2016). Social and economic impacts of climate. *Science, 353*(6304), aad9837. doi:10.1126cience.aad9837 PMID:27609899

Chang, C., & Liberty, P. L. (2017). The state of climate change education – reflections from a selection of studies around the world. *International Research in Geographical and Environmental Education, 26*(3), 177–179. doi:10.1080/10382046.2017.1331569

Chang, C., & Pascua, L. (2017). The state of climate change education – reflections from a selection of studies around the world. *International Research in Geographical and Environmental Education.*

Chang, C. H. (2015). Teaching Climate Change – A Fad or a Necessity? *International Research in Geographical and Environmental Education, 24*(3), 181–183. doi:10.1080/10382046.2015.1043763

Chang, C. H., & Pascua, L. (2016). Singapore students' misconceptions of climate change. *International Research in Geographical and Environmental Education, 25*(1), 84–96. doi:10.1080/10382046.2015.1106206

Cutter-Mackenzie, A., & Rousell, D. (2019). Education for what? Shaping the field of climate change education with children and young people as co-researchers. *Children's Geographies, 17*(1), 90–104. doi:10.1080/14733285.2018.1467556

Desjean-Perrotta, B., Moseley, C., & Cantu, L. E. (2008). Preservice Teachers' Perceptions of the Environment: Does Ethnicity or Dominant Residential Experience Matter? *The Journal of Environmental Education, 39*(2), 21–32. doi:10.3200/JOEE.39.2.21-32

Dillon, J. (2003). On Learners and Learning in Environmental Education: Missing Theories, Ignored Communities. *Environmental Education Research, 9*(2), 215–226. doi:10.1080/13504620303480

Dillon, J., Stevenson, R., & Wals, A. (2016). Introduction to special section: Moving from citizen to civic science to address wicked conservation problems. *Conservation Biology, 30*(3), 450–455. doi:10.1111/cobi.12689 PMID:27153527

Dragoni, L., Oh, I. S., Vankatwyk, P., & Tesluk, P. E. (2011). Developing executive leader: The relative contribution of cognitive ability, personality, and the accumulation of work experience in predicting strategic thinking competency. *Personnel Psychology*, *64*(4), 829–864. doi:10.1111/j.1744-6570.2011.01229.x

Fahey, S. J. (2012). Curriculum change and climate change: Inside outside pressures in higher education. *Journal of Curriculum Studies*, *44*(5), 703–722. doi:10.1080/00220272.2012.679011

Faria, F., Kelly, K., Daniel, P., & Azevedo, I. M. L. (2015). A New Approach of Science, Technology, Engineering, and Mathematics Outreach in Climate Change, Energy, and Environmental Decision Making." Sustainability. *The Journal of Record*, *8*(5), 261–271.

Feola, G. (2015). Societal transformation in response to global environmental change: A review of emerging concepts. *Ambio*, *44*(5), 376–390. doi:10.100713280-014-0582-z PMID:25431335

Field, E., Schwartzberg, P., & Berger, P. (2019). *Formal Report for Learning for a Sustainable Future*. http://www.LSF-LST.ca/cc-survey

Flora, J., Saphir, M., Lappé, M., Roser-Renouf, C., Maibach, E. W., & Leiserowitz, A. A. (2014). Evaluation of a National High School Entertainment Education Program: The Alliance for Climate Education. *Climatic Change*, *127*(3–4), 419–434. doi:10.100710584-014-1274-1

Fuertes, M., Andrés, S., Corrochano, D., Delgado, L., Herrero-Teijón, P., Ballegeer, A. M., Ferrari-Lagos, E., Fernández, R., & Ruiz, C. (2020). A proposal of a Category-Based Tool for Curriculum Analysis to Achieve the Climate Competence. *Educ. Knowl. Soc*, *21*, 13. doi:10.14201/eks.22823

Gardner, G. T., & Stern, P. C. (2008). *The Short List: The Most Effective Actions U.S. Households Can Take to Curb Climate Change*. Taylor and Francis Group. Available http://www.environmentmagazine.org/Archives/Back%20Issues/SeptemberOctober%202008/gardner-stern-full.html

Glasser, H. (2007). Minding the gap: The role of social learning in linking our stated desire for a more sustainable world to our everyday actions and policies. In A. E. J. Wals (Ed.), *Social learning: Toward a more sustainable world* (pp. 35–61). Wageningen Academic Publishers.

Greene, J. A., & Yu, S. B. (2016). Educating critical thinkers: The role of epistemic cognition. *Policy Insights from the Behavioral and Brain Sciences*, *3*(1), 45–53. doi:10.1177/2372732215622223

Hallar, A. G., McCubbin, I. B., & Wright, J. M. (2011). CHANGE: A Place-Based Curriculum for Understanding Climate Change at Storm Peak Laboratory, Colorado. *Bulletin of the American Meteorological Society*, *92*(7), 909–918. doi:10.1175/2011BAMS3026.1

Halstead, F., Parsons, L. R., Dunhill, A., & Parsons, K. (2021). A Journey of Emotions from a Young Environmental Activist. *Area*, *53*(4), 708–717. Advance online publication. doi:10.1111/area.12745

Halstead, J. M., & Taylor, M. J. (1996). *Values in education and education in values*. London: Falmer Press. Figueiredo, P., & Perkins, P. E. (2013). Women and Water Management in Times of Climate Change: Participatory and Inclusive Processes. *Journal of Cleaner Production*, *60*, 188–194.

Hannula, K. A. (2022). Education about climate change. *Journal of Geoscience Education*, *70*(4), 423. doi:10.1080/10899995.2022.2115763

Hofer, B. K. (2000). Dimensionality and disciplinary differences in personal epistemology. *Contemporary Educational Psychology*, *25*(4), 378–405. doi:10.1006/ceps.1999.1026 PMID:11001783

Irwin, R. (2008). *Heidegger, politics and climate change*. Continuum.

Kagawa, F., & Selby, D. (2010). Introduction. In F. Kagawa & D. Selby (Eds.), *Education and climate change: Living and learning in interesting times*. Taylor & Francis. doi:10.4324/9780203866399

Kagawa, F., & Selby, D. (2012). Ready for the storm: Education for disaster risk reduction and climate change adaptation and mitigation. *Journal of Education for Sustainable Development*, *6*(2), 207–217. doi:10.1177/0973408212475200

Kaplan, S., & Kaplan, R. (1982). *Cognition and Environment: Functioning in an Uncertain World*. Praeger.

Keller, L., Stötter, J., Oberrauch, A., Kuthe, A., Körfgen, A., & Hüfner, K. (2019). Article. *Changing Climate Change Education*, *28*(1), 35–43.

Kimberly, A. H. (2022). Education about climate change. *Journal of Geoscience Education*, *70*(4), 423–423. doi:10.1080/10899995.2022.2115763

Kolb, D. A. (1984). *Experiential Learning: Experience as the Source of Learning and Development* (Vol. 1). Prentice-Hall.

Læssøe, J., Schnac, K., Breiting, S., & Rolls, S. (2009). *Climate Change and Sustainable Development.* The Response from Education International Alliance of Leading Educational Institutes.

Lambert, D., & Hopkin, J. (2014). A possibilist analysis of the geography national curriculum in England. *International Research in Geographical and Environmental Education, 23*(1), 64–78. doi:10.1080/10382046.2013.858446

Lehtonen, A., Salonen, A. O., & Cantell, H. (2019). Climate change education: A new approach for a world of wicked problems. In J. W. Cook (Ed.), *Sustainability, human well-being, and the future of education* (pp. 339–374). Springer International Publishing. doi:10.1007/978-3-319-78580-6_11

Leigh, K. (2009). Energy Busters: Norfolk Schools Fight Climate Change. *Environmental Education, 91*, 13–14.

Leiserowitz, A., & Smith, N. (2010). *Knowledge of Climate Change across Global Warming's Six Americas.* Yale Project on Climate Change Communication. Available http://www.nationalacademies.org/bose/Climate_Change_Education_Leiserowitz_ Workshop_paper.pdf

Lotz-Sisitka, H. (2010). Climate injustice: How should education respond? In F. Kagawa & D. Selby (Eds.), *Education and climate change: Learning and living in interesting times* (pp. 71–88). Routledge.

McKeown, R., & Hopkins, C. (2010). Rethinking climate change education. *Green Teacher (Toronto), 89*, 17.

Meymaris, K., Henderson, S., Alaback, P., & Havens, K. (2008). *Project BudBurst: Citizen Science for All Seasons.* Paper presented at the AGU Fall Meeting.

Mochizuki, Y., & Bryan, A. (2015). Climate Change Education in the Context of Education for Sustainable Development: Rationale and Principles. *Journal of Education for Sustainable Development, 9*(1), 4–26. doi:10.1177/0973408215569109

Monroe, M., Plate, R. R., Oxarart, A., Bowers, A., & Chaves, W. A. (2017). Identifying effective climate change education strategies: A systematic review of the research. *Environmental Education Research, 25*(6), 791–812. doi:10.1080/13504622.2017.1360842

National Research Council. (2011). *Climate Change Education: Goals, Audiences, and Strategies. A Workshop Summary.* Washington, DC: The National Academies Press.

Nelson, D. R., Adger, W. N., & Brown, K. (2007). Adaptation to environmental change: Contributions of a resilience framework. *Annual Review of Environment and Resources*, *32*(1), 395–419. doi:10.1146/annurev.energy.32.051807.090348

Nicholls, J. (2016). *Understanding how Queensland teachers' views on climate change and climate change education shape their reported practices* [Unpublished doctoral dissertation]. James Cook University.

Olmos-Gómez, M. D. C., Estrada-Vidal, L. I., Ruiz-Garzón, F., López-Cordero, R., & Mohamed-Mohand, L. (2019). Making Future Teachers More Aware of Issues Related to Sustainability: An Assessment of Best Practices. *Sustainability*, *11*(24), 7222. doi:10.3390u11247222

Pogacnik, M., Znidarcic, D., & Strgar, J. (2014). A school garden in biotechnical education. *Archives of Biological Sciences*, *66*(1), 393–400. doi:10.2298/ABS1401393P

Pogaˇcnik, M., Žnidarˇciˇc, D., & Strgar, J. A. (2014). *School garden in biotechnical education*. Academic Press.

Powell, K. C., & Kalina, C. J. (2009). Cognitive and Social Constructivism: Developing Tools for an Effective Classroom. *Education*, *130*(2), 241–250.

Priestley, M., & Biesta, G. (2013). Introduction: The new curriculum. In M. Priestley & G. Biesta (Eds.), *Reinventing the curriculum: New trends in curriculum policy and practice* (pp. 229–236). Bloomsbury Academic Bloomsbury Collections. doi:10.5040/9781472553195.ch-001

Priestley, M., Biesta, G. J. J., & Robinson, S. (2015). Teacher agency: What is it and why does it matter? In R. Kneyber & J. Evers (Eds.), *Flip the system: Changing education from the bottom up* (pp. 134–148). Routledge. doi:10.4324/9781315678573-15

Pruneau, D., André, D., Liette, V., Joanne, L., Eileen, O., Elizabeth, M., Gilles, M., & Gaston, B. (2006). When Teachers Adopt Environmental Behaviors in the Aim of Protecting the Climate. *The Journal of Environmental Education*, *37*(3), 3–12. doi:10.3200/JOEE.37.3.3-12

Ratinen, I. (2016). *Primary student teachers' climate change conceptualization and implementation on inquiry-based and communicative science teaching: A design research*. University of Jyvaskyla, Jyvaskyla Studies in Education, Psychology and Social Research.

Robinson, Z. (2011). *Teaching climate change in higher education: Barriers and opportunities*.

Ross, A. (2000). *Curriculum: Construction and critique*. Taylor & Francis Group.

Salter, P., & Maxwell, J. (2016). The inherent vulnerability of the Australian curriculum cross-curriculum priorities. *Critical Studies in Education, 57*(3), 296–312. doi:10.1080/17508487.2015.1070363

Seowa, T., & Ho, L. (2016). Singapore teachers' beliefs about the purpose of climate change education and student readiness to handle controversy. *International Research in Geographical and Environmental Education, 25*(4), 358371. doi:10.1080/10382046.2016.1207993

Stapleton, S. R. (2015). Environmental Identity Development through Social Interactions, Action, and Recognition. *The Journal of Environmental Education, 46*(2), 94–113. doi:10.1080/00958964.2014.1000813

Stenhouse, L. (1975). *An Introduction to Curriculum Research and Development*. Heinemann.

Sterling, S. (2011). Transformative Learning and Sustainability: Sketching the Conceptual Ground. *Learning and Teaching in Higher Education, 5*(11), 17–33.

Stevenson, R. B. (2011). Environmental learning and agency in diverse educational and cultural contexts. In R. B. Stevenson & J. Dillon (Eds.), *Engaging environmental education: Learning, culture and agency* (pp. 219–238). Sense Publishers.

Stevenson, R. B., Nicholls, J., & Whitehouse, H. (2017). What Is Climate Change Education? *Curriculum Perspectives, 37*(1), 67–71. doi:10.100741297-017-0015-9

Teixeira, J., Silva, C., & Neves, J. V. (2018). School mobility management case study: German School of Oporto (Deutsche Schule zu Porto). *Case Stud. Transp. Policy, 7*(1), 13–21. doi:10.1016/j.cstp.2018.11.002

Theobald, E. J., Alison, C., Janneke, H., Mary, P. W., & Scott, F. (2015). Women Learn More from Local than Global Examples of the Biological Impacts of Climate Change. *Frontiers in Ecology and the Environment, 13*(3), 132–137. doi:10.1890/140261

Tolppanen, S., Aarnio-Linnanvuori, E., Cantell, H., & Lehtonen, A. (2017). Dealing with a Wicked Problem – A model for holistic climate change education. *Kasvatus, 48*(5), 456–468. https://erepo.uef.fi/handle/123456789/8285

Trott, C. D. (2022). Climate change education for transformation: Exploring the affective and attitudinal dimensions of children's learning and action. *Environmental Education Research, 28*(7), 1023–1042. doi:10.1080/13504622.2021.2007223

UKStudentClimateNetwork.(2019).AManifestoforTacklingtheClimateChangeCrisis. *The Guardian*. https://www.theguardian.com/commentisfree/2019/mar/15/uk-student-climate-network-manifestodeclare-emergency

UNESCO. (1997). Adult Environmental Education: Awareness and Environmental Action. UNESCO.

UNESCO. (2009). *Report of the International Seminar on Climate Change Education*. UNESCO.

UNESCO. (2017). Education for sustainable development goals: Learning objectives. UNESCO.

UNESCO & UNEP. (2011). Climate change starter's guidebook: An issues guide for educational planners and practitioners. Author.

UNESCO & UNFCCC. (2016). *Action for Climate Empowerment: Guidelines for Accelerating Solutions through Education, Training and Public*. https://unfccc.int/sites/default/files/action_for_climate_empowerment_guidelines.pdf

United Nations Framework Convention on Climate Change (UNFCC). May 9, 1992, S. Treaty Doc No. 102-38, 1771.

Valderrama-Hernández, R., Rubio, L. A., Sánchez-Carracedo, F., Franco, D. C., González, S. S., Gil-Doménech, D., Vidal-Raméntol, S., & Miñano, R. (2019). Does the Spanish university system teach sustainability? Perception of the students of four universities. *Visión Alumn. Cuatro Univ*, *23*, 221–245.

Varela-Losada, M., Arias-Correa, A., Pérez-Rodríguez, U., & Vega-Marcote, P. (2019). How Can Teachers Be Encouraged to Commit to Sustainability? Evaluation of a Teacher-Training Experience in Spain. *Sustainability*, *11*(16), 4309. doi:10.3390u11164309

W., Tolppanen, S., Gokpinar, T., Goldman, D., & Ben-Zvi ,A. O. (2021, June). *A multi-country comparison of climate change curricula in secondary schools*. Paper presented at the Annual Australasian Science Education Research Association ASERA Conference, South Australia, University of South Australia.

Wals, A. E. (2011). Learning Our Way to Sustainability. *Journal of Education for Sustainable Development*, *5*(2), 177–186. doi:10.1177/097340821100500208

Wenger, E. (2011). *Communities of practice: A brief introduction*. Retrieved April 27, 2023, from https://www.ohr.wisc.edu/cop/articles/communities_practice_intro_wenger.pdf

Yli-Panula, E., Jeronen, E., Koskinen, S., & Mäki, S. (2022). Finnish University Students' Views on Climate Change Education and Their Own Ability to Act as Climate Educators. *Education in Science*, *12*(3), 169. doi:10.3390/educsci12030169

Pedagogy of Climate Change: An Introduction. (2011). Higher Education Academy.

Young, M. (2014). Curriculum theory: What it is and why it is important. *Cadernos de Pesquisas*, *44*(151), 191–201. doi:10.1590/198053142851

ADDITIONAL READING

Fahey, S. J., Labadie, J. R., & Meyers, N. (2014). Turning the titanic: Inertia and the drivers of climate change education. *Journal of Applied Research in Higher Education*, *6*(1), 44–62. doi:10.1108/JARHE-01-2013-0003

Harland, C. M., Lamming, R. C., Zheng, J., Phillips, W. E., Caldwell, N. D., Johnsen, T. E., Knight, L. A., & Zheng, J. (2006). Supply management: Is it a discipline? *International Journal of Operations & Production Management*, *26*(7), 730–753. doi:10.1108/01443570610672211

Harris, R., & Burn, K. (2011). Curriculum theory, curriculum policy and the problem of ill-disciplined thinking. *Journal of Education Policy, 26*(2), 245–261. .498902 doi:10.1080/02680939.2010

Lee, J. J., Ceyhan, P., Jordan-Cooley, W., & Sung, W. (2013). Greenify: A Real-world Action Game for Climate Change Education. *Simulation & Gaming*, *44*(2–3), 349–365. doi:10.1177/1046878112470539

Margaret, E. M., Cathy, M., Jonathan, M., & Ackerman, S.A. (2022). The Demise of the Knowledge–Action Gap in Climate Change Education. *American meteorological society*, doi:10.1175/BAMS-D-21-0256

McKernan, J. (2007). *Curriculum and Imagination. Process theory, pedagogy and action research*. Routledge. doi:10.4324/9780203946930

Mockler, N. (2011). Becoming and 'Being' a Teacher: Understanding teacher professional identity. In N. Mockler & J. Sachs (Eds.), *Rethinking educational practice through reflexive inquiry, professional learning and development in schools and higher education* (pp. 123–149). Springer. doi:10.1007/978-94-007-0805-1_9

Morote, Á. F., Hernández, M., & Olcina, J. (2021). Are Future School Teachers Qualified to Teach Flood Risk? An Approach from the Geography Discipline in the Context of Climate Change. *Sustainability*, *13*(15), 8560. doi:10.3390u13158560

Sharplin, E. D. (2014). Reconceptualising out-of-field teaching: Experiences of rural teachers in Western Australia. *Educational Research, 56*(1), 97–110. .874160 doi:10.1080/00131881.2013

United Nations. (2015). Transforming Our World: The 2030 Agenda for Sustainable Development. Author.

Wise, S. B. (2010). Climate Change in the Classroom: Patterns, Motivations, and Barriers to Instruction among Colorado Science Teachers. *Journal of Geoscience Education*, *58*(5), 297–309. doi:10.5408/1.3559695

Young, M. (2013). Overcoming the crisis in curriculum theory: A knowledge-based approach. *Journal of Curriculum Studies*, *45*(2), 101–118. doi:10.1080/00220272 .2013.764505

KEY TERMS AND DEFINITIONS

Climate Change: It refers to the changes in climatic conditions experienced across the globe.

Climate Change Education: It refers to the knowledge about the climatic changes which aims at imparting skills of either adapting or mitigating the changes.

Incorporating: It refers to the process of combining sustainability education into the teaching curricula.

Pedagogical Strategies: It refers to a teaching method that can influence instruction.

Skills: They refer to the ability of the students to do some activities well especially if well guided.

Sustainability: It refers to the ability to maintain a certain level of climatic conditions.

Values: It refers to one's judgement of what is important in life.

Chapter 11
Being a Rural Woman During Climate Change:
An Analysis Based on the Concept of Ubuntu

Bilge Abukan
https://orcid.org/0000-0002-6690-9351
Ondokuz Mayıs University, Turkey

Kübranur Görmüş
https://orcid.org/0000-0001-9832-4430
Maltepe University, Turkey

Filiz Yildirim
Ankara University, Turkey

ABSTRACT

Climate change affects all living things, and especially women in rural areas. Focusing on human-environment interaction, ubuntu offers a holistic perspective to interventions to reduce the destructive effects of not only the social environment but also the physical environment. The main purpose of the present study is to evaluate the impact of climate change on the lives of rural women with a bio-psycho-social perspective, in the focus of ubuntu philosophy, and offer social work intervention suggestions.

DOI: 10.4018/978-1-6684-9099-0.ch011

INTRODUCTION

Posing an increasingly serious global threat, climate change is aggravating destructive environmental crises such as drought, desertification, hurricanes, excessive precipitation, melting of glaciers, forest fires, and erosion. In addition, it causes social problems such as poverty, loss of property and life, adverse effects on agricultural activities, the spread of infectious diseases, and forced migration (Kılıç, 2009). While these problems and changes were seen as natural phenomena in the past, they are now viewed as a result of human impact, and thus related to anthropogenic factors (Intergovernmental Panel on Climate Change [IPCC], 2001). In particular, human efforts to dominate the nature and the resulting destructive activities affect the natural habitats of other living things and cause deterioration in biodiversity (United Nations [UN], 2021). This deterioration has now reached to the point of "climate crisis", turning into a common problem faced by all the nations (UN, 2020).

Rural areas can be defined as out-of-town areas with low population density, widespread agricultural activities, and unique social structure (Abukan & Yıldırım, 2019). Being dependent on activities based on natural resources in these regions poses a threat to food security, livelihoods, health and physical infrastructure (Akinbami, 2021). The Food and Agriculture Organization (FAO) underlines that the severity of this impact differs significantly according to gender and reports that men migrate to urban areas with the desire to find alternative livelihoods (FAO, 2011), which increases the burden on women in rural areas both inside and outside the home and puts them in harsher conditions than men (Meyiwa, Maseti, Ngubane et al., 2014). In recent years, these problems have been discussed in the context of climate change, regarding social work practices (Dominelli, 2013; Gray, Coates & Hetherington, 2012; Nyahunda, 2021).

Ecological injustice created by environmental degradation and unjust distribution of natural resources, together with unsustainable development models, exacerbate structural inequalities and affect low-income individuals/communities (Dominelli, 2013). The inequalities and injustices caused by these effects, especially in Africa and poor countries with similar characteristics, are at the center of the *ubuntu* philosophy. *Ubuntu* highlights many basic values and principles of social work, such as social justice, human rights, equality, ecological justice, solidarity, spirituality, social change, human dignity, and respect for all (Chigangaidze, 2021; Chigangaidze, 2022; Mugumbate & Nyanguru, 2013). In addition to these values, *ubuntu* focuses on the interaction between humans and the environment (Chibvongodze, 2016; Murphy, 1994). Thus, *ubuntu* offers a holistic perspective for practices to include groups that are socio-economically disadvantaged due to the destructive effects of their physical environment (Chigangaidze, 2021; Chigangaidze, 2022).

Although all living things are affected by the negative effects of climate change, the poor, especially women, the elderly and children, who are relatively vulnerable in the society, experience these effects more intensely (UN Women, 2016). Although these groups are relatively less adaptable to climate change, they draw attention because they play a minimal role in changing the climate (Milne, 2005). On the other hand, women in rural areas are affected more strongly by poverty, which increases as climate change is associated with poverty in rural areas (Ar, Kadim & Gülçubuk, 2012; Dominelli, 2018; FAO, 2011; IPCC, 2014; Mitchell, Tanner & Lussier, 2007). The aim of the current study is to analyze the impact of climate change on the lives of rural women with a bio-psycho-social approach, from the lens of ubuntu. Such an analysis is expected to contribute to interventions aimed at improving the well-being of rural women in the context of knowledge, skills, and values by taking into account the traditional and emancipatory values of the social work.

BACKGROUND

Climate is one of the systems necessary for the formation of an ecological balance on earth and the continuity of living things (Özçağ, 2004). The deterioration of this ecological balance at certain times has caused various changes in the global climate. Climate change (Kılıç, 2009), which generally refers to high temperature values and long-term changes in weather events, emerged as a result of the natural interactions between the sun, atmosphere, and earth in the past (Öztürk, 2002). Today, these changes are reported to occur due to human influence (i.e. anthropogenic). Especially with the industrial revolution, the accumulation of carbon dioxide and greenhouse gases in the atmosphere has accelerated this process. Industrial Revolution has also led to the increased use of fossil fuels such as oil and coal, and caused the release of greenhouse gases such as carbon dioxide and methane, which trap solar heat and increase temperatures. In addition to these, the decrease in green areas, human activities such as farming and mining, as well as rapid economic and population growth have further accelerated this process. Greenhouse gases accumulating in the atmosphere have increased the temperature and negatively affected the climate (United Nations Framework Convention on Climate Change [UNFCCC], 2003). The average global temperature is 1.1 °C higher today than the temperatures measured between 1850 and 1950 and is projected to reach 3.2 °C towards the end of the 21st century (UN, 2021).

According to the World Meteorological Organization (WMO), greenhouse gas concentrations, sea level rise, ocean temperature, and ocean acidification are the four key indicators of climate change. With harmful and long-lasting effects on land, ocean and atmosphere for sustainable development and ecosystem, these indicators prove the severity of the global-scale impact caused by human activities (WMO, 2022). Climate change, extreme temperatures, melting of glaciers, rise in sea level, flood, drought, desertification, landslides, heat and cold waves, storms and similar events cause great risks for all living things and pose a grave threat to ecosystems (Dankelman, 2010). Such events cause adverse effects in terms of biodiversity, water resources, human health, residential areas and migration patterns, energy, transportation, industry, agriculture and food security (UN WomenWatch, 2009). These negative effects are observed in almost every country, including developed countries. In fact, climate change is considered to be the biggest ecological problem faced by Western countries (McCright, 2010). However, vulnerable countries are affected more severely by climate change in terms of inadequate livelihoods, hunger and poverty, social inequalities (including gender inequalities), and environmental degradation (United Nations Development Programme [UNDP], 2006).

The negative effects of climate change are more clearly observed in rural areas through lower agricultural production. This is due to the fact that livelihoods in rural areas rely heavily on the natural environment and, consequently, agricultural production (UN Women, 2016), which makes these areas become more vulnerable to environmental degradation and disasters. The regions and communities that are dependent on the natural environment in terms of food security, economic activities, livelihoods, health and physical infrastructure are more severely affected by the climate change. In particular, the increase in air temperatures, decrease in precipitation, insufficient soil moisture, difficulties in growing agricultural products, and decrease in productivity are among these effects. This problem is predicted to increase in the future and if the necessary preventive measures are not taken, both the environmental balance and the economic structure will be adversely affected. Agriculture, food, livestock, tourism, and energy sectors are particularly likely to be significantly affected by these changes (Edame, Ekpenyong, Fonta et al., 2011; FAO, 2008; Kumar & Sharma, 2013). In addition, adverse weather events due to climate change reduce agricultural productivity, cause product shortages and low income, resulting in rural poverty (Adak, 2010; Kızmaz, 2021).

The Impact of Cllmate Change on the Llves of Rural Women

Factors such as gender inequality, poverty, lack of access to education and other services, lack of employment options and low socio-economic level in the climate change process further increase the vulnerability of disadvantaged groups in rural areas. Especially women, children and the elderly are exposed to the threats posed by climate change, and they are more susceptible to malnutrition, increasing diseases, and difficulties in accessing clean water (Kızmaz, 2021). Some studies (Harding, 2004; Kaijer & Kronsell, 2014; Kara, 2021) explain this through the concept of *intersectionality*. In other words, social categories such as gender and socio-economic status are intertwined with many factors such as institutional arrangements and cultural practices; and the negative effects of climate change can only be understood through the lens of intersectionality.

In addition, the sector most affected by climate change is agriculture where women are heavily employed, and women living in rural areas directly interact with their environment (Kovancı, 2017; Öcal Kara, Eren Yalçın, Işgın et al., 2016). Further, men migrate to different cities and countries to find a job due to the gradual decrease of natural resources and insufficient opportunities in rural areas. As a result, women, the elderly and children are the ones living in the countryside. Since the majority of agricultural activities are carried out by women, "feminization of agriculture" is on the agenda all over the world (International Fund for Agricultural Development [IFAD], 2014b; Kadim, Ar & Gülçubuk, 2014), which causes women in rural areas to interact directly with nature and to be more affected by any disruption or disaster in the nature (Centre for Science and Environment [CSE], 1985; Dankelman, 2002).

In the 1970s, using the concept of "ecofeminism," Francoise D'Eubonne noted a parallelism between women and environmental exploitation, and related human oppression on the nature and man's oppression of women (Lorentzen & Eaton, 2002; Spretnak, 1990). "Ecofeminism" argues that the source of women and environmental problems is the patriarchal structure, and aims to eliminate gender inequalities, hierarchies and domination based on power relations (Buckingham, 2004; Warren, 1990). Looking at the level of harm done by environmental problems identified by the studies carried out in the 1980s, we can see that the gender dimension is strongly highlighted (Alston, 2015; Dankelman & Jansen, 2013; Meyiwa et al., 2014). Climate change is also exposing people to some physical, mental and psychological risks, leading to an increasing concern that it will pose a global health threat. While the risks are especially threatening for rural areas (Berry, Waite, Dear et al., 2018; Pal & Eltahir, 2015; Parry, Radel, Adamo et al., 2019) the bio-psycho-social impacts of climate change on rural women are of particular concern, as explained below.

Biological Impact

Human health is affected by ecological deterioration due to climate change (Costello, Abbas, Allen et al., 2009; Deschenes, 2014) because climate change increases exposure to extreme weather events, accelerates the spread of infectious diseases, affects population movements, reduces livelihoods and negatively affects mental health (Crimmins, Balbus, Gamble et al., 2016; Parry et al., 2019; Smith, Woodward, Campbell-Lendrum et al., 2014), which reveals climate-related health risks (Berry et al., 2014). As such, the biological effects of climate change is expected to increase the pressure on health care systems, and this pressure is predicted to worsen inequalities in society and affect disadvantaged individuals more severely (Dominelli, 2013; Romanello, McGushin, Di Napoli et al., 2021; Reisinger, Kitching, Chiew et al., 2014). Women's death rates increase while their life expectancy decreases due to climate change (Sorensen, Murray, Lemery et al., 2018), suggesting that the biological effects of climate change should not be overlooked.

Although mother-child health comes to mind when women's health is mentioned, it refers to a broad concept that impacts the whole life of a person (Alexander, La Rosa, Bader et al., 2020; Örüklü & Çakmak, 2021). Considering that women are heavily affected by biological and social factors, it is important that they have access to sustainable health care systems. Although improvements have been made in women's health in recent years, various health needs still remain unmet because inequalities based on gender roles still exist in many societies (Langer, Meleis, Knaul et al., 2015). Consequently, especially low-income women experience difficulties in maintaining their biological health or accessing treatment services (Şener & Başer, 2019). Poor women living in rural areas are more exposed to these effects (Dündar & Özsoy, 2020; UN Women, 2016).

With the effect of poverty, the number of mother and infant deaths is higher than the average, and various diseases occur (United Nations International Children's Emergency Fund [UNICEF], 2015). While women aged 60 and older in developed countries generally die of chronic diseases such as coronary failures and cancer, women in low-income countries die younger as a result of maternal and perinatal risk infections and infectious diseases (World Health Organization [WHO], 2014a). Women of reproductive age have been reported to die mostly due to pregnancy-related deaths, cervical cancer, AIDS, tuberculosis, depression, violence, and suicide. Especially in 2017, approximately 808 women lost their lives in low-income regions every day due to pregnancy and childbirth complications (WHO, 2014b; WHO, 2022). The changes in the climate and the destructions they bring along lead to a decrease in life expectancy, especially in women, and women face a greater risk of catching diseases (Neumayer & Plümper, 2007).

Risks such as death and illness have been reported to be very common in women's biological reproductive processes. The highest fertility rates in the world are in the African continent, and poor African women in rural areas have the highest number of children (Barrett, 1997; UNDP, 1995). In addition, climate change takes its toll on women's reproductive health (pregnancy, menopause, infertility) (Dündar Özsoy, 2020). On the other hand, the climate injustice that emerges as a result of these devastating effects of climate change can lead to problems such as an increase in women and mother-infant deaths, early marriage of girls, trafficking in women, prostitution, sexual assaults, and the inability of girls to receive education (Tatgın, 2019).

Psychological Impact

However, due to climate change, various disasters such as tsunamis, floods, landslides, forest fires, and avalanches are observed, and the harmful effects of these events are observed more in communities living in ecologically sensitive areas and people with genetic predisposition (Bourque & Willox, 2014). This causes psychological disorders such as depression, panic disorder, post-traumatic stress disorder, and anxiety in the society, and the probability of developing psychological disorders due to natural disasters may be higher in people with traumatic childhood experiences (Cankardaş & Sofuoğlu, 2021; Kar, 2009).

Coêlho, Adair and Mocellin (2004) state that the psychological effects experienced as a result of climate events manifest themselves differently in men and women, noting that after the drought in the Northeast region of Brazil, women became more anxious and stressed than men. This is interpreted as a result of women's simultaneous taking on roles as producers and livelihoods in household management. In another study (Mitchell et al., 2007), similar findings were obtained in the Ganges river basin of India. Women have been shown to be affected by the psychosocial effects of a flood disaster, such as losing their social support systems and caring for other family members (as cited in Goh, 2012). After the St. Helens volcanic eruption in Washington, the most common disorders in rural settlements in the region were found to be major depression, anxiety and post-traumatic stress disorder, and women showed higher baseline levels for these three types of disorders than men. However, men were reported to have a high baseline level only for anxiety disorder (Coêlho et al., 2004; Shore, Tatum & Vollmer, 1986a; Shore et al., 1986b).

Social Impact

The convergence of multiple inequalities, including limited resource access, constrained participation in decision-making processes, local organizing challenges, undervalued or unpaid labor, disparate caregiving responsibilities, cultural norms, and restricted freedom of movement, collectively render rural women more vulnerable to the impacts of climate change when compared to men (Alston, 2015; Dankelman, 2010). Poverty, undereducation, unpaid workforce, household management and inequality in access to technology are some striking characteristics of rural areas where patriarchal domination predominates. These inequalities bring about the restriction of the potential of these women to cope with the problems they experience and to improve themselves (Akinbami, Olawoye, Adesina et al., 2019). However, the traditional roles, cultural practices, and socio-economic conditions imposed on women in developing countries, such as their responsibilities for childcare and elderly care, often hinder their ability to proactively safeguard themselves before, during, and after natural disasters (UNDP, 2010). For example, a higher number of women than children and men were reported to have lost their lives during the Asian Tsunami, and one of the most important reasons for this was gender inequality. This consequence is associated with women's prioritizing their children and assets over their own lives (Goh, 2012; United Nations Office for Disaster Risk Reduction [UNISDR], 2008). Depending on socio-economic factors, women's levels of exposure to climate change may differ, and especially lower socio-economic groups may experience difficulties in accessing assistance, services, and information (WHO, 2014a). Pan-American Health Organization (2002) reports that relief efforts after natural disasters are aimed at the entire population in a disaster-affected region; however, women's access to these resources is limited, which is explained by the existing resource distribution mechanisms that reflect the patriarchal structure of the society (Neumayer & Plumper, 2007).

In addition, the income loss experienced by a household can cause food insecurity, poorer health, and limited water and energy resources, which reduce overall well-being. Inability to cover children's school expenses contributes to school dropouts, and when men migrate in search of alternative employment, it often places the entire responsibility and burden on women (Goh, 2012). In rural areas, women also undertake roles such as growing and processing food, carrying water from fountains or similar sources when there is no tap water at home, cooking by the fire for long hours, providing and burning fuel for heating in cold weather. In addition to these roles, if there is an elderly or disabled person at home with the child, they have to take on a lot of responsibilities in and out of the home in providing their care (Mies & Shiva, 1993; Özyol, 2013).

Cllmate Change, Soclal Work With Women and Ubuntu

Social work is a practice-based profession and academic discipline that aims at social change-development, social cohesion, empowerment and emancipation of people. Professional basic principles of social work such as social justice, respect for differences, human rights and collective responsibility (International Federation of Social Workers [IFSW], 2015) enable social workers to approach environmental crises from a more equal and fair lens (Dominelli, 2013). In recent years, researchers in the field of social work have emphasized that ecological problems and climate change-based disasters should be prioritized in social work practices (Dominelli, 2013; Gray et al., 2012; Nyahunda, 2021). In addition, as an important group that experiences the devastating effects of climate change the most, rural women can take on a social role in the work to be carried out against climate change (Brody, Demetriades & Esplen, 2008; UN Women, 2016). Thus, women should be prepared for the risks of climate change and environmental crises, and they should take an active role in the prevention and policy making efforts.

Social work is a profession and academic discipline that emphasizes the bio-psycho-social approach, and person-in-the-environment approach. It is important for social work practitioners to benefit from the ubuntu philosophy while evaluating human welfare holistically (Chigangaidze, 2021) because ubuntu's integration of the basic principles and values of social work with the emphasis on the natural environment (Chibvongodze, 2016) can enable this philosophical teaching to become an umbrella concept for all collaborative work while offering a holistic perspective.

Ubuntu Philosophy in Social Work Practices

In the face of the global spread of climate change and its detrimental impacts, it is now universally recognized as a shared challenge. The United Nations, along with its member states, has acknowledged this issue as a global concern, incorporating it into the framework of sustainable development goals. These goals present a plan of action that includes a global partnership to achieve peace and prosperity for people and the planet (UN, 2020). This plan is aimed at coordinating strategies to end poverty and deprivation, combating climate change, protecting the nature, improving health and education, reducing inequality, and fostering economic growth (UN, 2020; UN, 2021; UNDP, 2021). Although all these efforts are valuable, humanity fails to change the life practices adopted for modern life. Alternative efforts, rapid sanctions and transformation are needed to prevent excessive consumption of ecological resources with the focus of "human benefit". In today's world, where ecological degradation has reached critical levels, even minor interventions in nature can yield significant outcomes.

Terblanché-Greeff (2019) states that implementation of adaptation strategies is valuable in minimizing the vulnerability of communities to the negative effects of climate change; however, these strategies are characteristically based on the sustainable development paradigm, which adopts a more Western socio-economic-environmental approach. For example, Africa emits a less volume of greenhouse gases into the atmosphere compared to other developed and industrialized continents. However, despite this lower contribution to emissions, Africa is one of the continents most severely impacted by the detrimental effects of climate change. This situation underscores the necessity of addressing adaptation and coping strategies aligned with sustainable development goals. Consequently, concerns arise that the adaptation strategies devised and executed often neglect the cultural values of local communities (IPCC, 2012).

The strategies outlined within the framework of sustainable development goals, which center on the potential adverse impacts of climate change, particularly in economic terms, fall short of comprehensively addressing the challenges faced by impoverished nations like Africa. In this context, the significance of local knowledge and cultural values should be underscored. Exploring degrowth-oriented approaches as a viable alternative to growth-oriented strategies might yield more fruitful insights into the matter (Terblanché-Greeff, 2019). Every continent has distinct socio-economic-environmental characteristics, cultural and moral values, and the ubuntu philosophy, which reflects Africa's values, should be seen as an alternative approach in reducing the negative effects of climate change (Ngcoya, 2009; Terblanché-Greeff, 2019). Learning from local and traditional knowledge, particularly in rural areas, can be regarded as a crucial tool for adapting to climate change. This is because the communities relying on natural resources have a rich history of experiences in dealing with shifting social and ecological conditions.

The indigenous knowledge from these past experiences is neglected in policy and research. However, integrating this information with scientific knowledge can provide the evidence base for effective adaptation interventions (IPCC, 2012). Considering that the lifestyles in rural areas also vary depending on the specific characteristics of their natural environment, it is valuable to know the important natural disasters experienced by the communities in these areas in the past and to understand how they survived these disasters. The emphasis should be on uncovering the robust coping strategies embedded within the solutions crafted by these communities, rather than solely focusing on the occurrence of similar events in the past or present. However, such an assessment can also offer insights into the methods that may prove ineffective for these communities.

Philosophers adopting the ecofeminist thought state that poor people in Africa, like all people, may be responsible for events such as drought, desertification, climate change, biodiversity crisis, poverty, exploitation of the environment, and deforestation. However, they argue that disadvantaged people in Africa, especially women and children, are exposed to more exploitation, discrimination, oppression, social inequality and injustice due to philosophies/practices in which patriarchal domination is all-pervasive (Chemhuru, 2019; Warren, 1990). As such, Ubuntu can provide an important basis for social and ecological justice, human rights and social development, as it focuses not only on the relationship between human groups, but also on the relationship between humans and the natural environment (Chigangaidze, 2022). Furthermore, Ubuntu focuses on values such as communal values, human relations, social justice, human rights, social cohesion, human dignity, sharing, teamwork, and respect for everyone (Mugumbate & Nyanguru, 2013). Thus, Ubuntu values are similar to the fundamental principles of the social work profession (Chigangaidze, 2022). It is worth noting that while the philosophy and fundamental principles of Ubuntu originated from the values of African society, the core principles it emphasizes are not significantly distinct from those at the heart of the social work profession and its practices.

Ubuntu is also a humanist philosophical teaching that signifies the concepts of love, solidarity, compassion, harmony, consensus, hospitality, and sharing with everyone (Chigangaidze, 2021. In other words, it focuses on subjectivity rather than objectivity (Jones-Smith, 2012; Winston, 2015), placing an emphasis on the human experience. This shows why it is important for the social work profession to have a place for ubuntu (Chigangaidze, 2021; Chigangaidze, 2022). A code of ethics and behavior specific to African culture, ubuntu is described with the following: "I am a person because you are, I am because I share and participate" and "I am because of others." (Chigangaidze, 2021). This code emphasizes humanism by focusing on the need for everyone (or society) to exist for a person to exist. In other words, as Weiss-Gal (2008) stated, ubuntu is based on the idea that a person exists through relationships with people. In this regard, Mugumbate and Chereni (2019) draw attention to the importance of the concept by arguing that individuals are part of the family, families are part of the community and communities are part of the natural environment (Van Breda, 2019). In addition to the values that Ubuntu adopts, the natural environment, cultural values, local knowledge, and spiritualism it embraces offer a holistic and humanistic perspective. Therefore, ubuntu offers a broader perspective for social work practices.

Social Work Practices and Ubuntu for Rural Women

While addressing the problems experienced by rural women, the possibilities of the geographical region, the relations of the community, its characteristics, cultural values and gender roles emerge as the key aspects. As a matter of fact, the five main dimensions that Hailey (2008) defines within the scope of ubuntu can reveal some effective strategies to solve the climate change-related problems of rural women, which are:

- Valuing individual identity and community
- Role in community building
- Fostering collective work and reconciliation
- Conflict mediation and its potential role in reconciliation
- Supporting organizational effectiveness and efficiency.

Nyahunda (2021) focuses on the components of ecofeminism theory in her empowerment-based model to increase the participation of rural women in the climate change activities. This theory associates women's vulnerabilities during the climate change with poverty, inequality, discrimination and gender roles (Mies &

Shiva, 1993; Tuana, 2013). The key concepts of the model are ecological justice, gender blindness, gender mainstreaming, and social justice. In the model, a process is carried out in five steps as shown in Table 1.

Table 1. Empowerment guidelines for social workers

Steps		Descriptions
Step 1	Increasing climate change awareness of social workers	The reason for the passive participation of social workers in climate change problems is that they have insufficient knowledge and lack of education on this subject.
Step 2	Organizing campaigns and climate change trainings to strengthen gender awareness for rural women	The purpose is to ensure that awareness and empowerment campaigns are carried out on the problems faced by rural women in their communities, and to increase the skills of women in rural areas to understand and interpret the causes and consequences of climate change.
Step 3	Designing, creating, and implementing gender-based policy lobbies	There is no program or policy prepared by social workers for climate change practices. Therefore, women's voices need to be integrated into public planning and decision-making processes.
Step 4	Supporting good practice of rural women	Although it is important to support the best practices of women, the intermediary role of social workers is crucial in this process in bringing women together with resources.
Step 5	Monitoring- Evaluation	The aim of this step is to ensure that a monitoring system is established for the programs to be planned and to check whether there is a change in gender inequality indicators.

In the table above, an empowerment-oriented guide model for the problems of rural women, which has become increasingly evident in the process of climate change, is presented by Nyahunda (2021). This guide aims to increase the knowledge and awareness level of women in rural areas, to create programs and policies, to participate in decision-making mechanisms, and to have an equal position in access to resources like every individual, and a series of stages are specified for these purposes. This model can provide a general framework and be a good starting point for the applications to be carried out to address the problems of rural women in the climate change process with an ubuntu focus. Thus, the integration of the model with the core values of both the bio-psycho-social approach and ubuntu can make social work practices more effective.

Mayaka and Truel (2021) draw attention to the fact that ubuntu expresses a different philosophy from the form established in the social work profession in many countries and that a different meaning can be attributed to it in every culture. It especially emphasizes that ubuntu's focus on the cultural value will provide local practitioners with a perspective on how they can expand their own work in sustainable development and understanding the cultural resistance of regions or communities against their interventions. However, considering that the patriarchal structure may be dominant in the culture of these communities, it is important to interpret the community values and to find solutions to its problems.

The following example, published on the Thomson Reuters Foundation on May 18, 2022, points out to women's problems that have become increasingly evident and increasing in severity in the climate change process (Zami, 2022). Ubuntu can be an important starting point for solving these problems. The values put forward by ubuntu are important in ensuring equality for rural women. Because ubuntu aims to both strengthen local values and leave no one behind. Thus, increasing the level of knowledge and awareness of women in rural areas on climate change and creating local programs and policies can be expected to be effective.

Table 2. Case example

Bangladesh, one of the countries in South Asia, is geographically lower than other countries. With the effect of global temperatures, it becomes more vulnerable to severe storms, severe monsoon floods, and extreme sea level rise. Especially families in rural areas take various measures to adapt to such effects of climate changes and to reduce the possible risks. among these measures are raising the floors of the houses above the flood-water level, building shelters, and planting trees to ensure the safety of their animals. According to the results of the research conducted by the International Institute for Environment and Development (IIED) in 2019, families in Bangladesh spend 12 times more than the country's income from foreign aid to prepare for and take measures against the negative effects of climate change. In this regard, it was reported that the losses and damages in urban areas should also be determined, and it is appropriate to make an assessment on a wider scale in climate-related expenditures, with particular emphasis on poor households in urban areas. In another study conducted by IIED, the share of female-directed household expenditures in Bangladesh is noted to be higher than that of male-led household expenditures, which is associated with the fact that women have lower incomes than men. IIED, Kingston University London, and the UN Development Programme collaborated to conduct a survey on 3,094 rural households in 10 districts of Bangladesh to find out how various socioeconomic and gender-related factors affect family spending during disasters such as storms, floods, and droughts. They found that 43 percent of households were exposed to floods, 41 percent to storms, and 83 percent to long-term stresses such as drought or salinity. In Bangladesh, mostly men living in the northwestern regions migrate to other places as seasonal workers, and women bear the sole responsibility of household management. The IIED study explains that climate change also shows differences between women and men. With social norms, women are perceived only as those who provide food, water, care, and other basic needs. They usually do this kind of work for free or receive very little wages. Therefore, although women play a significant role in adapting to climate change, they tend to be ignored.

SOLUTIONS AND RECOMMENDATIONS

With climate change, women experience various inequalities and are affected bio-psycho-socially. Social injustices, violations of rights, ecological injustices, discrimination, social and structural inequalities, poverty, and gender inequality stand out at the center of the problems experienced by rural women. Considering these effects, the principles adopted by the ubuntu philosophy in social work interventions for the needs of these women are important in terms of ensuring the localization of the intervention. Hence, the active involvement of social workers in interventions addressing the biological, social, and psychological impacts of climate change is essential to enable rural women to adapt to its consequences.

Considering the social and psychological effects of climate change on rural women, one of the most important ways to achieve ecological justice is to ensure gender equality (Özcan Elçi, 2022; Özyol, 2013; Tatgın, 2019). As such, climate change adaptation policies should be created by considering gender equality (Gökalp Alica, 2017). However, the basic indicators regarding gender equality, which is one of the 17 global development goals, are insufficient in the case of Turkey (Özcan Elçi, 2021), which is not very different at the international level. Policies sensitive to women should be more extensively included into the international, national and local action policies regarding climate change (Talu, 2017).

Considering the gender-based problems of rural women with climate change, social workers should have sufficient knowledge and equipment to empower and liberate them. However, as Nyahunda (2021) points out, social workers may not be aware of their roles and responsibilities in this field. Therefore, subjects such as climate change, the natural environment, ecological justice, and sustainability should be integrated into the curriculum of existing social work undergraduate programs focused on women's well-being. It should be especially noted that it is important for social workers to actively participate in practices in this field to transform their knowledge on these subjects into skills and to be aware of ethical principles and values. In addition, before starting to practice with women in rural areas, the geographical and demographic characteristics of the region in question and the social, cultural, political, and economic characteristics of the community living there must be considered. Trying to understand the relationship-power dynamics can guide the solutions that need to be developed. This can also provide information about the resistance structures that these communities can display during the interventions. With the recognition of the region where the intervention will be made, how women and their families in rural areas make sense of this phenomenon can be learned by reconsidering the information they have about climate change and its negative effects. They can also be informed about the concepts and how they relate these concepts to their own life experiences. Taking into account the values and principles of solidarity and spirituality, conducting interviews with women in rural areas becomes imperative. These interviews can help uncover past successful practices and grasp the underlying value and belief systems. By seeking to comprehend the favorable impacts of these practices on women, a deeper understanding can be gained for the effective implementation of such practices.

Women's awareness and sensitivity towards the problem of climate change has been reported to be high (Besnili Memiş, 2019). Therefore, professional practices for women's organizations can be helpful. Women can be important stakeholders and actors in ensuring climate justice and solving ecological problems. In addition, creating environments where women come together frequently, ensuring their active participation and supporting their participation in decision-making mechanisms are critical. Nonetheless, relying solely on activities involving women may not prove to be sufficient. It is equally important to engage men within the community in these activities as well. Particularly, successful practices carried out by women in the past can be exemplified for them. Focusing on concepts such as division of labor, equality, and fairness in the implementations in rural areas is also important. The participation of rural women in program-policy and decision-making processes can be realized by ubuntu practitioners with exemplary practices. Hence, the development and implementation of gender-sensitive policies should be approached as an ongoing process throughout all stages, rather than as a standalone phase. It is advisable to tailor the subsequent stages based on this foundation, as the necessary steps may vary depending on the unique characteristics of each community.

Finally, when considering the effects of climate change on the health of rural women, local leaders such as *mukhtars* (neighborhood representatives) are also important since they are the representatives that communicate directly with the relevant institutions of the state to find solutions for the negative conditions caused by climate change in rural areas (Ekinci & Ökde, 2021). However, *mukhtars* may not be aware of the negative effects of climate change on rural women, their needs, and their own responsibilities regarding climate change. Thus, it is critical to increase their awareness about climate change so that rural women can access adequate resources and services to meet their needs.

FUTURE RESEARCH DIRECTIONS

Further research is needed to provide evidence base for interventions to improve the well-being of rural women in the climate change process. Thus, the Ubuntu philosophy can be taken as a basis in reducing the biopsychosocial effects of climate change in rural areas and the inequalities caused by these effects. However, discussion of the Ubuntu philosophy in social work is still new. For this reason, research that will provide a knowledge base on the role of this philosophical approach in the lives of rural women in the climate change process can be recommended. Conducting studies that will contribute to the development of policies and intervention methods to increase the welfare of these women is also important since the number of studies on these issues in the literature is quite low.

Exploring the Ubuntu philosophy becomes particularly pertinent within the context of the biological impacts of climate change. This exploration can center on rural women's efforts to safeguard their biological well-being and secure access to essential healthcare services. In addition, studies for the adoption of ubuntu can be functional in reducing the gender-based inequalities experienced by rural women and their greater involvement in social life. Thus, it is important to develop interventions based on local knowledge. Similarly, studies focusing on family and society dynamics specific to rural life can be carried out. In particular, the psychological effects of climate change on rural women can be investigated.

CONCLUSION

The consequences of unjust distribution of natural resources, climate change and environmental degradation increase inequalities in the social structure. The poor, especially women, the elderly, and children, who are relatively vulnerable, are affected more intensely by these inequalities. Rural women directly experience the effects of climate change related to rural poverty. Especially low-income women have difficulties in maintaining their biological health or accessing treatment services. The psychological effects of climate-related events are observed to manifest differently in men and women. Reports indicate that women tend to experience higher levels of anxiety and stress compared to men in such situations. However, rural women face numerous inequalities, including limited access to resources, minimal participation in decision-making mechanisms, challenges in local organizing, undercompensated or unpaid labor, imbalanced caregiving responsibilities, cultural norms, and restrictions on their freedom of movement. These inequalities are exacerbated for women during the climate change. These inequalities are central to the philosophy of ubuntu. Ubuntu offers a holistic view of applications for disadvantaged groups who are exposed to the destructive effects of their physical environment. In this respect, social justice embraced by social work covers many basic values and principles such as human rights, equality, ecological justice, solidarity, spirituality, social change, human dignity, and respect for everyone. The Ubuntu philosophy embodies liberating values that play a vital role in mitigating the adverse impacts of biological, psychological, and social challenges faced by rural women due to climate change. Furthermore, it holds the potential to address the existing inequalities. Thus, incorporating this philosophy into interventions for rural women can play a pivotal role in enhancing their well-being.

REFERENCES

Abukan, B., & Yıldırım, F. (2019). Kırsal alanda toplum kalkınması için sosyal hizmette kültürel yetkinliğin önemi [The importance of cultural competence in social work for community development in rural area]. *Journal Of History School (Johs)*, *43*, 1666–1681. doi:10.29228/joh.24748

Adak, N. (2010). *Kırsal kadın ve çevre* [Paper presentation]. Türkiye IX. Tarım Ekonomisi Kongresi, Şanlıurfa, Turkey.

Akinbami, C. A. O. (2021). Climatepreneurship: Adaptation strategy for climate change impacts on rural women entrepreneurship development in Nigeria. African Handbook of Climate Change Adaptation, 1-27. doi:10.1007/978-3-030-42091-8_191-1

Akinbami, C. A. O., Olawoye, J. E., Adesina, F. A., & Nelson, V. (2019). Exploring potential climate-related entrepreneurship opportunities and challenges for rural Nigerian women. *Journal of Global Entrepreneurship Research*, *9*(1), 1–28. doi:10.118640497-018-0141-3

Alexander, L. L., La Rosa, J. H., Bader, H., & Garfield, S. (2020). *New dimensions in women's health*. Jones & Bartlett Publishers.

Alston, M. (2015). Women and climate change in Bangladesh. Routledge.

Ar, H., Kadim, F., & Gülçubuk, B. (2012). *İklim değişikliği ve yoksulluğun en çok etkilenenleri: Kırsal alandaki kadınlar* [Paper presentation]. Türkiye X. Tarım Ekonomisi Kongresi.

Barrett, H. R. (1997). Women, occupation and health in rural Africa: Adaptation to a changing socioeconomic climate. *Journal of Occupational Science*, *4*(3), 93–105. doi:10.1080/14427591.1997.9686425

Berry, H. L., Waite, T. D., Dear, K. B., Capon, A. G., & Murray, V. (2018). The case for systems thinking about climate change and mental health. *Nature Climate Change*, *8*(4), 282–290. doi:10.103841558-018-0102-4

Berry, P., Clarke, K.-L., Fleury, M. D., & Parker, S. (2014). Human health. In F. J. Warren & D. S. Lemmen (Eds.), *Canada in changing climate: Sector perspectives on impacts and adaptation* (pp. 191–232). Government of Canada.

Besnili Memiş, O. (2019). Kadınların iklim değişikliği ile ilgili algılarının belirlenmesi [Determination of women's perceptions related to climate change]. *Journal of Academic Value Studies*, *5*(4), 700–718. doi:10.23929/javs.948

Bourque, F., & Cunsolo Willox, A. (2014). Climate change: The next challenge for public mental health? *International Review of Psychiatry (Abingdon, England), 26*(4), 415–422. doi:10.3109/09540261.2014.925851 PMID:25137107

Brody, A., Demetriades, J., & Esplen, E. (2008). Gender and climate change: Mapping the linkages- A scoping study on knowledge and gaps. Institute of Development Studies, 1-27.

Buckingham, S. (2004). Ecofeminism in the twenty-first century. *The Geographical Journal, 170*(2), 146–154. doi:10.1111/j.0016-7398.2004.00116.x

Cankardaş, S., & Sofuoğlu, Z. (2021). İklim değişikliği ve birey üzerindeki etkilerinin gözden geçirilmesi. *Nesne Psikoloji Dergisi, 9*(19), 139–146. doi:10.7816/nesne-09-19-11

Centre for Science and Environment. (1985). The state of India's environment 1984-1985: The second citizens' report. Author.

Chemhuru, M. (2019). Interpreting ecofeminist environmentalism in African communitarian philosophy and Ubuntu: An alternative to anthropocentrism. *Philosophical Papers, 48*(2), 241-264. . doi:10.1080/05568641.2018.1450643

Chibvongodze, D. T. (2016). Ubuntu is not only about the human! An analysis of the role of African philosophy and ethics in environment management. *Journal of Human Ecology (Delhi, India), 53*(2), 157–166. doi:10.1080/09709274.2016.11906968

Chigangaidze, R. K. (2021). An exposition of humanistic- existential social work in light of ubuntu philosophy: Towards theorizing ubuntu in social work practice. *Journal of Religion & Spirituality in Social Work, 40*(2), 146–165. doi:10.1080/15426432.2020.1859431

Chigangaidze, R. K. (2022). The environment has rights: Eco-spiritual social work through ubuntu philosophy and Pachamama: A commentary. *International Social Work.* Advance online publication. doi:10.1177/00208728211073382

Coêlho, A. E., Adair, J. G., & Mocellin, J. S. (2004). Psychological responses to drought in northeastern Brazil. *Revista Interamericana de Psicologia. Interamerican Journal of Psychology, 38*(1).

Costello, A., Abbas, M., Allen, A., Ball, S., Bellamy, R., Friel, S., Groce, N., Johnson, A., Kett, M., Lee, M., Levy, C., Maslin, M., McCoy, D., McGuire, B., Montgomery, H., Napier, D., Pagel, C., Patel, J., ... Patterson, C. (2009). Managing the health effects of climate change. *Lancet, 373*(9676), 1693–1733. doi:10.1016/S0140-6736(09)60935-1 PMID:19447250

Council Conclusion on Policy Coherence for Development. (2009). In *2974th External relations council meeting*. Council of European Union.

Crimmins, A., Balbus, J., Gamble, J. L., Beard, C. B., Bell, J. E., Dodgen, D., Eisen, R. J., Fann, N., Hawkins, M. D., Herring, S. C., Jantarasami, L., Mills, D. M., Saha, S., Sarofim, M. C., Trtanj, J., & Ziska, L. (2016). *The Impacts of Climate Change on Human Health in the United States: A Scientific Assessment*. Executive Summary. U.S. Global Change Research Program. doi:10.7930/J00P0WXS

Dankelman, I. (2002). Climate change: Learning from gender analysis and women's experiences of organising for sustainable development. *Gender and Development, 10*(2), 21–29. doi:10.1080/13552070215899

Dankelman, I. (2010). Climate Change, Human Security and Gender. In Gender and climate change: An introduction. Routledge.

Dankelman, I., & Jansen, W. (2013). Gender, environment and climate change: understanding the linkages. In *Gender and climate change: An introduction* (pp. 49–82). Routledge.

Deschenes, O. (2014). Temperature, human health, and adaptation: A review of the empirical literature. *Energy Economics, 46*, 606–619. doi:10.1016/j.eneco.2013.10.013

Dominelli, L. (2013). Environmental justice at the heart of social work practice: Greening the profession. *International Journal of Social Welfare, 22*(4), 431–439. doi:10.1111/ijsw.12024

Dominelli, L. (Ed.). (2018). *The Routledge handbook of green social work*. Routledge. doi:10.4324/9781315183213

Dündar, T., & Özsoy, S. (2020). İklim değişikliğinin kadın üreme sağlığına etkileri [Climate change effects on female reproductive health]. *Arşiv Kaynak Tarama Dergisi, 29*(3), 190–198.

Edame, G. E., Ekpenyong, A. B., Fonta, W. M., & Duru, E. J. C. (2011). Climate change, food security and agricultural productivity in Africa: Issues and policy directions. *International Journal of Humanities and Social Science, 1*(21), 205–223.

Ekinci, E., & Ökde, F. (2021). İklim değişikliğinin kırsal alanlara etkisinin muhtarlar üzerinden değerlendirilmesi: Yozgat köyleri örneğinde. *Afet ve Risk Dergisi*, *4*(2), 300–318. doi:10.35341/afet.993555

Food And Agriculture Organization. (2008). *Climate change, bioenergy and food security: Options for decision makers identified by expert meeting*. Prepared for the high-level conference World Food Security: The Challenges of Climate Change and Bioenergy, Rome, Italy.

Food And Agriculture Organization. (2011). *The global bioenergy partnership sustainability indicators for bioenergy (GBEP). Executive Summary*. https://www.fao.org/publications/card/en/c/939f849c-043f-5674-bf0c-fe4a488065ca/

Goh, A. H. X. (2012). *A literature review of the gender-differentiated impacts of climate change on women's and men's assets and well-being in developing countries*. International Food Policy Research Institute. doi:10.2499/CAPRiWP106

Gökalp Alıca, S.S (2017). Kadın ve Çevre. *Tbb Dergisi, Özel Sayı*, 151-172.

Gray, M., Coates, J., & Hetherington, T. (Eds.). (2012). *Environmental social work*. Routledge. doi:10.4324/9780203095300

Hailey, J. (2008). *Ubuntu: A literature review*. Paper Document. Tutu Foundation.

Harding, S. (2004). *The feminist standpoint theory reader: Intellectual and political controversies*. Routledge.

Intergovernmental Panel on Climate Change. (2001). *Climate change: The scientific basis. Contribution of working group I to the third assessment report of the Intergovernmental Panel on Climate Change*. Cambridge University Press.

Intergovernmental Panel on Climate Change. (2012). Glossary of Terms. In Managing the Risks of Extreme Events and Disasters to Advance Climate Change. A Special Report of Working Groups I and II of the Intergovernmental Panel on Climate Change. Cambridge University Press.

Intergovernmental Panel on Climate Change. (2014). *Climate Change: Impact, Adaptation and Vulnerability-Part A: Global and Sectoral Aspects*. Fifth Assessment Report. https://www.ipcc.ch/site/assets/uploads/2018/02/WGIIAR5-PartA_FINAL.pdf

International Federation of Social Workers. (2015). *IFSW's Policies 2015*. IFSW.

International Fund for Agricultural Development. (2014b). *Annual Report 2014*. https://www.ifad.org/documents/38714170/40252580/AR_print_2014.pdf/868aaf81-0ef8-492c-80b0-f76b6034068c

International Fund for Agricultural Development. (2016). *Rural development report 2016: Fostering inclusive rural transformation*. https://www.ifad.org/documents/38714170/39155702/Rural+development+report+2016.pdf/347402dd-a37f-41b7-9990-aa745dc113b9

Jones-Smith, E. (2012). Theories of counseling and psychotherapy: An integrative approach. *Sage (Atlanta, Ga.)*.

Kadim, F., Ar, H., & Gülçubuk, B. (2014). *Gıda güvenilirliği açısından kırsal alandaki kadınlar ve aile çiftçliğinin önemi* [Paper presentation]. Aile Çiftçiliği Sempozyumu.

Kaijer, A., & Kronsell, A. (2014). Climate change through the lens of intersectionality. *Environmental Politics*, *23*(3), 417–433. doi:10.1080/09644016.2013.835203

Kar, N. (2009). Psychological impact of disasters on children: Review of assessment and interventions. *World Journal of Pediatrics*, *5*(1), 5–11. doi:10.100712519-009-0001-x PMID:19172325

Kara, Y. (2021). Kesişimsellik iklim krizinin neresinde? [Where is the intersectionality of climate crises?]. *International Journal of Cultural and Social Studies*, *7*(1), 1–16. doi:10.46442/intjcss.729485

Kılıç, C. (2009). Küresel iklim değişikliği çerçevesinde sürdürülebilir kalkınma çabaları ve Türkiye [Sustainable development efforts in the frame of global climate change and Turkey]. *C.Ü. İktisadi ve İdari Bilimler Dergisi, 10*(2), 19-41.

Kızmaz, Z. (2021). İklim değişikliğinin kırsal alandaki etkisi ve alternatif arayışlar: Sosyolojik bir yaklaşım [The impact of climate change in rural area and alternative searches: A sociological approach]. *Fırat Üniversitesi Sosyal Bilimler Dergisi*, *31*(1), 431–453. doi:10.18069/firatsbed.827702

Kovancı, E. (2017). Kronik: Kadınların sürdürülebilir gelişme için örgütlenme deneyimlerinden ve toplumsal cinsiyet analizinden iklim değişikliğini anlamak [Climate change: Learning from gender analysis and women's experiences of organising for sustainable development]. *Ankara Üniversitesi Sbf Dergisi*, *72*(2), 493-503. (Original work published 2002) doi:10.1501/SBFder_0000002456

Kumar, A., & Sharma, P. (2013). Impact of climate variation on agricultural productivity and food security in rural India. In *Economics Discussion Papers*. Kiel Institute for the World Economy. http://www.economics-ejournal.org/dataset/PDFs/discussionpapers_2013-43.pdf

Langer, A., Meleis, A., Knaul, F. M., Atun, R., Aran, M., Arreola-Ornelas, H., Bhutta, A. Z., Binagwaho, A., Bonita, R., Caglia, J. M., Claeson, M., Davies, J., Donnay, F. A., Gausman, J. M., Glickman, C., Kearns, A. D., Kendall, T., Lozano, R., Seboni, N., ... Frenk, J. (2015). Women and health: The key for sustainable development. *Lancet*, *386*(9999), 1165–1210. doi:10.1016/S0140-6736(15)60497-4 PMID:26051370

Lorentzen, L. A., & Eaton, H. (2002). Ecofeminism: An overview. In The Forum on Religion and Ecology at Yale (pp. 1-2). Academic Press.

Mayaka, B., & Truell, R. (2021). Ubuntu and its potential impact on the international social work profession. *International Social Work*, *64*(5), 649–662. doi:10.1177/00208728211022787

McCright, A. M. (2010). The effects of gender on climate change knowledge and concern in the American public. *Population and Environment*, *32*(1), 66–87. doi:10.100711111-010-0113-1

Meyiwa, T., Maseti, T., Ngubane, S., Letsekha, T., & Rozani, C. (2014). Women in selected rural municipalities: Resilience and agency against vulnerabilities to climate change. *Agenda (Durban, South Africa)*, *28*(3), 102–114. doi:10.1080/10130950.2014.955686

Mies, M., & Shiva, V. (1993). *Ecofeminism*. Zed Books. doi:10.5040/9781350219786

Milne, W. (2005). Changing climate, uncertain future: Considering rural women in climate changes policies and strategies. *Canadian Women's Studies*, *24*(4), 49–54.

Mitchell, T., Tanner, T., & Lussier, K. (2007). Action Aid: 'We know what we need': South Asian women speak out on climate change adaptation. Institute of Development Studies, 1-24.

Mugumbate, J., & Chereni, A. (2019). Using African ubuntu theory in social work with children in Zimbabwe. *African Journal of Social Work*, *9*(1).

Mugumbate, J., & Nyanguru, A. (2013). Exploring African philosophy: The value of ubuntu in social work. *African Journal of Social Work*, *3*(1), 82–100.

Murphy, R. (1994). *Rationality and Nature: A Sociological Inquiry into a Changing Relationship*. Westview Press.

Neumayer, E., & Plümper, T. (2007). The Gendered nature of natural disasters: The Impact of catastrophic events on the gender gap in life expectancy, 1981–2002. *Annals of the Association of American Geographers*, *97*(3), 551–556. doi:10.1111/j.1467-8306.2007.00563.x

Ngcoya, M. (2009). *Ubuntu: Globalization, accommodation, and contestation in South Africa*. American University.

Nyahunda, L. (2021). Social work empowerment model for mainstreaming the participation of rural women in the climate change discourse. *Journal of Human Rights and Social Work*, *6*(2), 120–129. doi:10.100741134-020-00148-8

Öcal Kara, F., Eren Yalcın, G., Işgın, T., & Özel, R. (2016). İklim değişikliğinde kadının rolü ve tarımsal yayım [Paper presentation]. 7. Tarım Ekonomi Kongresi, Isparta.

Örüklü, C., & Çakmak, S. (2021). Sürdürülebilir kalkınma hedefleri çerçevesinde kadın sağlığı [Women's health within the framework of sustainable development goals]. *Journal of Human Sciences*, *18*(3), 364–373. doi:10.14687/jhs.v18i3.6118

Özçağ, M. (2004). *Sürdürülebilir kalkınma sürecinde iklim değişikliği ve Türkiye analizi* [Doctoral dissertation]. Adnan Menderes Üniversitesi.

Özcan-Elçi, D. (2022). İklim değişikliği, toplumsal cinsiyet ve kadın temalı çalışmalara dair bir literatür taraması. *Proceedings of the International (Bio) Climate Change Symposium*.

Öztürk, K. (2002). Küresel İklim Değişikliği ve Türkiye'ye Olası Etkileri [Global Climatic Changes and Their Probable Effect upon Turkey]. *Gazi Üniversitesi Gazi Eğitim Fakültesi Dergisi*, *22*(1), 47–65.

Özyol, A., & Çobanoğlu, N. (2013). *Yeşil kadın* [Green women]. https://www.academia.edu/17311834/YEŞİL_KADIN

Pal, J. S., & Eltahir, E. A. (2015). Future temperature in southwest Asia projected to exceed a threshold for human adaptability. *Nature Climate Change*, *6*(2), 197–200. doi:10.1038/nclimate2833

Pan-American Health Organization. (2002). *Gender and natural disasters*. Author.

Parry, L., Radel, C., Adamo, S. B., Clark, N., Counterman, M., Flores-Yeffal, N., Pons, D., Romero-Lankao, P., & Vargo, J. (2019). The (in) visible health risks of climate change. *Social Science & Medicine*, *241*, 112448. doi:10.1016/j.socscimed.2019.112448 PMID:31481245

Reisinger, A., Kitching, R. L., Chiew, F., Hughes, L., Newton, P. C. D., Schuster, S. S., Tait, A., & Whetton, P. (2014). Climate change vulnerability in adaptation in rural areas. In B. Fitzharris & D. Karoly (Eds.), *Climate change 2014: Impacts, adaptation, and vulnerability. Part B: Regional aspects. Contribution of working group II to the fifth assessment report of the intergovernmental panel on climate Change* (pp. 1371–1438). Cambridge University Press.

Romanello, M., McGushin, A., Di Napoli, C., Drummond, P., Hughes, N., Jamart, L., Kennard, H., Lampard, P., Rodriguez, B. S., Arnell, R., Ayeb-Karlsson, S., Belesova, K., Cai, W., Campbell-Lendrum, D., Chambers, J., Chu, L., Ciampi, L., Dalin, C., & Hamilton, I. (2021). The 2021 report of the Lancet Countdown on health and climate change: Code red for a healthy future. *Lancet*, *398*(10311), 1619–1662. doi:10.1016/S0140-6736(21)01787-6 PMID:34687662

Şener, Y., & Başer, M. (2019). Küreselleşme ve yoksulluğun kadın sağlığı üzerine etkileri [The Effects of Globalization and Poverty on Women's Health]. *ERÜ Sağlık Bilimleri Fakültesi Dergisi*, *6*(2), 6–10.

Shore, J. H., Tatum, E. L., & Vollmer, W. M. (1986a). Evaluation of mental effects of disaster, Mount St. Helens eruption. *American Journal of Public Health*, *76*(3, Suppl), 76–83. doi:10.2105/AJPH.76.Suppl.76 PMID:3946730

Shore, J. H., Tatum, E. L., & Vollmer, W. M. (1986b). Psychiatric reactions to disaster: The Mount St. Helens experience. *The American Journal of Psychiatry*, *143*(5), 590–595. doi:10.1176/ajp.143.5.590 PMID:3963245

Smith, K. R., Woodward, A., Campbell-Lendrum, D., Chadee, D. D., Honda, Y., Liu, Q., Olwoch, J. M., Revich, B., & Sauerborn, R. (2014). Human health: impacts, adaptation, and co-benefits. In Climate Change: Impacts, Adaptation, and Vulnerability. Part A: Global and Sectoral Aspects. Contribution of Working Group II to the Fifth Assessment. Report of the Intergovernmental Panel on Climate Change (pp. 709-754). Cambridge University Press.

Sorensen, C., Murray, V., Lemery, J., & Balbus, J. (2018). Climate change and women's health: Impacts and policy directions. *PLoS Medicine*, *15*(7), 1–10. doi:10.1371/journal.pmed.1002603 PMID:29990343

Spretnak, C. (1990). Ecofeminism: Our roots and flowering. Reweaving the world: The emergence of ecofeminism. *The Institute for Ecosophical Studies*, *3*(2), 1–10.

Talu, N. (2017). İklim değişikliği ve toplumsal cinsiyet- Politika belirleme süreçleri [Climate change and gender- Policy making processes]. *Yasama Dergisi*, *33*, 68–87.

Tatgın, E. (2019). İklim adaleti kapsamında toplumsal cinsiyet tartışması. *Kent ve Çevre Araştırmaları Dergisi, 1*(1), 104–118.

Terblanché-Greeff, A. C. (2019). Ubuntu and environmental ethics: The West can learn from Africa when faced with climate change. In M. Chemhuru (Ed.), *African environmental ethics* (pp. 93–109). The International Library of Environmental, Agricultural and Food Ethics. doi:10.1007/978-3-030-18807-8_7

Tuana, N. (2013). Gendering climate knowledge for justice: Catalyzing a new research agenda. In M. Alston & K. Whittenbury (Eds.), *Research, action and policy: Addressing the gendered impacts of climate change* (pp. 17–31). Springer. doi:10.1007/978-94-007-5518-5_2

United Nations. (2020). *Sustainable development goals - Guidelines for the use of the SDG logo including the colour wheel, and 17 Icons*. https://www.un.org/sustainabledevelopment/wp-content/uploads/2019/01/SDG_Guidelines_AUG_2019_Final.pdf

United Nations. (2021). *What is climate change?* https://www.un.org/en/climatechange/what-is-climate-change#

United Nations Development Programme. (1995). *Human development report*. Oxford University Press.

United Nations Development Programme. (2006). *Global partnership for development.* Annual Report. https://www.undp.org/sites/g/files/zskgke326/files/publications/UNDP-in-action-2006-en.pdf

United Nations Development Programme. (2010). *Designing climate change adaptation initiatives: A UNDP toolkit for practitioners*. UNDP Bureau for Development Policy.

United Nations Framework Convention on Climate Change. (2003). *İklime özen göstermek: İklim değişikliği sözleşmesi ve Kyoto Protokolü için kılavuz.* https://unfccc.int/resource/docs/publications/caring_trk.pdf

United Nations International Children's Emergency Fund. (2015). *Level and Trends in Child Mortality Report*. United Nations Children's Fund.

United Nations Office for Disaster Risk Reduction. (2008). *Gender perspectives: Integrating disaster risk reduction into climate change adaptation.* United Nations International Strategy for Disaster Reduction. Available at: http://www.unisdr.org/preventionweb/files/3391_GenderPerspectivesIntegratingDRRCCGood20Practices.pdf

United Nations Women. (2016). *Rural women in climate change in Jordan. In Brief Report.* https://jordan.unwomen.org/sites/default/files/Field%20Office%20 Jordan/Attachments/publications/2018/UN%20Women-RuralWOmenJordan-Brief-WEB%20EN.pdf

United Nations WomenWatch. (2009). *Fact Sheet. Women, Gender equality and climate change.* https://www.un.org/womenwatch/feature/climate_change/

Van Breda, A. D. (2019). Developing the notion of Ubuntu as African theory for social work practice. *Social Work, 55*(4), 439–450. doi:10.15270/52-2-762

Warren, K. J. (1990). The power and the promise of ecological feminism. *Environmental Ethics, 12*(2), 125–146. doi:10.5840/enviroethics199012221

Weiss-Gal, I. (2008). The person-in-environment approach: Professional ideology and practice of social workers in Israel. *Social Work, 53*(1), 65–75. doi:10.1093w/53.1.65 PMID:18610822

Winston, C. N. (2015). Points of convergence and divergence between existential and humanistic psychology: A few observations. *The Humanistic Psychologist, 43*(1), 40–53. Advance online publication. doi:10.1080/08873267.2014.993067

World Health Organization. (2014a). *Gender, climate change and health.* https://apps.who.int/iris/bitstream/handle/10665/144781/9789241508186_eng.pdf

World Health Organization. (2014b). *Quantitative Risk assessment of the effects of climate change on selected causes of death, 2030s and 2050s.* Author.

World Health Organization. (2022). *Maternal and reproductive health.* https://www.who.int/data/gho/data/themes/maternal-and-reproductive-health

World Meteorological Organization. (2022). *State of the global climate 2021.* https://public.wmo.int/en/media/press-release/four-key-climate-change-indicators-break-records-2021

Zani, M. T. (2022, May 18). *Women in rural Bangladesh pay more for rising cost of climate disasters.* Thomson Reuters Foundation. https://news.trust.org/item/20220518143335-m17kx

ADDITIONAL READING

Abdelwahed, N. A. A. (2022). Developing Entrepreneurial Sustainability among Saudi Arabia's University Students. *Sustainability*, *14*(19), 118–190. doi:10.3390u141911890

Dlamini, S., Block, E. K., & Mathibela, N. (2022). Knowledge of and attitudes towards Sustainable Development Goals among adults in South Africa. *The South African Geographical Journal*, 1–18.

McCauleyP.MathurK.CohenN.Henriques-ThompsonK.GopinathJ.MiabR.K. (2023). Environmental sustainability, healthcare workers and healthcare students: A literature review of attitudes and learning approaches. Research Square, 1-18. doi:10.21203/rs.3.rs-2783033/v1

Powers, M. C., & Engstrom, S. (2020). Radical Self-Care for Social Workers in the Global Climate Crisis. *Social Work*, *65*(1), 29–37. doi:10.1093wwz043 PMID:31828329

Rambaree, K. (2013). Social work and sustainable development: Local voices from Mauritius. *Australian Social Work*, *66*(2), 261–276. doi:10.1080/0312407X.2013.784793

Ramsay, S., & Boddy, J. (2017). Environmental social work: A concept analyses. *British Journal of Social Work*, *47*(1), 68–86. doi:10.1093/bjsw/bcw078

Rinkel, M., & Powers, M. (Eds.). (2019). Social work promoting community and environmental sustainability: A workbook for global social workers and educators (Volume 3). International Federation of Social Work (IFSW).

Thysell, M., & Cuadra, C. B. (2022). Imagining the ecosocial within social work. *International Journal of Social Welfare*, ijsw.12571. Advance online publication. doi:10.1111/ijsw.12571

KEY TERMS AND DEFINITIONS

Bio-Psycho-Social Approach: It is an approach that focuses on the interrelationships and interactions of biological, psychological, and social effects in the analysis and resolution of a situation/problem, enabling these effects to be addressed holistically.

Climate Change: Refers to the continuing increase in global average temperature and its effects on the Earth's climate system.

Rural Women: Refers to women who lead their lives and livelihoods in rural areas, contribute to production in rural areas, and generally engage in agriculture or animal husbandry.

Social Work: A practice-based professional and academic discipline that promotes social change and development, social integration, empowerment and emancipation.

Social Work With Women: It is a specific application area of social work that focuses on the position and relations of women in society, points out structural inequalities, produces information, services and policies for increasing women's welfare and solving women's problems.

Ubuntu: It is a philosophical teaching that emerged in Africa, emphasizes being "us" instead of "me", aims to leave no one behind, expresses the concepts of love, solidarity, compassion, harmony, consensus, hospitality and sharing with everyone.

Chapter 12
Best Practice in the Use of the Water Sources of Portoviejo Canton, Ecuador, and Its Parishes

Arturo Luque González
https://orcid.org/0000-0002-4872-891X
Universidad Tecnica de Manabí, Ecuador & Observatorio Euromediterráneo de Espacio Público y Democracia, URJC, Spain

José Humberto Ponce Ponce
Universidad Tecnica de Manabí, Ecuador & Observatorio Euromediterráneo de Espacio Público y Democracia, URJC, Spain

Cristina Raluca Gh. Popescu
https://orcid.org/0000-0002-5876-0550
University of Bucharest, Romania & The Bucharest University of Economic Studies, Romania

ABSTRACT

Water is a fundamental element for the sustenance of all living beings in nature, hence the importance of its conservation and the correct management of its sources. In the case of the Portoviejo River, water is used as a means of irrigation in plantations, a drinking water supply to the city, and a space for developing tourism. However, new settlements along the banks of the Portoviejo River have led to a rise in pollution levels and other misuses of the water. Owing to these circumstances, there was a clear need to evaluate the historical background and conservation mechanisms of water sources used by the canton authorities and its control bodies. The main objective

DOI: 10.4018/978-1-6684-9099-0.ch012

of this study was to carry out an exhaustive analysis of good practices in the use of water sources within the Portoviejo Canton. The study used a hermeneutic research approach, which allows the researcher to perform a qualitative analysis of various interpretations of different texts through a dialectical process, which supports the investigation and substantiates the results obtained.

INTRODUCTION

Natural resources are of great importance because of the value they represent for human subsistence. According to the United Nations (UN) (2023), "water is at the epicenter of sustainable development and is fundamental for socioeconomic development, energy, food production, ecosystems and for the survival of human beings". The care of these resources means the guarantee of survival on this planet. Nonetheless, currently 2.2 billion people lack access to safely managed drinking water services (World Health Organization (WHO), 2022a) and 4.2 billion people do not have safely managed sanitation services, while 3 billion lack basic hand-washing facilities (UNICEF, 2019). Water is the key to the life of all species of plants and animals, so its care should not be neglected anywhere in the world (Krause & Strang, 2016). The planet's water is maintained in different reserves: oceans, lakes, rivers, groundwater, glaciers, the atmosphere, among others. Of these reservoirs, the oceans are the most important, storing 97% of the total water resource; however, this saltwater is of limited direct use. The availability of fresh water is therefore restricted and its distribution in the world is variable, which is why its care is essential (IICA, 2008).

Water sources offer a large number of benefits to social groups living along the banks of rivers or lakes, in addition to contributing to the local economy. Water is an economic resource for which there is a limited supply and a growing demand, one which tends to exceed the availability of the resource (Ortiz et al., 2017). Natural resources must be used responsibly and even more so when it comes to water for human consumption. A large part of the weight of this responsibility falls on public institutions, which are responsible for the proper administration and distribution of drinking water, as well as its treatment and distribution in optimal conditions for human consumption. However, all citizens have a duty to make good use of the resource that is being provided, either for the daily consumption of drinking water or for agricultural purposes.

According to the United Nations World Water Development Report 2023, cooperation and partnerships represent the key to fulfilling the water related aims (United Nations (UN), 2023). Also, according to specialists, access to clean water and sanitation represents a vital part in accomplishing the human rights in terms of a successful sustainable water management (World Water Week (WWW), 2022).

Pollution processes are an important issue in environmental sustainability and a challenge for the world, imposing themselves on government agendas at the global level (Bordeleau et al., 2008; Xue et al., 2012; Zhang et al., 2018; Fadhullah et al., 2020). In 2020, 45% of the domestic wastewater generated in the world was discharged without applying safe treatment, and it is estimated that at least 10% of the world's population consumes food irrigated with wastewater (WHO, 2022b). The contamination of water sources is an issue of special relevance and delicacy since it is common for factories to dump contaminated water into all types of waterways, which are then left untreated. This represents a double problem from both the action of the contaminators and through the omission of responsibility by the public authorities (Halder & Islam, 2015; Luque & Jiménez, 2019).

In the past, little importance was given to the treatment of contaminated water by industries, causing irreparable consequences, in addition to entailing a high economic cost in finding a solution. According to Afroz & Rahman (2017), water pollution is a serious problem in Malaysia and has an adverse impact on the sustainability of its water resources, on plants and living organisms and on the health and economy of the population. The total availability of water is significantly reduced since the cost of treating contaminated water is too high and, in some cases, the contaminated water is not suitable for consumption. In India, Goldar & Banerjee (2004) show a significant negative correlation between the level of industrialization, irrigation intensity and fertilizer use in a given district and the water quality at monitoring points. In China, according to Zhang et al. (2021, p.10), with the increase in GDP, the use of nitrogen fertilizers, sewage discharge and nitrate concentration also increased, indicating that human activity aggravates nitrate pollution. Nitrate in surface waters in most provinces of China poses a medium risk to human health, especially among children, who have a higher risk of non-carcinogenic health issues. Increasing nitrate concentration also has a serious effect on marine ecosystems, leading to eutrophication, toxic algal blooms and hypoxia (Carey et al., 2011; Hoagland et al., 2019).

THE CASE OF ECUADOR

In Ecuador, the problem lies in focusing the responsibility for water resources. In this matter, reputed specialists highlighted that "the drinking water and sanitation management boards in Ecuador need to have a legal status and must affiliate their staff to social security" (United Nations (UN), 2023, p.125). That is why the government dedicates financial resources specifically for its care, which together with The Nature Conservancy[1] (2019), constitutes the so-called "Water Fund", a budget that seeks to improve the care of water sources in the main cities of the country. These funds allow both public institutions and private companies to establish strategies and undertake projects in populated areas and in the agricultural sectors. In the province of Manabí, this has been characterized by bottom-up planning[2], which seeks to provide a total or partial solution to water demands, and goes hand-in-hand with the socioeconomic development of the province. With the passage of time, the demand for water increases along with population growth, forcing the province to establish policies based on well-defined objectives and strategies to be able to allocate resources adequately according to the new demand (Comprehensive Development Plan, 2020).

The communities that come under the Development Plan are encouraged to carry out suitable management and conservation of waterways and improve the quality of life through rational agricultural production based on irrigation. Portoviejo canton is dominated by the agricultural sector and a large amount of water is destined for use in irrigation. According to Martín de Santa Olalla, López Fuster & Calera Belmonte, (2005), "the efficient management of irrigation water is a key factor in the management of the resource, since crop irrigation accounts for between 60 and 80% of global water uses and applications. The measurement and monitoring of water in the soil is an essential aspect in the efficient management of irrigation water."

For the purposes of this study, it was necessary to analyze policies and strategies aimed at the good administration of water resources in the canton and verify that these are being effectively applied. To this end, according to the National Development Plan (2017-2021), the primary tool is the Decentralized System of Participatory Planning (SNDPP), having as its main objectives: 1) guaranteeing constitutional rights, 2) implementing the development regime and the provisions of the development regime (through the implementation of public policies) and 3) undertaking programs, projects and interventions, resting on 2 pillars: environmental sustainability and equitable territorial development. Based on the experiences of the last 10 years, the SNDPP highlights the existence of new undertakings, which fall into 3 categories: a) rights for all throughout life, b) economy at the service of society and c) greater importance for society and less for the state.

It should be noted that these objectives are based on the international commitments of global development. According to Salas (2018), this global perspective works together with the community in the planting of trees along the banks of the Portoviejo River with the purpose of protecting the riverbeds. The river is the main water source that provides water to the canton and its parishes and so taking care of the source and preventing contamination is of paramount importance. In addition, floods must be prevented in the rainy season, so that contamination is not spread to surrounding areas. The guiding principles and approaches of the development plan of the Portoviejo canton, according to the local government administration (2015), is also to prioritize the river due to its geographical location, its great potential in terms of agricultural and marine development, its biodiversity, and above all its great quantity of fresh water. The fulfilment of these objectives aim to make Portoviejo a sustainable city and a pilot center of local productive development for the region and the country.

HISTORICAL BACKGROUND OF THE WATER SOURCES OF THE PORTOVIEJO CANTON

The municipal authority of Portoviejo canton made an agreement with the OIKOS Corporation (1998) to carry out the Environmental Training and Pollution Prevention Program, which is the parent of the EDUCAR Project sponsored by the United States Agency for International Development (USAID). This corporation carried out an exhaustive environmental diagnostic in three stages: 1) it described the situation of the natural resources by collecting secondary and primary sources, 2) it identified, prioritized and analyzed environmental problems and 3) it predicted the main trends of environmental problems and their future dynamics. This diagnostic had 3 main objectives: 1) to contribute to the process of decentralization of local governments underway in Ecuador by raising the level of technical capabilities and local policies in the field of environmental management; 2) to contribute to the strengthening of the processes of education, participation and community interest in environmental management in order to improve understanding of the socio-environmental reality and increase the quality of life of the local population; 3) design and apply a portfolio of methodological instruments and procedures to be used by municipal officials and community leaders to find viable and co-responsible solutions to the problems that affect them.

This diagnostic was first organized by making an approximation to the situation that encompassed the problem, advantages and disadvantages of the water and natural resources of the Portoviejo canton. All the information that was collected was analyzed and systematized in the process of use, conservation and preservation of the water resources. Air, soil, flora and fauna, energy and climate were all taken into consideration (Beiswenger, 1997). In the same way, bad practices and use of water sources within the Municipality of Portoviejo was analyzed and described, developing an instrument that marks these problems on a scale of importance to be solved (INEFAN, 2017). At that time the Municipality of Portoviejo, along with its collaborators, did not have the wherewithal to solve problems of a natural origin, let alone the management, conservation, care and preservation of the water sources of the Portoviejo River in relation to pollution and excessive population growth. The increase in settlements on the banks of the river left little space for its overflow especially in the rainy season (Mirandiña & Endocara, 2019). It was possible to appreciate that there was no type of regulation that would allow the municipality to control the use of water and the management of discharges. This evidenced the partial or total absence of programs that provided education, training or instruction on the use of this resource to citizens who settled on the banks of the Portoviejo River (Noguera, 2020). This situation denotes the lack of a plan in the face of the projected development and growth of the city, which should have been the responsibility of the municipality, the National Council of Water Resources (CNRH) and the Regulatory Corporation of Water Management of Manabí. This problem is linked to the low level of information and knowledge that citizens have about the practices and use of water sources, environmental pollution, as well as the lack of regulations that must be followed by the inhabitants settled along the banks of the rivers (Intriago & Quiroz, 2021).

The use of water sources is limited to the amount of water contained in each source, which in turn determines its quality; standards are checked by comparing the physical and chemical characteristics of the water according to established guidelines. (UN, 2015). Olguín et al. (2010) state that physical, chemical and microbiological aspects should be considered when measuring quality by quantitative methods. These offer sufficient proof to make the appropriate decisions and targeted actions for the mitigation of any inadequacies, thus guaranteeing the safety of water for human consumption. Alternatively, Torres et al., (2009) recommend that the ICA water quality index method be applied as a statistical tool to make adequate and timely assessments of water quality. This is corroborated by Espinoza & Rodríguez (2016), who note that the ICA method provides guidelines to classify water quality on a scale of 0 to 100, according to the constituents evaluated; this, in turn, makes it possible to rate the water according to its condition as excellent, good, average, poor or very poor, as shown in Table No. 1.

Table 1. ICA classification range according to the general criteria

Class	Quality	Index of Quality	Meaning
I	Excellent	91-100	Very clean water
II	Good	71-90	Slightly polluted water
III	Average	51-70	Moderately polluted water
IV	Poor	26-50	Very polluted water
V	Very poor	0-25	Heavily polluted water

Source: Quiroz and Menéndez (2016)

According to Giler et al. (2020), the tributaries that flow into the Portoviejo River are derived from the cantons of Portoviejo, Rocafuerte, Santa Ana and a part of the 24 de Mayo canton, with a total catchment area of 2,105.07 km, as detailed in Table No. 2.

Table 2. Main morphological data of the Portoviejo river basin

Description	Quantity
Area (Km2)	2.105.07
Perimeter (km)	293.58
Length (km)	52.02
Breadth (km)	36.30

Source: Giler et al. (2020)

Within the area that comprises the Portoviejo River and the populations settled along its banks, agriculture is one of the main uses given to these tributaries, including crops of different varieties, as well as pastures and wooded areas. The areas covered with water amount to 714 hectares (Montilla & Pacheco, 2017). The new population settlements close to the banks of the Portoviejo River, and the economic activities they carry out, has led to an increase in the pollution and misuse of the water source. This study aimed to carry out an exhaustive analysis of good practices in the use of water sources that flow into the Portoviejo canton and its parishes. The investigation used hermeneutic research methodologies, which provide the opportunity to perform an analysis of various interpretations from different primary and secondary sources.

THEORETICAL FRAMEWORK

Water Sources

A water source includes bodies of water such as rivers, streams, lakes, reservoirs, springs and groundwater, from which drinking water systems take their supplies to provide water in conditions suitable for human consumption. This is the principal justification for the protection and conservation of these water sources by reducing the risks of contamination, especially where large industries are located (EPA, 2021). Water sources are of great importance for the survival of all living beings, and their contamination would trigger a problem of great magnitude incompatible with all types of life. According to Correa et al. (2020), for some time, serious environmental crises have derived from the damage caused to water sources, often due to the lack of awareness of local communities, the lack of collective commitment, or limited willingness to adopt measures that favor the environment. However, in general communities demonstrate both a low level of knowledge at the same time as a positive attitude toward water preservation.

Importance of Taking Care of Water Sources

Water is a renewable natural resource of great importance for processes such as energy generation and economic activity; rivers, streams and other sources are commonly used for these purposes[3] in areas where the population makes exclusive use of groundwater for the supply of drinking water (Costa et al. 2012). Groundwater represents more than 95% of the world's reserves of fresh water; however, despite its importance for human consumption, it is often not given appropriate treatment to avoid adverse effects such as the multiplication of water-borne diseases. According to Cortés (2015), one of the main problems leading to the degradation of water is the expansion of the population, with the establishment of new settlements and large industries that unintentionally or willfully carry out an inadequate management of water sources. According to article 318 of the constitution of Ecuador, the government, through the Single Water Authority, is responsible for the planning and management of water resources, including river banks, that are destined for human consumption and irrigation, thus guaranteeing food sovereignty and related productive activities (Organic Law on Water Resources, 2014).

All water resources need good planning processes for suitably comprehensive management. This can be achieved by establishing water-use policies and creating projects that facilitate the conservation and preservation of this resource; for example, by building dams and wells for induction and water collection. An adequate pollution control process is essential, one that provides information on the return of water to the ecosystem, and that facilitates adequate monitoring of the quantity and quality of water being used. To this end, sufficient economic resources must be provisioned to deploy control measures of the deterioration of these resources. Adequate management of basins must also be ensured, including reforestation projects, especially on the banks of the rivers, with a view to protecting the soil. On the other hand, it is important to note that the major sources of contamination are in large urban centers; the Ministry of the Environment (2012) states about one million six hundred thousand tons of pollutants find their way into wastewater each year, which significantly decreases the amount of water available for human consumption.

In all human activity, responsible care must be given to water and its sources. This priority should be present in the daily lives of each person since the care given to water sources is vital for human survival and the misuse of water or mismanagement of the soil, albeit unconsciously, of those people who live along the river banks contributes directly to the reduction of the water available (Faldez, 2020). This also will have a negative impact on future generations, especially when compounded by the effects of climatic change on rainfall, with the amount of water required to cover all the needs of an area under increasing pressure (García, Carvajal & Jiménez, 2016).

Sustainable Use of Water

The protection of water resources is a vital component of the circular economy[4] through the sustainability of this type of resources. With respect to this, Cruz & Bodnar (2011) point out that the adoption of interdisciplinary legal measures from different thinkers allows a glimpse of the path to achieve water sustainability and confirms sustainability as a new paradigm of law. The sustainability of water can be understood as an engine that drives a whole process of consolidation based on the axiology of law. It is therefore essential to recognize the various dimensions of sustainability in conjunction with the development of people's livelihoods and changing habits, considering that human activity tends to interfere with the natural cycles of water. In the view of Voulvoulis (2018), the sustainable use of water resources is directly associated with progress and improvement, yet in general, as humanity has expanded, environmental protection has decreased. Human activity has led little by little to the destruction of the sustainability of water, which puts the future of new generations at risk. It should be noted that water is at the center of the processes of sustainable development and is fundamental to socio-economic development. Humanity must guarantee healthy ecosystems that allow and ensure the survival of all living beings. The quality of water sources is vital to reduce the level of disease worldwide, to improve health, to protect nature and to conserve the land that provides benefits for all (UN-DASE, 2015).

Water Sources in Ecuador

Ecuador has four times the world average per capita surface water. These water sources are derived mainly from the rainy season, which contributes to the underground reserves as well as the surface runoff of rivers. The water sources are often contaminated during the process of distribution, affecting both their quality and quantity. According to Camacho & Astudillo (2021), Ecuador consumes approximately 9,700 Hm3 between irrigation, domestic use and the industrial sector. The development of the country's hydrographic system is determined by the Andes mountain range, which crosses the country from north to south and is considered an area of great environmental heterogeneity and biological diversity. The physiographic complexity of this mountain range affects the movement of air and conditions the climate, which directly shapes the hydrological regimes of its rivers (Guambo et al., 2022). The geographical location of the Andes also determines the three main natural regions of the country: coast, highlands and the Amazon region (the fourth region is the Galapagos Archipelago in the Pacific Ocean) (Galárraga, 2020). Within the territory of Ecuador, there are 31 hydrographic systems, made up of 79 basins; these systems come from springs in the Andes, 24 of which empty their waters in the Pacific Ocean and 7 to the East and ultimately the Atlantic. The Coast is considered a region with sui generis characteristics, and the central highland zone constitutes a hydrogeological unit of exceptional characteristics with the largest number of wells nationwide (Burbano et al., 2015).

Water Sources in the Portoviejo Canton, Manabí

The hydrographic system of Manabí consists mainly of the rivers of the Portoviejo, Chone, Jama, Pedernales and Jipijapa basins, which supply water to the surrounding communities, both for purification for human consumption and for irrigation (Burbano et al., 2011). Manabi province is unique in mainland Ecuador since it does not obtain its water from the Andes Mountains but from the Coastal mountain range. The basins are formed from the runoff of the Coastal range of Chongón-Colonche, which is located almost in the center of the province and runs in a north-south direction, separating the small basins of the coastal rivers to the west, from the large basins of the eastern rivers (Guambo et al. 2022). It should be noted that the area of Manabí has a great wealth of water resources, and their care is of great importance for the development of the population in general, as well as in the activities of the agricultural sectors. This duty falls to the local water authorities to establish measures and budget as necessary for their care.

In the Portoviejo canton, an adequate supply of water is essential to the improvement in the social, industrial, agricultural and health sectors, and especially in the quality of life of people. Conversely, a lack of water sources in Portoviejo represents a slide into poverty, the slowing of the economy, the increase in the presence of diseases and a reduction of the quality of life of its inhabitants. Therefore, the supply of water is fundamental to the development of the canton and its inhabitants (Camacho & Astudillo, 2021). Good practices in the use of water sources are all aimed at satisfying the different demands of a system, often over and above the actual availability of water. This generates a reduction in the level of satisfaction of these demands, as well as a reduction in the environmental conditions necessary to achieve a good ecological balance. In the long term, these problems may be intensified as a result of climate change. Satisfying the demands of a system without affecting the environment is one of the main objectives of effective management: to care for and preserve the water sources that supply all of the parishes in the Portoviejo canton.

METHODOLOGY

The present study used a variety of methods to access the precise information required for the object of study. The scope of the research was at the descriptive level, since it sought to identify the characteristics of the good use of water sources; Arias (2012) indicates that descriptive research consists of the characterization of a fact, phenomenon, individual or group, in order to establish its structure or behavior, that is, this type of research covers a certain depth of knowledge in accordance with the case study. At the same time, the research design was documentary, which according to Arias (2012), is a process based on the search, retrieval, analysis, criticism and interpretation of secondary data, that is, those obtained and recorded by other researchers in documentary sources. A field research method was also applied, which in this case consisted of semi-structured interviews, characterized by a conversation or informal interrogation. A series of questions were formulated designed to elicit the information required. In this case, the interviewees were Jean Carlos Giler Álava (JCG for the purposes of analysis), manager of the Municipal Company of Drinking Water and Sanitation of Portoviejo, PORTOAGUAS, EP; José Gabriel Ávila Navarrete (JGA), head of customer service at PORTOAGUAS, EP; and Geovanny Xavier Vera Zambrano (GXV), legal counsel of PORTOAGUAS, EP. These were selected in order to obtain the appropriate information for the case study by grouping the results from the social, legal, political and ethical and moral dimensions (Capdevielle, 2011). The different research approaches could then be used to analyze the use of water sources in the Portoviejo canton, together with the principal state plans that protect and manage the water resources of this area.

ANALYSIS AND RESULTS

The proper use of water resources is a responsibility of all citizens, with special responsibility falling to the public authorities. The state plays an important role in the implementation of policies, as well as projects whose aim is to reduce water pollution. There is also a need to assess and quantify risks to human health in order to create certainty (Jia et al., 2018). According to Maestre, the ecologist and director of the Laboratory of Ecology of Arid Zones and Global Change at the University of Alicante, Spain (2023) "the biggest cause of desertification is the misuse of soil and water, not drought." Table No.3, Interviews part I, shows a conceptual summary of the in-depth interviews of experts in the area, grouped by social and legal dimensions. This evidences the need to establish processes of certainty through public policies.

Table 3. Interviews part I

<table>
<tr><td rowspan="9">**Social**</td><td rowspan="3">What is the role of the state in guaranteeing the reduction of pollution in water sources in the Portoviejo canton?</td><td>JCG- A leading role in the creation of public policies</td></tr>
<tr><td>JGA- To guarantee public policies in conjunction with Portoviejo local authority</td></tr>
<tr><td>GXV- To guarantee the reduction of environmental pollution that affects water sources through public policies</td></tr>
<tr><td rowspan="3">What measures does the municipality of Portoviejo propose for the treatment of wastewater in water sources in the Portoviejo canton?</td><td>JCG- Water source pollution mitigation plans and new investments</td></tr>
<tr><td>JGA- Expand the water treatment networks and the construction of a new treatment plant</td></tr>
<tr><td>GXV-of a technical nature for waste water treatment [sic]</td></tr>
<tr><td rowspan="3">Does the municipality employ the necessary mechanisms for wastewater reuse as an alternative water source?</td><td>JCG- We do not have a system that allows the reuse of wastewater that has had effective treatment</td></tr>
<tr><td>JGA- Yes, since once treated, the water returns to the Portoviejo River</td></tr>
<tr><td>GXV- In the not too distant future, it should inevitably constitute an obligation for the local authority's reuse of wastewater that has had an effective treatment; however, today, no mechanism has been put into practice</td></tr>
<tr><td rowspan="12">**Legal**</td><td rowspan="3">Which prevention measures fail to prevent further pollution of water sources?</td><td>JCG- lack of control; irresponsible conduct of citizens—perpetrated crimes; improve environmental practices.</td></tr>
<tr><td>JGA- Lack of control of organizations; greater environmental training</td></tr>
<tr><td>GXV- lack of control</td></tr>
<tr><td rowspan="3">Is Ecuadorian legislation sufficient to ensure the protection and conservation of water sources</td><td>JCG- Yes, but the powers have not been not correctly assumed by the public bodies involved. Normative regulations must be established that avoid the disarticulation that currently exists between the bodies of the central government and the local authorities.</td></tr>
<tr><td>JGA- No, the regulation only exists on paper.</td></tr>
<tr><td>GXV- Yes, but the powers are not correctly assumed by the public bodies involved</td></tr>
<tr><td rowspan="3">What is the problem of pollution of the Portoviejo River?</td><td>JCG- Wastewater discharges; the deposit of materials such as debris; tree waste; remains of organic matter and plastic objects and other objects.</td></tr>
<tr><td>JGA- Deforestation, industrial, agricultural and livestock activities, as well as domestic wastewater dumping.</td></tr>
<tr><td>GXV- Wastewater discharges; the deposit of materials such as debris; tree waste; remains of organic matter and plastic objects.</td></tr>
<tr><td rowspan="3">Do you believe that state sanctions should be more rigorous for those who commit acts that harm the quality of water from water sources in the canton?</td><td>JCG- The sanctions that the central government could impose must be framed in attacking the lack of control over actions and the urgent adoption in the face of such facts, since if the fulfillment of the existing sanctioning powers is not viable, it will be useless to define legal consequences that are stricter.</td></tr>
<tr><td>JGA- Yes</td></tr>
<tr><td>GXV- The structural lack of control of actions must be addressed, and their urgent adoption in the face of such facts, since, if compliance with the existing sanctioning powers is not feasible, it will be useless to define stricter legal consequences.</td></tr>
<tr><td colspan="3">Source: compiled by the authors</td></tr>
</table>

In the case of Portoviejo, the municipal authority has the obligation to allocate projects and budget for the conservation and care of the water resources of the canton. The most important water source is the Portoviejo River, since drinking water is obtained from this to supply the city and its surroundings. For this reason, the municipal authority, in conjunction with PORTOAGUAS EP, implements various measures for the correct treatment of wastewater that flows into the river and the adequate maintenance of the treatment systems. According to articles 12, 313 and 318 of the constitution of Ecuador (2008), water is a strategic national resource, in the public domain, and the inalienable, imprescriptible and unseizable property of the state; it constitutes a vital element for the natural environment and for the subsistence of human beings; it is solely the state's right and obligation to administer, regulate, control and manage strategic resources, in accordance with the principles of environmental sustainability, precaution, prevention and efficiency. But even with all the measures proposed by the state, there is a disarticulation in the actions taken by the local authorities of the Portoviejo Canton with those proposed by the central government, which causes many projects and their budget allocations to be neglected or measures to be insufficient to counteract the factors that harm water sources. In this way, budgets are dedicated to projects that are considered of greater importance, leaving on hold smaller yet vital projects that are aimed at conserving the good treatment of water resources.

The pollution of the Portoviejo River not only occurs within the canton, but extends to the basins upstream as far as the Santa Ana canton; therefore, the study of the polluting factors must not only fall to the authorities of Portoviejo, but also to the administrations of other cantons within the catchment area (mainly Santa Ana and 24 de Mayo), especially since in the rural areas of these cantons the inhabitants have free and uncontrolled access to the water of the river. Similarly, in places where the population uses the water for domestic purposes, polluted wastewaters are dumped into the same river. According to Guambo et al., (2020), the inhabitants of Picoazá, on the banks of the river, use these waters to bathe, wash their clothes, as well as for consumption. Local residents report that tankers come every afternoon to extract water for sale in other areas. In addition, one of the biggest problems arises from the inadequate management of soils and crops, and from agrochemical waste and fertilizers together with water erosion, contributing contaminants ranging from sediments to untreated sewage. For these reasons, the river is significantly polluted before it reaches the Portoviejo canton, where the same pattern is repeated and where the greater population increases the discharges of wastewater. Greater quantities of sediment also enter the river in the lower reaches of the basin and, closer to the sea, there is the added pollutant of wastewater from shrimp farms, as shown in Table No.4:

Table 4. Main pollutants of the Portoviejo River

Main Pollutants of the Portoviejo River		
Pollutants	**Origin**	**Prevention Measures**
Wastewater	Wastewater in most cases comes from local inhabitants who in their domestic context flush sewage into the river. On the other hand, there are wastewater products of the industrial sector which tend to pollute in greater excess due to their amount of chemicals.	The Portoviejo local authority, through PORTOAGUAS EP, implements technical measures in order to promote the correct treatment of wastewater in water sources within the coverage area of the water treatment systems that it operates and manages.
Debris	In many areas it has been shown that people irresponsibly dump debris from building sites, which they eject onto the riverbank and this contaminates the river little by little.	Those who dump debris in public areas are punished with a fine equivalent to one month's minimum wage in the hope that people will become more aware of the damage they are causing.
Tree waste	Seasonally, fallen twigs and branches accumulate on the banks of the river, despite being a type of natural pollution, it is vitally important that the authorities keep the banks of the river clean.	The municipal authority has the obligation to allocate a budget for cleaning of tree waste in the Portoviejo River.
Remains of organic matter	This corresponds to the waste that people throw on the banks of the river, which then accumulates and enters the river, thus polluting it even more...	This type of dumping is sanctioned with a fine of 15% of one month's minimum wage in the hope that people will become more aware of the damage they are causing.
Plastic waste	The plastic waste that is thrown into the river is mixed together with organic waste; this type of waste is then dragged downstream and spreads for many kilometers until it reaches the sea.	In 2021, the AZURE project was inaugurated, which facilitates the extraction of up to 80 tons of plastic waste per day from the Portoviejo River. The river was expressly chosen to start the project, which has the support of the Provincial Government of Manabí. (Manabí, 2021)
Agricultural pollutants	The agricultural sector contributes to the pollution of water sources, albeit by smaller measure, by chemical waste from agricultural work.	The municipal authority is obliged to give prevention talks to farmers to encourage the proper management of soil and water and thus avoid environmental pollution.

Source: Compiled by the authors from Guambo et al. (2022) and Manabí (2021)

According to the National Institute of Statistics and Censuses of Ecuador (INEC) (2016), issues with water managemet persist. The UN (2016) states that more than 80% of wastewater products from human activities such as domestic work are discharged into rivers or oceans without receiving any treatment. Thus, water pollution is on a large scale, which, if not treated in time, will eventually cause irreparable damage to the environment. This pollution produces disease in many parts of the world, with children under 5 being the most affected; indeed, there is a direct correlation with increased infant mortality, with more than 800 children dying daily due to waterborne diseases. Additionally, high nitrate concentration in drinking water increases the risk of diseases such as methemoglobinemia, diabetes, spontaneous abortion, thyroid disease, and gastric cancer (Havelaar & Melse, 2003; Hord, 2011; Sanchez et al., 2017; Danni et al., 2019a, 2019b). As the amount of wastewater increases, the negative impacts on health multiply (Dahan & Kashiwase, 2016).

The use of water resources mainly for domestic purposes will increase worldwide in the coming years due to the increase in the world population; it is expected that by 2050 the domestic use of water will increase its demand by between 60% and 80%. Therefore, if strict measures have not been taken for its conservation before that time, it is expected that there will be a shortage of water. Water will become the "gold" of the next generations as its supply becomes more limited and fewer people have easy access to its supply; if disinterest in this matter continues, a worldwide social crisis will likely emerge. The effects of climate change will only aggravate the future situation, making it increasingly urgent that the quality of water conservation systems be addressed. As an example of a successful model of environmental management that might be followed elsewhere, the restoration of the Escher River in the Ruhr basin in Germany demonstrates the opportunities that arise from the modernization of the wastewater treatment infrastructure of a region (Guambo et al., 2022).

Table No.5, Interviews Part II, provides a conceptual summary grouped by political and ethical dimensions on the in-depth interviews of the experts in the area of water management. It can be seen that budgetary concerns are prioritized, and that factors lying outside of the legislative framework must also be addressed. The commitment and responsibility of citizens is essential to success.

Table 5. Interviews Part II

Politics	Do you believe that local officials need to do better in the face of constant complaints about poor water quality?	JCG- Public services in all their dimensions and contexts must be aimed at the commitment and responsibility of public officials to address the needs of citizens regarding the provision of poor-quality drinking water. Administrative actions must be immediately undertaken to solve such a problem, channeling the official bodies to identify the neuralgic cause [sic] of the problem and determine the respective sanction for the offending person or persons.
		JGA- Of course, since they are also citizens who receive that public service.
		GXV- There must be greater commitment and responsibility on the part of public officials in response to the disconformity of citizens related to the provision of poor-quality drinking water. Administrative actions must immediately be undertaken to solve such a problem, channeling official bodies to denounce and identify the root cause of the problem and determine the respective sanction to the offending person or persons.
Ethics and morals	Do you consider that the economic resources allocated for the protection and conservation of the natural environment are sufficient to solve the problem?	JCG- It is preponderant and a reality that the lack of planning and incorrect allocation of economic resources has always been a neuralgic [sic] problem in public administrations. That is why the authorities must allocate sufficient resources to solve these problems and that they always move towards a sustainable development of our canton.
		JGA- It is not enough, since budgets are often allocated to other projects.
		GXV- The lack of planning and incorrect allocation of economic resources has always been a constant in public administrations. Therefore, based on the objective criteria of the authorities, resources must be allocated that are sufficient to solve these structural problems.
Source: Compiled by the authors		

In the Portoviejo canton, the state must play an important role in the establishment of public policies, and these must effectively be aimed at guaranteeing the reduction of environmental pollution of water sources. The Ecuadorian legal system responds to the requirements for the protection and conservation of water resources, but the allocated powers have not been adequately assumed by the pertinent public bodies. Stricter, clearer and more accessible regulations must be established to avoid the disarticulation that currently exists between central government and local authorities. A greater sense of commitment and responsibility must be taken on by public officials in response to the provision of poor-quality drinking water, and the current administrations must immediately endeavor to solve the issues; for example, by identifying the root cause of the problem and sanctioning those discovered to be responsible. The lack of planning and incorrect provision of economic resources is a constant theme in public administration in Ecuador; therefore, sufficient resources must be allocated to solve these structural problems.

CONCLUSION

The contamination of water sources and, indeed all natural sources worldwide, is an issue that generates much controversy, concern and confrontation. While it is true that international organizations fight for the conservation of the natural environment, creating laws and agreements between states with the aim of minimizing pollution, these efforts have hitherto been insufficient. The high level of pollution and neglect inherent in water-source regulation contribute to environmental degeneration and negatively influence the processes of global warming. Certain international organizations create awareness campaigns to protect natural sources of water (for example by educating citizens on domestic water consumption), but this is not the solution to the problem in its entirety since many households lack basic services. In addition, wastewater is not treated properly and frequently empties into rivers and their tributaries that then flow into the sea. If water quality is not improved, a greater problem will emerge combining both shortages and an increase in diseases and leading to more poverty, inequalities or armed conflict (as in Syria, originating in drought). This situation is aggravated by the net increase in the global population up to 2050, which will put the greatest pressure on the urban areas of low-income countries.

The impact of the contamination of water sources in Ecuador is most keenly felt in areas of limited economic resources, where the lack of sewage systems leads to the dumping of wastewater. One way to address this is by forcing companies to have an environmental certification regarding the proper use of their wastewater, thus minimizing the risk of toxic substances reaching water basins and the population centers they support. This study has shown that water sources are of vital importance within Portoviejo canton and the different settlements that benefit from the tributaries of the Portoviejo River, both for agricultural use and drinking water for households. The importance of the state in creating public policies that mitigate water pollution is abundantly clear. The Water Law enacted in Ecuador is an important step in the conservation of these natural sources of water and shows that it is possible to aspire toward the correct use of water sources.

At present, water sources in the Portoviejo canton are not of good quality despite the laws created by the Ecuadorian state and awareness campaigns in the area. Both central and local government do not fully fulfill their functions in this area and do not give sufficient importance to the issue despite knowing that the quality of the natural environment is key to a better quality of life for citizens. Minimizing the environmental impact on and raising public awareness about the use of these natural resources has become urgent. However, the current market-led economic model of accumulation is incompatible with this need, as the natural world reaches its limits and life itself is under threat.

REFERENCES

Afroz, R., & Rahman, A. (2017). Health impact of river water pollution in Malaysia. *International Journal of Advanced and Applied Sciences, 4*(5), 78–85. doi:10.21833/ijaas.2017.05.014

Arias, F. (2012). *El proyecto de investigacion*. Episteme.

Beiswenger, H. (2020). *Proyecto de saneamiento de las fuentes hídricas del río Portoviejo. Informe final. Municipio de Portoviejo*. Portoviejo - Ecuador.

Bordeleau, G., Savard, M. M., Martel, R., Ampleman, G., & Thiboutot, S. (2008). Determination of the origin of groundwater nitrate at an air weapons range using the dual isotope approach. *Journal of Contaminant Hydrology, 98*(3), 97–105. doi:10.1016/j.jconhyd.2008.03.004 PMID:18499297

Burbano, N., Becerra, S., & Pasquel, E. (2011). *Caracterización hidrogeolóica de las Cuencas Portoviejo - Chone*. Retrieved from INAMHI: https://issuu.com/inamhi/docs/caract._hidrogeologica-manabi/7

Burbano, S., Becerra, P., Efrén, S., & Pérez, L. (2015). Introducción a la Hidrología del Ecuador. Academic Press.

Camacho, L., & Astudillo, F. (2021). Evaluación del aprovechamiento empírico del agua subterránea somera del cantón Morona-Ecuador. *Rev. Dominio las Ciencias, 6*(2), 460–473. https://dominiodelasciencias.com/ojs/index.php/es/article/view/1179/html

Carey, R. O., Migliaccio, K. W., & Brown, M. T. (2011). Nutrient discharges to Biscayne Bay, Florida: Trends, loads, and a pollutant index. *The Science of the Total Environment, 409*(3), 530–539. doi:10.1016/j.scitotenv.2010.10.029 PMID:21111457

Constitución de la República del Ecuador. (2008). *Constitucion de la Republica del Ecuador*. Montecristi.

Correa, L., López, L., & Vergara, M. (2020). Investigaciones de representaciones sociales sobre fuentes hídricas y condiciones para la preservación hídrica (2013-2019). *Revista Espacios, 41*(29), 329–344.

Cortés, I., Cortés, M., & Cisneros, H. (2015). Recursos hídricos. *Rev. Cultura científica y tecnológica, 31*(6). Retrieved from http://erevistas.uacj.mx/ojs/index.php/culcyt/article/view/344

Costa, A., Teixeira, C., Do, N., Oliveira, M., Oliveira, Q., & Silva, J. (2012). Recursos hídricos. Cuadernos de Ciencias exactas y tecnología. *UNIT-SERGIPE, 1*(1), 67-73. Retrieved from https://periodicos.set.edu.br/cadernoexatas/article/view/201/126

Cruz, P. M., & Bodnar, Z. (2011). El nuevo paradigma del derecho en la posmodernidad. Revista de Estudios Constitucionales, Hermenéutica y teoría del derecho, 3, 75-83. doi:10.4013/rechtd.2011.31.08

Dahan, S., & Kashiwase, H. (2016). El *ODS 6 sobre agua y saneamiento es esencial para el desarrollo sostenible.* Obtenido de Banco Mundial BLOGS. Retrieved from https://blogs.worldbank.org/es/opendata/el-ods-6-sobre-agua-y-saneamiento-es-esencial-para-el-desarrollo-sostenible

Danni, S. O., Bouchaou, L., Elmouden, A., Brahim, Y. A., & N'da, B. (2019). Assessment of water quality and nitrate source in the Massa catchment (Morocco) using δ15N and δ18O tracers. *Applied Radiation and Isotopes*, *154*, 108859. doi:10.1016/j.apradiso.2019.108859 PMID:31539708

EPA. (2021). *Información sobre la protección de las fuentes de agua.* Retrieved from Agencia de Protección Ambiental de Estados Unidos. Retrieved from https://bit.ly/3O9d9Xm

Espinoza, T., & Rodríguez, C. (2016). Notas técnicas: Índice de calidad de agua ICA, de los ríos Aroa Yaracuy de los estados de Falcón y Yaracuy. *Rev. Ingeniería UC.*, *23*(3), 381–386.

Fadhullah, W., Yaccob, N. S., Syakir, M. I., Muhammad, S. A., Yue, F.-J., & Li, S.-L. (2020). Nitrate sources and processes in the surface water of a tropical reservoir by stable isotopes and mixing model. *The Science of the Total Environment*, *700*(134517), 134517. doi:10.1016/j.scitotenv.2019.134517 PMID:31629263

Faldez, R. L. (2020). *Conservación del agua: importancia y técnicas.* Retrieved from Ecología verde: https://bit.ly/3I5WJuU

Galárraga, S. (2020). *Estado de los recursos hídricos en Ecuador.* Retrieved from HidroRed. Retrieved from http://tierra.rediris.es/hidrored/basededatos/docu1.html

García, G. M., Carvajal, E. Y., & Jiménez, H. (2016). La gestión integrada de los recursos hídricos como estrategia de adaptación al cambio climático. *Rev. Ingeniería competitive, 9*(1), 19-29. Retrieved from https://www.ecologiaverde.com/conservacion-del-agua-importancia-y-tecnicas-2195.html

Giler, A., Donoso, P., & Zaldumbide, D. (2020). Manejo sostenible de inundaciones, cuencas hidrográficas y riberas en la provincia de Manabí. *Rev. La Técnica.*, *1*(5), 55–72.

Gobierno autónomo Descentralizado. (2015). *Propuesta de Desarrollo estratégico del cantón Portoviejo*. Retrieved from Gobierno Autónomo Descentralizado del cantón Portoviejo. Retrieved from https://bit.ly/41veXNr

Goldar, B., & Banerjee, N. (2004). Impact of informal regulation of pollution on water quality in rivers in India. *Journal of Environmental Management*, *73*(2), 117–130. doi:10.1016/j.jenvman.2004.06.008 PMID:15380317

Guambo, G., Torres, J., & Quiroz, S. (2022). Análisis histórico de la contaminación hídrica en el río Portoviejo. *Minerva Journal, 3*(8), 54-60. Retrieved from https://minerva.autanabooks.com/index.php/Minerva/article/view/64/211

Halder, J. N. & Islam, M. N. (2015). Water pollution and its impact on the human health. *Journal of Environment and Human, 2*(1), 36-46. doi:10.15764/EH.2015.01005

Havelaar, A. H. & Melse, J. M. (2003). *Quantifying public health risk in the WHO guidelines for drinking-water quality: a burden of disease approach*. Academic Press.

Hoagland, B., Schmidt, C., Russo, T. A., Adams, R., & Kaye, J. (2019). Controls on nitrogen transformation rates on restored floodplains along the Cosumnes River. California. *The Science of the Total Environment*, *649*, 979–994. doi:10.1016/j.scitotenv.2018.08.379 PMID:30179826

Hord, N. G. (2011). Dietary nitrates, nitrites, and cardiovascular disease. *Current Atherosclerosis Reports*, *13*(6), 484–492. doi:10.100711883-011-0209-9 PMID:21968645

IICA. (2008). *Buenas prácticas agrícolas. Guía para pequeños y medianos agroempresarios*. Instituto Interamericano de Cooperación para la Agricultura. Retrieved from http://repiica.iica.int/docs/B0737e/B0737e.PDF

INEC. (2016). *Indicadores ODS de agua, saneamiento e higiene en Ecuador*. Retrieved from https://www.ecuadorencifras.gob.ec/documentos/web-inec/Bibliotecas/Libros/Diagnostico_ASH_pobreza_INEC_BM.pdf

INEFAN. (2017). *Plan de manejo de área de boques y vegetación protectores de las colinas que circundan la ciuda de Portoviejo*. Instituto Ecuatoriano Forestal de Áreas Naturales y Vida Silvestre.

Intriago, F. J., & Quiroz, F. L. (2021). Calidad del agua de la cuenca media del río Portoviejo. Estrategias para mitigar la contaminación. *Rev. Polo del conocimiento, 6*(6), 1144-1171. PMID:33414495

Jia, Z.M., Li, S.Y., & Wang, L., (2018). Assessment of soil heavy metals for eco-environment and human health in a rapidly urbanization area of the upper Yangtze Basin. *Scientific Reports*, *8*(1), 1–14. https://www.nature.com/articles/s41598-018-21569-6 PMID:33414495

Krause, F., & Strang, V. (2016). Thinking relationships through water. *Society & Natural Resources*, *29*(6), 633–638. doi:10.1080/08941920.2016.1151714

Ley Orgánica de recursos hídricos. (2014). *Usos y aprovechamiento del agua.* Retrieved from https://bit.ly/3W4QFZB

Luque, A. & Jiménez, A. (2019). Procesos de sostenibilidad textil: éxito, o nueva forma de contaminar bajo un paradigma amable. *Revista Inclusiones, 6*, 259-288.

Maestre, T. (2023). *La mayor causa de desertificación es el mal uso del suelo y del agua, no la sequía.* Retrieved from https://www.eldiario.es/ballenablanca/economia/fernando-t-maestre-ecologo-mayor-causa-desertificacion-mal-suelo-agua-no-sequia_128_10157041.html

Manabí, P. (2021). *Predectura de Manabi, Desarrollo y Equidad.* Retrieved from https://bit.ly/3W9pBIE

Martín de Santa Olalla, F., López Fuster, P. & Belmonte, C. (2005). *Agua y agronomía.* Madrid: Universidad de Castilla.

Ministerio del Ambiente. (2012). *Contaminación de fuentes hídricas.* Retrieved from https://www.youtube.com/watch?v=w60Z2z5PF9w

Mirandiña, A., & Endocara, M. (2019). *Evaluación de necesidades tecnológicas para el manejo de la oferta hírica en cantidad y calidad.* Manabí - Ecuador.

Montilla, A., & Pacheco, H. (2017). Comportamiento temporal y espacial del bosque ribereño en el curso bajo del rio Portoviejo y la quebrada Chilán, provincia de Manabí, Ecuador. *Rev. Internacional de Contaminación Ambiental.*, *33*(1), 21–35. doi:10.20937/RICA.2017.33.01.02

Noguera, C. (2020). *Desarrollo urbano sostenible. Sociología del medio ambiente.* Retrieved from www.ua.es/personal/antonio.aledo/alumnos/nogera.doc

OIKOS. (1998). *Diagnóstico Ambiental del cantón Portoviejo. Corporación de Gestión Tecnológica y Científica sobre el Ambiente.* Proyecto Educar.

Olguín, E., González, R., Sánchez, G., & Zamora, J. (2010). Contaminación de ríos urbanos: El caso de la subcencia del río Sordo en Xalapa, Veracrua, México. *Rev. Latinoam Biiotecnol Amb Algal, 1*(2), 178–190.

ONU-DASES. (2015). *Decenio Internacional para la acción "El agua fuente de vida"*. Departamento de Asuntos Económicos y Sociales de Naciones Unidas: Retrieved from https://bit.ly/3Mpm85c

Organización de Naciones Unidas. (2015). *Decenio Internacional para la acción. El agua como fuente de vida. Departamento de asuntos económicos y sociales de las naciones undas.* Organización de las Naciones Unidas. Obtenido de Departamento de asuntos económicos y sociales de las Naciones Unidas. Retrieved from https://www.un.org/spanish/waterforlifedecade/

Organización de Naciones Unidas. (2016). *Agua limpia y saneamiento; Por que es importante.* Retrieved from https://bit.ly/3W6prlk

Organización de Naciones Unidas. (2023). *Desafíos globales del agua.* Retrieved from https://www.un.org/es/global-issues/water

Organización Mundial de la Salud. (2022a). *Agua para consumo humano.* Retrieved from https://bit.ly/2HxnryQ

Organización Mundial de la Salud. (2022b). *Saneamiento.* Retrieved from https://bit.ly/2UYncQK

Ortiz, A. A., Ruiz, O. M., & Rodríguez, M. J. (2017). Planificación y gestión de los recursos hídricos: Una revisión de la importancia de la viariabilidad climática. *Rev. Logos Ciencia y Tecnología, 9*(1), 100–105.

Plan Integral de Desarrollo. (2020). *Plan integral de desarrollo de los recursos hídricos de la provincia de Manabí.* Secretaría ejecutiva para asuntos económicos y sociales. República del Ecuador. Retrieved from https://bit.ly/42VmXbG

Plan Nacional de Desarrollo. (2017-2021). *Toda una vida. Ecuador.* Observatorio Regional de Planificación para el desarrollo de América Latina y el Caribe. Retrieved from https://bit.ly/2t3yGFG

Quiroz, E. & Menéndez, C. (2016). Modelación matemática de la capacidad de autodepuración de corrientes superficiales. Caso de estudio Río Portoviejo. *Rev. Cubana de ingeniería, 6*(2), 64-70.

Salas, Q. N. (2018). *Siembra de árboles en la rivera del río Portoviejo.* Retrieved from https://bit.ly/3Oh9ruy

Sanchez, D. A., Szynkiewicz, A., & Faiia, A. M. (2017). Determining sources of nitrate in the semi-arid Rio Grande using nitrogen and oxygen isotopes. *Applied Geochemistry, 86*, 59–69. doi:10.1016/j.apgeochem.2017.09.012

Shiva, V. (2005). *Las mujeres de Kerala contra Coca-Cola.* Retrieved from https://mondiplo.com/las-mujeres-de-kerala-contra-coca-cola

The Nature Conservancy. (2019). Estrategia hídrica en Ecuador. *Nature*. Retrieved from https://www.nature.org/es-us/sobre-tnc/donde-trabajamos/tnc-en-latinoamerica/ecuador/estrategia-hidrica

Torres, P., Cruz, C., & Patiño, P. (2009). Índices de calidad de agua en fuentes superficiales utilizadas en la producción de agua para consumo humano: Una revisión crítica. *Revista Ingenierías Universidad de Medellín*, *8*(15), 79–94.

UNICEF. (2019). *1 de cada 3 personas en el mundo no tiene acceso a agua potable*. Retrieved from https://uni.cf/2LaVcqR

United Nations (UN). (2023). *The United Nations World Water Development Report 2023: Partnerships and Cooperation for Water*. UNESCO. https://www.unesco.org/reports/wwdr/2023/en/download

Voulvoulis, N. (2018). W*ater reuse from a circular economy perspectiva and potential risks from an unregulated approach. Retrieved from Current Opinion in Environmental Science & Healh*, *2*(3), 32–45. doi:10.1016/j.coesh.2018.01.005

World Water Week (WWW). (2022). *A Rising Tide: Shared Vision for Women in Water Diplomacy.* Online & On-site session, 30 August 2022. worldwaterweek.org/event/10314-a-rising-tide-shared-vision-for-womenin-water-diplomacy.

Xue, D. M., Baets, B. D., Cleemput, O. V., Hennessy, C., Berglund, M., & Boeckx, P. (2012). Use of a Bayesian isotopemixingmodel to estimate proportional contributions of multiple nitrate sources in surface water. *J. Environ. Pollut.*, *161*, 43–49. doi:10.1016/j.envpol.2011.09.033 PMID:22230066

Zhang, Y., Shi, P., Li, F. D., Wei, A. L., Song, J. X., & Ma, J. J. (2018). Quantification of nitrate sources and fates in rivers in an irrigated agricultural area using environmental isotopes and a Bayesian isotope mixing model. *The Journal of Chemical Thermodynamics*, *208*, 493–501. doi:10.1016/j.chemosphere.2018.05.164 PMID:29886338

KEY TERMS AND DEFINITIONS

Climate Change: Means a change of climate which is attributed directly or indirectly to human activity that alters the composition of the global atmosphere and which is in addition to natural climate variability observed over comparable time periods.

Economic Globalization: This is a phenomenon in expansion that causes profound changes on the world stage. It revolves around trade, the flow of investment, financial capital, division of labor and specialization. The concept is not limited only to economic variables since its effects extend to individuals, society to the state. Developing countries are experiencing stagnation in the face of their inability to cope with globalization, which is compounded by poor management of their financial markets, leading to an increase in the income inequality gap. Economic globalization brings with it the mobilization of goods and capital, reduces distance between borders and energizes international trade with some alterations to sovereignty.

Globalization: Concept used to describe the growing interdependence of the world's economies, cultures, and populations, brought about by cross-border trade in goods and services, technology, knowledge and flows of investment, people, and information. It also generates large asymmetries.

Partnerships and Cooperation for Water: This refers to understanding the critical role of water related aims and targets at an international level, in a world that is constantly facing important negative effects from pollution and global changes in terms of biodiversity, ecosystem, and climate.

Public Policy: This refers to decisions and actions that a government takes when addressing public or collective issues.

Resilience: Transformations within a complex system related to the capacity for self-organization while maintaining internal structure, together with the ability to create adaptive responses, generate knowledge, experience, and learning. Resilience and sustainability are directly related to changes within societies, economies, and the human system as a whole. The transformation of systems is inevitable since it allows systems to strengthen.

Water: Is at the core of sustainable development and is critical for socio-economic development, energy and food production, healthy ecosystems and for human survival itself.

ENDNOTES

1 In Ecuador, The Nature Conservancy (TNC) (2019), and its partners are helping cities protect their water sources through the creation of Water Funds, innovative financial mechanisms whose goal is to secure long-term financing for the conservation of key water sources.

2 Bottom-up planning: This takes into account the microeconomics in this case of each parish of the canton, whose objectives are framed in covering the

current and future demands of each sector; in addition, it does not limit its representatives in abstract and intangible objectives.

3 After being expelled by the government in 1977, Coca-Cola re-established itself in India on October 23, 1993, at the same time that the other American dinks multinational, Pepsi Cola, settled there. Both companies currently have 90 "bottling plants" that are, in fact, "extraction plants": 52 units belong to Coca-Cola and 38 to Pepsi Cola. Each of them extracts between 1 and 1.5 million liters of water per day, Shiva, (2005).

4 A tool and structure that has emerged as an alignment of economic and environmental outcomes.

Chapter 13
Dynamics of Teaching in the Era of Climate Change Education

John M. Onyango Yambo
Kisii University, Kenya

ABSTRACT

This chapter addresses the dynamics of teaching in the era of climate change education. It equally looked at encounters of mobile schools and teaching and facilities in mobile schools in mitigating climate change. It necessitates that the education researchers engage in inquiry and involvement of other stakeholders. Two methods were used, partial relevant literature review and qualitative assessment of teaching methods. The study targeted policy makers, curriculum and instructional designers, faculty, administrators, teacher educators, and other stakeholders interested in averting the impact of climate change and promoting sustainable development in Kenya and beyond.

INTRODUCTION

Ordinarily, Education for Sustainable Development (ESD) is all about encouraging behavior that promotes a sustainable future in the era of climate change. According to UNESCO (2009), ESD is based on five types of learning: learning to know; learning to do; learning to live together; learning to be and learning to transform oneself and society in the context of climate change. When dealing with ideologies of practice and teacher actions that influence effective teaching of numeracy, the work of Muir (2008) linked the dynamics of teaching in the era of climate change

DOI: 10.4018/978-1-6684-9099-0.ch013

education to the observable performance of teaching acts, involving interacting effectively with students through questions and probes, answers and reactions, and praise and criticism that would be an ideal for this era of climate change crisis.

In schools and teaching, it is more of behavior change. In this regard, Mbah, Shingruf and Molthan-Hill (2022) opined that in the case of climate change, the relationship between variables influencing pro-climatic behavior and the role of knowledge seem to be even more complex and multi-faceted. In the same vain, analyses by Lehnert, Fiedor, Frajer, Hercik and Jurek (2020) confirmed that there is a weak or no correlation between climate change awareness/knowledge and the willingness to act in favor of climate protection. Interestingly, other studies by Stevenson, Nils-Peterson and Bondell (2018) found the link between climate knowledge and behavior to be significant. This therefore, calls for a pro-active teaching and equipping teachers and educators with dynamics of teaching in the era of climate change education.

Since in our schools and educational institutions normally have both gender that should be made more aware of climate change education, some educative information by Kolenatý, Kroufek and Cinˇcera (2022) revealed that the analyses of survey data showed that girls exhibited significantly higher levels of climate change concern and willingness to act than boys (in both pre-test and post-test), but there was no significant difference between girls and boys with regard to climate change knowledge and self-efficacy. This gap on climate change knowledge and self-efficacy is explored in this chapter.

The chapter proposes recommendations for effective teaching using modern technology that would lead to increased student achievement and mastery and comprehension of course material provided. This sentiment is supported by Barrick (2011) who opined that when the subject matter to be learned possesses meaning, organization and structure which is clear, then learning proceeds more rapidly and what is eventually learnt is retained longer by students. On the contrary, teachers would strive to do their best to teach, however, the key problem is that the higher authorities of the schools have to decide what the schools need to achieve. This could be possible by arranging plans for schoolwork which will eventually attain the desired ends. More importantly, teachers should be involved in making some of these decisions. In dynamics of teaching in the era of climate change, more participants should be reached with climate change informative education. This is because early 'information-deficit' models assumed that if people were well-informed about the causes and consequences of environmental problems, which are likely to be caused by climate change, then they would adopt pro-environmental behavior to remedy the situation effectively (Kolenatý et al. 2022).

Due to the fact that change is a process, dynamics of teaching in the era of climate change education cannot be realized in isolation. Involving other ardent education stakeholders is key. Studies by Farrell, Devlin and James (2017) in Australia emphasized that in order to achieve climate change education, members of the community have to be effectively included by initiating and implementing constructive change in their communities. This should include their professions and workplaces. These members of the community need to have excellent interpersonal and decision-making skills, including an awareness of personal strengths and limitations, mentor future generations of learners in the community; engage in meaningful public discourse and creative participation with a profound awareness of their needs (Davies & Devlin, 2007).

In Africa, it has been equally realized that climate change has a direct impact on education. A typical example has been pointed out in the Zimbabwe Human Development Report (2017) that climate change has a direct impact on education outcome. The primary impacts of climate change on education, according to this report, arise from the effects of extreme weather events, such as heavy rains accompanied by flash floods, strong winds and hailstorms with short and long-term consequences. Moreover, drought and increasing temperatures lead to poor harvests and food scarcity which have negative impacts upon educational attainment. It has been realized that poorly fed learners with fear and uncertainty of weather, interfere most with academic achievement (ZHDR, 2017).

BACKGROUND

Theoretical Framework

This study was guided by the theory of change (TOC). It was propagated by Fulbright-Anderson, Kubisch and Connell (1998). Here, they propose a guiding framework (a pathway) to discuss the sequence of events that leads to a particular desired outcome which is associated with programme and project evaluation, and informed social action. It is meant for the evaluation of complex community initiatives. The TOC has been adopted by other scholars like Connolly and Seymour (2009) who applied it in assessing educational reform in a variety of contexts, such as Mathematics and Science; early childhood and community school linkages and district-wide and school reform initiatives (Fullan, 2006). In further research, Stein and Valter (2012) argued that the TOC helps in the planning and assessment of education reform initiatives in an urban environment. As applied to this study, it fosters the creation of collective ownership, which is necessary for dynamics of teaching in the era of climate change education. Having quality education in mobile and tent schools needs effective collaboration by various education stakeholders.

Despite its wide application, Connell and Klem (2000) observed that the TOC approach suffers from conceptual shortcomings, including the fact that it does not prescribe specific evaluative methods, or the way the theories of change are articulated. There are, however, certain points of connection and similarities between the varieties of TOC that have been developed. On the contrary, TOC is an approach used in international development which lays out a roadmap to achieve a long-term goal (Stein & Valter, 2012). Given its tested and proven potential to create a desirable impact, the DFID's guidelines provide a broader perspective, underscoring the need for implementers of the TOC to draw on objectives and frameworks that are owned nationally, supported with evidence that guarantees behavioral climate changes and economic transformation among others (Fullan, 2006). This therefore makes TOC relevant in this paper.

TEACHING IN THE ERA OF GLOBAL CLIMATE CHANGE

Ordinarily, teaching in this era of climate change needs teachers and educators who are conscious of the current global phenomena, including climate change and its related effects. According to Landorf, Doscher and Rocco (2008), the educator needs to be responsible for evaluating the contextual circumstances that will impact the students' well-being and provide a form of teaching that enables their students to understand their own capabilities so as to adapt (Bangay & Blum, 2010). It provides the prospective teachers with an opportunity of establishing an appropriate teacher-pupil relationship, a chance of evaluating the student potentials and the suitability in the teaching profession (Landorf et al. 2008).

From the Australian perspective, Dianne (2009) contended that climate change education is a relatively new field in the region. In July 2008, the Australian Association for Environmental Education (AAEE) applied pressure to the Federal Government to improve its practices to educate people about climate change, which was timely. Climate change education has been generally incorporated into ESD and not treated as a separate field (Elliott & Crosswell, 2002). In this regard, the National Solar School Initiative was a major ESD program that has climate change mitigation as an element and teachers were trained to embrace it effectively.

The work of Kim and Kim (2009) in South Korea asserted that climate change has been considered as a key theme of ESD and has been included in school topics and subjects, such as climate (geography), energy (science) and global warming (environment). However, a national curriculum or guideline for Climate Change Education (CCE) has not yet been fully established. Korea is striving to better inform and educate its public about global warming and climate change to achieve a national consensus on the country's commitment to international climate change efforts. Additionally, Kasim and Behcet (2015) argued that systemic educational programs on climate change inform the public about the adverse effects of climate change and propose preventive effective measures.

Through educating children and youth at their early stage about the significant role they can play in preserving the environment, they can carry into their adulthood a changed perspective and attitude to make a lasting and positive impact on future greenhouse gas reduction. Going forward, the government has promoted several activities in the education sector. The environment is being instituted as an independent subject in all secondary school curriculum to ensure systematized education concerning energy, climate change and other environmental issues. From the primary school level, the environment is introduced within each subject of study. The Korean government has also encouraged relevant agencies to modify or add entries to textbooks regarding climate change and green growth (Kim & Kim 2009).

Educators have noted that meaningful achievement of CCE can be effected in teacher training institutions then the knowledge can be cascaded down to learners. In this regard, Aglazor (2017) prescribed that the purpose of teacher training is to expose student teachers to the actual teaching and learning environment and the role of teaching practice in teacher education has been to equip them with the needed knowledge for environmental conservation, especially in this era of climate change. Normally, there are cardinal related issues to teaching in the era of global and environmental climate change as explained below.

Encounters of Mobile Schools

The concept of mobile schools has its origin from six objectives of EFA and other sections of the Dakar Framework of Action (UNEP, 2011) which focus on access and learning environments. According to Muhammad, Mussawar and Irfan (2017), it was established during the Participatory Action Research (PAR) project where seven mobile schools were established in seven nomadic Bede groups, which roam around the country in boats in Pakistan. Since these groups change around 90 spots in a year and do not stay more than 3-4 days in one spot, simultaneously, the schools also remain in constant move with them as they went. These mobile schools were found to be significant and transformative indeed. As noted by Maksud and Imtiaj (2006), they not only eradicated illiteracy but also proved that it is possible for a nomad to have access to modern education in real time.

An encounter in Nigeria estimated that 9.3 million nomadic pastoralists, a third of this population consists of children of school going age. Nomadic pastoralists have a literacy rate of 0.2 to 2 percent, and their formal and informal school participation is generally low (ADEA, 2000). This was attributed to the lifestyle which makes it extremely difficult to participate in school (UNEP 2011). In order to address this problem, the Nigerian government has undertaken measures to avail education to this nomadic group. A relevant curriculum suited to their lifestyles has been developed. Moreover, teacher training has been conducted to equip them with knowledge of the new curriculum. Additionally, flexible schedules and shifts were also adopted to give room for children learning in varied lifestyles. In this initiative, radio programs were designed and mobile classrooms availed to both the teachers and pastoralists. Later, the education underwent monitoring whereby performance indicators were developed. As a result, enrolment of nomadic pastoralists' children rose from 18,831 to 20,006 in 1998 (ADEA, 2001). Mobile schools finally improved the academic standards of the pastoralists significantly.

Comparatively, in Kenya, studies done by Okutu (2019) found that the main constraints on the education supply in pastoralist zones include vastness of area, long distances from schools and low population densities which make it difficult to attract numbers in classrooms. Furthermore, there are such other constraints as inadequacy of educational infrastructure; curricula not adapted to pastoral modes of life; scarcity of teachers and absenteeism among herders' children who are relied on heavily to work (MoE, 2007). This has been coupled with insecurity as a major threat to school attendance in both Turkana and West Pokot areas (Okutu, 2019). As indicated by Nyatuka (2023), a hostile school environment is by itself a hazard as it neglects the needs of its learners and teachers, especially the girls, children with disabilities, orphans and others with special needs who need more security.

Ideally, mobile schools would be the best alternative since they counter challenges and respond to lifestyle and needs of nomadic pastoralists' school going children. They equally facilitate continuous learning since even when nomadic pastoralists move from one place to another, the mobile schoolteacher moves with the learners accordingly. This makes EFA a reality for nomadic pastoralists' children, especially teaching them in the era of climate change education. The attempt to provide viable quality education to the nomadic pastoralists, innovations by local communities, Non-profit organizations, faith based organizations (FBOs) and the Ministry of Education have involved the support and establishment of mobile schools as well as promoting the use of mobile education and e-Learning (Yambo, 2022a; MoE, 2007). According to the United States AID Kenya country office report (USAID, 2013), the mobile school initiative aims to increase access to education and improve quality through the provision of culturally and religiously appropriate basic education to children who would otherwise find it difficult to access formal quality education (Yambo, 2022a).

Teaching and Facilities in Mobile Schools

Apparently, one of the solutions found in the provision of education for the children of nomadic pastoralists' children in the era of climate change has been mobile schools. Facilities and teaching process in mobile schools has been equally a matter of concern. According to Kasim and Behcet (2015), one of the most important factors affecting teaching and facilities in mobile schools has been culture-based pedagogy or culture responsiveness. Culturally responsive teaching is an endeavor to make learning activities more relevant with students and more effective for them considering cultural knowledge, past experience, reference frame and performance styles of students who are ethnically diverse (Bangay & Blum, 2010). In culturally responsive teaching strategies, Kasim and Behcet (2015) suggest that building a classroom atmosphere would encourage students make new learning significant, providing opportunities for students to search subjects they are interested in, creating a platform in the classroom, presenting new information with examples regarding students' lives, giving life examples from students' culture and incorporating people from students' lives into the learning process with the purpose of building a bridge between students' new learning plus their existent experiences (USAID, 2013).

Stable learners are receptive to learning. The United Nations International Children Emergency Fund (UNICEF, 2008) recognizes that quality education includes learners who are healthy, well-nourished and ready to participate, learn and supported in the learning process by their families and members of the community. The education should be offered in environments that are healthy, safe, gender sensitive, protective with adequate resources and facilities. According to UNESCO (2005), the teachers should be well trained and skillful in pupil assessment to facilitate learning and reduce disparities. According to USAID (2013), teachers are the single most important education resource in any country. One of the most important requirements for sustained progress towards better quality in education is an improved learning environment, encompassing the physical infrastructure, the learning process and the interaction between children and teachers (UNICEF, 2008).

As put forward by Muhammad et al. (2017), white tents were used as schools for tackling the issue of nomads' children education in Mongolia. These tents were made attractive to lure learners and some food was provided to make them friendlier. Teachers were also nomads for they moved along with the household, and the government paid them for their work hence were motivated to continue teaching (UNEP, 2011). Even in Somalia, the concept of Quranic schools brought revolutionary changes in the lives of nomads. This was single teacher schools, provided multiple entry points for sparse and remote nomadic communities as compared to national formal schooling (Maksud & Imtiaj, 2006). Studies by Koissaba (2017) also showed that pastoralists have poor access to education due to the fact that the physical facilities are in poor state with dilapidated buildings. The basic learning facilities are equally lacking. The classrooms are inadequate as a result of which some pupils learn under trees. The primary schools are few and far in between, while the secondary schools are even fewer.

POLICIES FOR CLIMATE CHANGE EDUCATION

Universally, policy makers maintain that education is the single strongest predictor of climate change awareness (World Bank, 2010). Mainstreaming climate education would change mindsets and behaviors on climate; not just for students but also for their parents, communities and governments altogether. This will eventually foster better preparedness and resilience to climate shocks among entire populations worldwide. Conversely, Wedell (2009) found that there is still a tendency for national educational change policy makers and planners in different parts of the world to ignore the human factors that strongly influence change processes. Instead, they view the change process as a purely linear rational-technical planning and legislative matter. Such an approach seems to assume that once the policy decision to initiate change has been taken, any necessary legislation passed and funding secured, successful change implementation is merely a matter of issuing clear instructions to those lower down the administrative ladder to introduce changes in the classroom from a given date (UNESCO, 2009). This approach does not seem to be a recipe for success and many educational change initiatives are yet to be adopted for successful implementation in the era of climate change education.

According to Mbah et al. (2022), the climate change policy in Bangladesh includes action points for disaster management, infrastructure and research and knowledge management. Within this policy context, research and knowledge management are intended to predict the likely scale and timing of climate change impacts on different sectors including education. Likewise, Sri Lankan government has declared that it is on a mission to develop a generation which is sensitive to the impact of climate change (Ajaps & Mbah, 2022). It intends to do so by promoting research and development on climate change issues at a countrywide level. Within its policy goals, Sri Lanka has an aim to educate and empower both local governments and academia to understand and implement adaptation practices in vulnerable areas of the country, especially in the education sector.

Further in Southern Asia, Pakistan supported the issue of CCE via the National Climate Change Policy (2021) where it was acknowledged that there is need to raise awareness of climate change on a much greater scale than is currently being carried out in the country. Particularly, it aims to develop a climate change awareness program which will involves communities and guides them on sustainable management practices for forests, water and energy. In this regard, the twenty-first century is therefore likely to be a period of great change in education systems internationally, as national policy makers try to understand and develop alternative means of providing their citizens with an education that will help them acquire the skills and understandings needed for the continuing, and also rapidly changing, challenges of techno-globalization (Mbah et al. 2022; Wedell, 2009). Certainly, educational change does not become visible in classrooms directly as a result of a written policy document, instead, whether it occurs, and what form it ultimately takes, directly depends on how people understand what is written down and how they behave in response to that particular understanding (Fullan, 2001; UNICEF, 2008).

In Kenya, a policy framework for nomadic education as put forward by Okutu (2019:123) states in part that: "Measures will be put in place to ensure that learning in nomadic counties is not interrupted by weather and constant threats/acts of insecurity. The Government shall put in place adequate contingent measures to alleviate the effects of any emergencies in nomadic regions; establish adequate early warning systems in all nomadic regions; provide adequate security in all nomadic regions with special attention to learning centers and institutions and institutionalize traditional conflict resolution and interventions for peaceful coexistence." Either, these measures have not been put in place or have not yielded desired results. Insecurity caused by cattle rustling is still a challenge in some parts of the county, which still adversely affect provision of quality education in some regions. Schools are often closed from time to time due to perennial insecurity. Teachers and their pupils are forced to flee from time to time to safer areas, hence, disrupting learning altogether (Ayiro & Sang, 2017).

LEARNING ENVIRONMENT AND CLIMATE CHANGE

One of the key proponents of dynamics of teaching has been an active learning environment. When dealing with the ten principles of effective teaching, Kerns et al. (2005:4) suggested that: "Active learning provides opportunities for students to talk and listen, read, write, and reflect as they approach course content through problem-solving exercises. Formation of small informal groups, simulations, case studies, role playing, in class questions and other activities, all of which require students to apply what they are learning and/or think about what they are learning as they are learning". In this regard, efforts should be made to create an active learning environment in order for students to gain knowledge from the presentation of information they should actively participate in their learning (Lubawy, 2003).

An active learning environment also involves adequate learning space, for example, blackboard use where a teacher provides case studies, quizzes, and problems for students to study, reflect upon and answer to their best ability. Furthermore, teachers endeavor to have students prepare a "minute paper" to answer "what was the most important thing learned in class?" and "what important question remains unanswered?" (Kerns et al. 2005:8). More interactive learning would be more beneficial. In addition, Lubawy (2003) opined that the students be led to paraphrase the same explanation to a completely different audience (parents, children, professionals, novices, experts) as this this would eventually make them internalize what they have learnt. Furthermore, teachers are required to assess paraphrases for accuracy and appropriateness; have students to research concepts relating to specific subjects on the internet or in the library and be made aware of and accountable for the proper procedure for reporting and annotating correctly (Kerns et al. 2005).

On culture and learning environment, Muhammad et al. (2017) opined that culturally responsive teaching is an endeavor to make learning activities more relevant with students and more effective for them considering cultural knowledge, past experience, reference frame and performance styles of students who are ethnically diverse and marginalized. In line with this, education plays a pivotal role in the development of society and decreases the risks of individual and group marginality or vulnerability by internalizing new skills and coping competencies.

The place where learning takes place is important. According to Wedell (2009), learners will continue to spend the bulk of their language-learning time in the learning environment that they are familiar with. It will be necessary for teachers to be able to explain what each type of learning is trying to achieve. In the medium term, it may be necessary to reconsider the mismatch between the two different learning environments. Consequently, the quality of the learning environment at the classroom level is influenced by the teaching methods and classroom practices used by teachers.

Active learning environment in schools should be a collective responsibility. According to Stevenson, Nicholls and Whitehouse (2017) at a school-wide level, administrative and teaching staff should work to ensure safety of school learning environment (Bangay & Blum, 2010). Schools should be familiar with potential local hazards, undertake risk assessments and improve/ maintain buildings where appropriate (Stevenson et al. (2017). Given that children are particularly vulnerable to the effects of natural disasters, school curricula that incorporate learning about local disaster risks and how to be prepared for the same and cope when disaster strikes can enhance young people's resilience effectively. Community based and child-centered disaster risk reduction would increasingly become daily part of teacher's work. Stakeholders and supporters of education on disaster risk reduction argue that students themselves should also be involved with risk reduction processes and be familiar with what to do in the event of an emergency (UNESCO &UNEP, 2011). A closer consideration of the role of education in disaster, risk reduction, Stevenson et al. (2017), for instance, argued that for learning that moves beyond a knowledge of natural hazards to learning that involves students themselves in exploring their local surroundings, assessing risks and speaking with community members about historical extreme weather events has been considered healthy enough.

EFFECTIVENESS OF ENVIRONMENTAL EDUCATION PROVISION

Education stakeholders have a responsibility to create an enabling environment whereby learners can employ local knowledge and propose actions for climate change adaptation effectively. Interest in education about climate change, according to UNESCO (2009), has increased in recent years, attributable in part to expanded funding and leadership for educational programs that address climate change. Addition of climate change lessons to educational curriculum guidelines has increased awareness of unusual weather patterns and the deepening concern of the likelihood of global environmental, social, and economic changes due to climate change (Wedell, 2009).

Provision of education in more challenged environment has been an effort worth appreciating. In this regard, Ngugi (2016) pointed out that reaching the nomads with formal education has been a major challenge as indicated by efforts meant to hook them into school with interventions such as free education, school feeding programmes, introduction of boarding schools, as well as provision of uniforms. Similarly, equipping and provision of books and stationery to pupils have remained futile. Retaining them in schools is problematic and dropping out appears to be the norm. Those who do not drop out are however, pushed out by early marriages and migration among other factors. To improve and increase community engagement in education projects, leaders have to play a significant role by creating an environment that allow and welcomes community members to be part of the project in all stages of life (MoE, 2007).

In order to realize the desired environmental education change, Wedell (2009) observes that the policy makers should not ignore existing local practices and beliefs when deciding on the content and process of change; it immediately makes it very unlikely that the change will ever reach the institutionalized stage of becoming an accepted and normal aspect of most classrooms. Then how can policy makers become aware of people's reality, and ideally also enable representatives of those who will be affected to appreciate their own thinking about, and hopes for, change? The answer seems obvious: by talking to them and closely observing their reality.

The work of Nyatuka (2023) opined that environmental impacts on education change considerably. According to Republic of Kenya (ROK, 2010), inclusive learning, which refers to the enrolment and teaching of all the children in formal or non-formal learning environments without regard to gender, physical, intellectual, social, emotional, ethnic, cultural, religious or other considerations should be considered. Inclusion equally means a philosophy focusing on the process of adjusting the home, school and society so that all the individuals, regardless of their differences, may have the opportunity to interact, play, learn, work and experience the feeling of belonging and to develop in accordance with their potentials and difficulties. Additionally, research indicates that hereditary factors, experience, the environment and their personalities influence the way the children learn and the need to use a variety of teaching methods and activities to meet their respective learning needs (Yambo, 2022b; ROK, 2010:32).

TEACHER COMMITMENT AND CLIMATE CHANGE EDUCATION PROVISION

Studies by Park (2005) reiterated that teacher commitment to school organization like an organizational commitment refers to agreed-on organizational values or goals and building strong staff unity. To increase teachers' engagement in their work, school improvement strategies have stressed on creating innovative working conditions such as participation in decision-making and continuous learning and timely incentives. Ordinarily, teacher commitment to the teaching profession as an occupational commitment is a positive affective attachment to one's occupation (Somech & Bogler, 2002). In order to realize students' achievement, Park (2005) enumerates three teacher commitment dimensions which are necessary to accomplish school goals: to improve teacher professionalism, to foster client-oriented teaching and being mindful of learning environment. However, they can also be in conflict when they are pursued simultaneously. For example, teachers who are more committed to the profession are less likely to be committed to school (Somech & Bogler, 2002; Park, 2005). Moreover, commitment to students may lead teachers to invest their efforts mostly in building relationships with students, independently of the needs of the school organization as a whole.

In teaching profession, both students' background and entry behavior plus environment variables affect teacher commitment greatly. Students' academic levels influence teacher commitment to student learning. Heretofore and Dannetta (2002) assert that academically advanced students are more likely to challenge teacher's teaching practice and commitment to their achievement. The socioeconomic status of students is equally strongly associated with teacher's job commitment. Other variables, such as gender and ethnicity are correlated with teacher's attitude toward teaching and students achievement. More variables at play in teachers' commitment, according to Elliott and Crosswell (2002) are school background variables, like whether the school is private or public, urban or rural, size and socioeconomic status. In general, private school teachers are more committed than public school teachers because the former stress a normative work orientation more than the latter ones do (Yambo, 2012; Reyes, 1990).

Teachers who are working at the urban and low socioeconomic status schools are more likely to leave the school and even the teaching profession (Elliott & Crosswell, 2002). School size as a work group scale influences teacher commitment while larger schools or districts would have less committed teachers (Park, 2005). According to Garrison and Liston (2004), teacher commitment is expected to achieve school goals, improve teacher professionalism, and foster client-oriented instruction. In addition, teacher commitment is assumed as a key to solving the teacher turnover problem, which is not limited to a school or a district but it occurs even in the profession. These considerations provide a strong necessity to reinforce teacher commitment in a systemic way. Studies by Yambo (2022b) found that teacher's commitment has been seen as key to learners' academic achievement in schools.

The work of Mart (2013) holds that those who have the desire to teach; feeling that teaching is a profoundly meaningful part of their life, definitely have a passion for teaching (Garrison &Liston, 2004). Teacher commitment has dimensions hence Eliot and Croswell (2001:2) state that teacher commitment is multidimensional and some external factors may influence commitment degree, including the school or organization, students, career continuance, professional knowledge base and the teaching profession.

Clearly, one of the most important factors in the development of teacher commitment is the passion for teaching. According to Rowe (2003), passionate teachers are fiercely devoted to their work and greatly inspire their students. The strength of any profession depends upon the degree of commitment of its members. Teaching is no exception here. It is widely accepted that a distinguishing feature that teachers have is, their dedication and commitment to the development of student achievement. Comparatively, there is a relationship between teacher commitment and student achievement. Mart (2013) supported this idea by stating that there is a strong connection between passionate teaching and the quality of student learning. Teachers with passion work with enthusiasm, their dedication and commitment increase, and they believe in the importance of their job (Rowe, 2003). Additionally, Ayiro and Sang (2017) concluded that teachers' commitment and dedication is depended on administrative implementation. Compatible administrative implementation strengthens teachers' commitment in school and class. Committed teachers are critical in promoting climate change education.

FUTURE RESEARCH DIRECTIONS

Having considered education as a tool to combat climate change, it should be noted that as the degree of the impact of climate change varies across nations and regions, if one wants to anchor education activities locally to facilitate adaptation processes, place-based specific conditions can play a key role (Ajaps & Mbah, 2022). Most of the schools in marginalized areas are also in dire state of neglect. Chances for students coming from these schools to proceed to universities are slim. In these regions, teachers are also few due to insecurity and hardship (Okutu, 2019). Among others, more studies on the teachers' security, incentives and motivation should be carried out to bolster the efforts being made to mainstream climate change education in the curricula. This will go a long way in realizing the ESD goals.

CONCLUSION

If well implemented, CCE has a huge potential in addressing numerous environmental and sustainability challenges being experienced today, both nationally as well as internationally. Among others, it will enable people make informed decisions on climate adaptation strategies. In this chapter, relevant examples including mobile schools and the National Solar School Initiative have been presented. Both of them form major components of the ESD program that has a climate change mitigation as an element. Since education plays a pivotal role in this endeavor, more mobile schools and trained teachers should be rolled out in various regions, especially those with more nomadic communities. Governments should budget for more funds to cater for education in marginalized areas worldwide. Proactive climate education change policies should also be implemented in such regions as well.

REFERENCES

Aglazor, G. (2017). The role of Teaching Practice in Teacher Education Programmes: Designing Framework for Best Practice. *Global Journal of Educational Research, 16*(4) 101-110. doi:10.4314/gjedr.v16i2.4

Ajaps, S., & Mbah, M. F. (2022). Towards a critical pedagogy of place for environmental conservation. *Environmental Education Research, 28*(4), 508–523. doi:10.1080/13504622.2022.2050889

Association for the Development of Education in Africa. (2001). What works and what is plans in Education: Africa Speaks. ADEA.

Ayiro, L. P., & Sang, J. K. (2017). Provision of Education to the 'Hard to Reach' Amidst Discontinuity in Nomadic Communities in Kenya. *FIRE: Forum for International Research in Education, 3*(3). Advance online publication. doi:10.18275/fire201603031070

Bangay, C., & Blum, N. (2010). Education responses to climate change and quality: Two parts of the same agenda? *International Journal of Educational Development, 30*(4), 359–368. doi:10.1016/j.ijedudev.2009.11.011

BarrickR. K. (2011). *MEAS Project*. creativecommons.org/licenses/by/3.0/

Connell, J. P., & Klem, A. K. (2000). You can get there from here: Using a theory of change approach to plan urban education reform. *Journal of Educational & Psychological Consultation, 11*(0), 93–120. doi:10.12071532768Xjepc1101_06

Connolly, M., & Seymour, E. (2009). *Why theories of change matter.* http://mobilizingstem. wceruw.org/documents/Why%20Theories%20of%20Change%20Matter.pdf

Dannetta, V. (2002). What factors influence a teacher's commitment to student learning? *Leadership and Policy in Schools, 1*(2), 144–171. doi:10.1076/lpos.1.2.144.5398

Davies, M., & Devlin, M. (2007). *Interdisciplinary higher education: Implications for teaching and learning*. Centre for the Study of Higher Education. University of Melbourne.

Dianne, F. (2009). Sustainable development: The response from education - Australian country report. Melbourne Graduate School of Education.

Elliott, B., & Crosswell, L. (2002). *Teacher commitment and engagement: The dimensions of ideology and practice associated with teacher commitment and engagement within an Australian perspective*. Available: https://www.aare.edu.au/02pap/cro02522.htm

Farrell, K., Devlin, M., & James, R. (2017). *Nine Principles Guiding Teaching and Learning*. University of Melbourne Centre for the Study of Higher Education.

Fulbright-Anderson, A., Kubisch, A. C., & Connell, J. P. (1998). *New approaches to evaluating community initiatives: Theory, measurement and analysis*. Aspen Institute.

Fullan, M. (2006). Change theory: A force for school improvement. Seminar Series No. 157. Jolimont, Australia: Centre for Strategic Education.

Fullan, M. G. (2001). *The New Meaning of Educational Change* (3rd ed.). Cassell. doi:10.4324/9780203986561

Garrison, J., & Liston, D. (2004). *Teaching, learning, and loving*. Teachers College Press.

Kasim, K., & Behcet, O. (2015). Teachers' Perceptions on Culturally Responsiveness in Education. *Journal of Ethnic and Cultural Studies, 2*(2), 47-57.

Kerns, B., Elhouar, S., Sterling, M. J., Grant, J. M., McGowan, M., Rubash, A., Neelly, K., & Wolffe, R. (2005). Ten Principles of Effective Teaching and Practical Examples for the Classroom and Blackboard. Office of Teaching Excellence and Faculty Development, Bradley University.

Kim, J. W., & Kim, C. (2009). *National Report for the project 'Climate Change and Sustainable Development – The Response from Education*. College of Education, Seoul National University.

Koissaba, B. R. (2017). *Education For All: Prospects and Challenges of Mobile Schools, Mobile Education, and E-Learning for the Nomadic Pastoralists in Kenya* (2nd Ed. 2017). Research Gate. https://www.researchgate.net/publication/319630142

Kolenatý, M., Kroufek, R., & Činčera, J. (2022). What Triggers Climate ˇ Action: The Impact of a Climate Change Education Program on Students' Climate Literacy and Their Willingness to Act. *Sustainability, 14*(16), 10365. doi:10.3390u141610365

Landorf, H., Doscher, S., & Rocco, T. (2008). Education for sustainable human development: Towards a definition. *Theory and Research in Education, 6*(0), 221–236. doi:10.1177/1477878508091114

Lehnert, M., Fiedor, D., Frajer, J., Hercik, J., & Jurek, M. (2020). Czech Students and Mitigation of Global Warming: Beliefs and Willingness to Take Action. *Environmental Education Research, 2020*(26), 864–889. doi:10.1080/13504622.2019.1694140

Lubawy, W. C. (2003). Evaluating teaching using the best practices model. *American Journal of Pharmaceutical Education, 67*(3), 87. doi:10.5688/aj670387

Maksud, A. K., & Imtiaj, R. (2006). The Nomadic Bede Community and their Mobile School Program. What works for the Poorest: knowledge policies and Practices. Academic Press.

Mart, Ç. T. (2013). A Passionate Teacher: Teacher Commitment and Dedication to Student Learning. *International Journal of Academic Research in Progressive Education and Development, 2*(1), 437-444.

Mbah, M. F., Shingruf, A., & Molthan-Hill, P. (2022). Policies and practices of climate change education in South Asia: Towards a support framework for an impactful climate change adaptation. *Climate Action, 1*(1), 28. doi:10.100744168-022-00028-z

MoE. (2007). *Economic survey. Kenya National Bureau of Statistics*. Government Printer.

Muhammad, S., Mussawar, S., & Irfan, U. (2017). Addressing the Issue of Nomadic Communities Children Educational Exclusion through Mobile Tent Schools in Malakand, Pakistan. *Imperial Journal of Interdisciplinary Research, 3*(3). http://www.onlinejournal.in

Muir, T. (2008). Principles of Practice and Teacher Actions: Influences on Effective Teaching of Numeracy. *Mathematics Education Research Journal, 20*(3), 78–101. doi:10.1007/BF03217531

National Climate Change Policy. (2021). *Government of Pakistan, Ministry of Climate Change*. Retrieved on 14 April 2023 from https://mocc.gov.pk/SiteImage/Policy/NCCP%20Report.pdf

Ngugi, M. (2016). Challenges Facing Mobile Schools among Nomadic Pastoralists: A Case Study of Turkana County, Kenya. *American Journal of Educational Research, 4*(1), 22–32. doi:10.12691/education-4-1-6

Nyatuka, B. O. (2023). Partnering to Bolster Student Achievement: A Case of the Child Friendly School Concept. In J. Keengwe (Ed.), *Handbook of Research on Race, Culture, and Student Achievement* (pp. 376–394). IGI Global. doi:10.4018/978-1-6684-5705-4.ch019

Okutu, A. A. (2019). Nomadism and Provision of Basic Education in West Pokot County, Kenya. *Journal of African Interdisciplinary Studies, 3*(6), 114–125.

Park, I. (2005). Teacher Commitment and its Effects on Student Achievement in American High Schools. *Educational Research and Evaluation, 11*(5), 461–485. doi:10.1080/13803610500146269

Republic of Kenya. (2010). *Child friendly School Manual*. Ministry of Education.

Republic of Kenya. (2012). *Sessional Paper No. 14 of 2012: Reforming education and training sectors in Kenya*. Ministry of Education and Ministry of Higher Education, Science and Technology.

Rowe, K. (2003). *The Importance of Teacher Quality as a Key Determinant of Students' Experiences and outcomes of Schooling*. Background paper to keynote address presented at the ACER Research Conference 2003, 19-21 October 2003.

Somech, A., & Bogler, R. (2002). Antecedents and consequences of teacher organizational and professional commitment. *Educational Administration Quarterly*, *38*(4), 555–577. doi:10.1177/001316102237672

Stein, D., & Valters, C. (2012). *Understanding theory of change in international development.* The Justice and Security Research Program. https://www.theoryofchange.org/wp-content/ uploads/toco_library/pdf/ UNDERSTANDINGTHEORYOFChangeSte inValtersPN.pdf

Stevenson, K. T., Nils-Peterson, M., & Bondell, H. D. (2018). Developing a Model of Climate Change Behavior among Adolescents. *Climatic Change*, *151*(3-4), 589–603. doi:10.100710584-018-2313-0

Stevenson, R. B., Nicholls, J., & Whitehouse, H. (2017). What Is Climate Change Education? *Curriculum Perspectives*, *37*(1), 67–71. Advance online publication. doi:10.100741297-017-0015-9

UNESCO. (2005). *Education for All: The Leap to Equality.* EFA Global Monitoring Report'03. portal.unesco.org/education/en/ev.phpurl_id=23023&url_do=do_topic&url_section=201.html

UNESCO. (2009). *Climate Change Education Inside and Outside the Classroom.* UNESCO.

UNESCO &UNEP. (2011). *Climate change starters guide book.* UNESCO.

UNICEF. (2008). *United Nations International Children Emergency Fund Curriculum report card.* Working Paper Series, Education Section, Programme Division.

UNICEF. (2011). *UNICEF'S Child friendly schools Kenya case study.* Nairobi: Management Project Office of CFS. Available at: http://cdpen. cbern.cn/cdp/tjwz /297.htm

USAID (2013). *Alternative Education: Mobile schools for Nomadic Pastoralists in North Eastern province.* Available from http//Kenya.Usaid.gov/programs/education. and – Yo

Wedell, M. (2009). *Planning for Educational Change: Putting people and their contexts first.* Continuum International Publishing Group.

World Bank. (2010). *Knowledge, Productivity and Innovation in Nigeria: Creating a New Economy.* The World Bank.

World Declaration on Education for All. (1990). *Education for all and Program of Action.* Common Wealth.

Yambo, J. M. O. (2012). *Determinants of KCSE Examination Performance in SDA Sponsored Schools: A Key to Academic Promotion to the next Level of Learning*. Lambert Academic Publishing.

Yambo, J. M. O. (2022a). Involvement of Government and Non-Governmental Organizations' Teamwork in the Advancement of Quality Education in Kisumu West Sub County, Kenya. *Journal of Advances in Education and Philosophy*, *6*(4), 194–202. doi:10.36348/jaep.2022.v06i04.001

Yambo, J. M. O. (2022b). Influence of the Principals Decision Making Skills on Students' Academic outcome in Public Teachers Training Colleges in Nyanza Region, Kenya. *Journal of Research Innovation and Implications in Education*, *6*(3), 136–144.

Zimbabwe Human Development Report. (2017). *Human Development: Towards Building a Climate Resilient Nation Climate Change and Education*. The Embassy of Sweden in Harare Zimbabwe.

KEY TERMS AND DEFINITIONS

Climate Change: Long term alteration in global weather patterns, especially increase in temperature and storm activity.

Climate Change Education: During the shift in global weather patterns, which brings about increase in temperature and stormy activities, learning still has to take place hence education and training committed to promoting learning on environmental issues (including climate change) has to be taught.

Community: This refers to such groups as businesses or corporations, service agencies, sports clubs as well as cultural and civic organizations that have interest in the provision of sound education to learners. It also includes the municipality as well as philanthropists.

Dynamic: Vigorous activity, energy and a sense of purpose to get things done like teaching learners despite hard climatic conditions.

Nomadic Pastoralist: Denotes members of ethnic groups that move with animals and children from place to place in search of water and pasture for their livestock.

Passion: Simply defined as a strong inclination or desire towards an activity that one likes and finds important.

Teaching and Learning Facilities: Means all materials designed, built, or installed in support of teaching and learning process.

Compilation of References

Abe, Y. (2018). Logistics Emissions and Energy Consumption Models. In A. Nagurney (Ed.), *Sustainable Supply Chains* (pp. 121–146). Springer.

Abtahi, Y., Gøtze, P., Steffensen, L., Hauge, H. K., & Barwell, R. (2017). *Teaching climate change in mathematics classrooms: An ethical responsibility.* https://www.researchgate.net/publication/343385757

Abukan, B., & Yıldırım, F. (2019). Kırsal alanda toplum kalkınması için sosyal hizmette kültürel yetkinliğin önemi [The importance of cultural competence in social work for community development in rural area]. *Journal Of History School (Johs), 43*, 1666–1681. doi:10.29228/joh.24748

Adak, N. (2010). *Kırsal kadın ve çevre* [Paper presentation]. Türkiye IX. Tarım Ekonomisi Kongresi, Şanlıurfa, Turkey.

Adams, W. M., & Hutton, J. (2007). People, parks, and poverty: Political ecology and biodiversity conservation. *Conservation & Society*, *5*(2), 147–183.

ADB. (2021). *Climate Risk Country Profile: Nepal.* Manila: Asian Development Bank. Retrieved from https://www.adb.org/sites/default/files/publication/677231/climate-risk-country-profile-nepal.pdf

Adger, W. N., Agrawala, S., Mirza, M. M. Q., Conde, C., O'Brien, K., Pulhin, J., Pulwarty, R., Smit, B., & Takahashi, K. (2007). Assessment of adaptation practices, options, constraints, and capacity. In M. L. Parry, O. F. Canziani, J. P. Palutikof, P. J. van der Linden, & C. E. Hanson (Eds.), *Climate Change 2007: Impacts, Adaptation, and Vulnerability. Contribution of Working Group II to the Fourth Assessment Report of the Intergovernmental Panel on Climate Change* (pp. 717–743). Cambridge University Press.

Afriyie, K., Ganle, J.K., & Santos, E. (2018). The floods came and we lost everything': Weather extremes and households' asset vulnerability and adaptation in rural Ghana. *Climate and Development, 10*(3), 259–274. doi:10.1080/17565529.2017.1291403

Afroz, R., & Rahman, A. (2017). Health impact of river water pollution in Malaysia. *International Journal of Advanced and Applied Sciences*, *4*(5), 78–85. doi:10.21833/ijaas.2017.05.014

Agarwal, B. (2009). Gender and forest conservation: The impact of women's participation in community forest governance. *Ecological Economics*, *69*(11), 2785–2799. doi:10.1016/j.ecolecon.2009.04.025

Aglazor, G. (2017). The role of Teaching Practice in Teacher Education Programmes: Designing Framework for Best Practice. *Global Journal of Educational Research, 16*(4) 101-110. doi:10.4314/gjedr.v16i2.4

Agrawal, A., & Gibson, C. C. (1999). Enchantment and disenchantment: The role of community in natural resource conservation. *World Development*, *27*(4), 629–649. doi:10.1016/S0305-750X(98)00161-2

Agyeman, J., Bullard, R. D., & Evans, B. (Eds.). (2003). *Just Sustainability's: Development in an Unequal World.* MIT Press.

Ajaps, S., & Mbah, M. F. (2022). Towards a critical pedagogy of place for environ- mental conservation. *Environmental Education Research*, *28*(4), 508–523. doi:10.1080/13504622.2022.2050889

Akinbami, C. A. O. (2021). Climatepreneurship: Adaptation strategy for climate change impacts on rural women entrepreneurship development in Nigeria. African Handbook of Climate Change Adaptation, 1-27. doi:10.1007/978-3-030-42091-8_191-1

Akinbami, C. A. O., Olawoye, J. E., Adesina, F. A., & Nelson, V. (2019). Exploring potential climate-related entrepreneurship opportunities and challenges for rural Nigerian women. *Journal of Global Entrepreneurship Research*, *9*(1), 1–28. doi:10.118640497-018-0141-3

Alberta Council for Environmental Education. (2017). *What is excellent climate change education?* Alberta Council for Environmental Education.

Alcázar, M., & Chavez, J. R. (2014). Educational Program for Promoting the Application of Bioclimatic and Sustainable Architecture in Elementary Schools. *Energy Procedia*, *57*, 999–1004. Advance online publication. doi:10.1016/j.egypro.2014.10.083

Alexander, L. L., La Rosa, J. H., Bader, H., & Garfield, S. (2020). *New dimensions in women's health.* Jones & Bartlett Publishers.

Ali, S., & Cheema, A. R. (2018). *Exploring the many barriers to a girl's education in Sindh.* LSE South Asia Center. Retrieved from http://blogs.lse.ac.uk/ southasia/2018/05/16/exploring-the-many-barriers-to-a-girls-education-in-sindhpakistan/

Alston, M. (2014). Gender mainstreaming and climate change. *Women's Studies International Forum, 47*(B), 287-294.

Alston, M. (2015). Women and climate change in Bangladesh. Routledge.

Althor, G., Watson, J. E., & Fuller, R. A. (2016). Global mismatch between greenhouse gas emissions and the burden of climate change. *Scientific Reports*, *6*(1), 20281. Advance online publication. doi:10.1038rep20281 PMID:26848052

Altomonte, S. (2009). Climate Change and Architecture: Mitigation and Adaptation Strategies for a Sustainable Development. *Journal of Sustainable Development, 1*(1), 1. Advance online publication. doi:10.5539/jsd.v1n1p97

Amaeshi, K., & Amao, O. O. (2012). Corporate Accountability and the Politics of CSR in a Developing Country. *Journal of Business Ethics, 117*(1), 173–190.

AMCEN. (2012). *The Africa Ministerial Conference on Environment (AMCEN)*. Retrieved from https://www.unep.org/roa/amcen/

AMCEN. (2017). *The Africa Ministerial Conference on Environment* (AMCEN). Retrieved from https://www.unep.org/roa/amcen/

Amer, N. (2019). Biomimetic Approach in Architectural Education: Case study of 'Biomimicry in Architecture' Course. *Ain Shams Engineering Journal, 10*(3), 3. Advance online publication. doi:10.1016/j.asej.2018.11.005

Anderson, A. (2012). Climate Change Education for Mitigation and Adaptation. *Journal of Education for Sustainable Development, 6*(2), 191–206. doi:10.1177/0973408212475199

Anderson, K., Stahl, C., & Utterwulghe, A. (2017). Resource Extraction and Indigenous Livelihoods: A Case Study in Papua New Guinea. *Ecological Economics, 138*, 66–73.

Anyanwu, J., & Njoku, C. (2023). Integrating Climate Change Education into Social Studies Curriculum at The Basic Education Level. *International Journal of Education, Learning and Development, 11*(2), 33–40. Advance online publication. doi:10.37745/ijeld.2013/vol11n23340

Apollo, A., & Mbah, M. F. (2021). Challenges and Opportunities for Climate Change Education (CCE) in East Africa: A Critical Review. *Climate (Basel), 9*(6), 93. doi:10.3390/cli9060093

Ardoin, N. M., Falk, J., Heimlich, J. E., & Foutz, S. (2009). Behavior-change theories and free-choice environmental learning. In J. H. Falk (Ed.), *Free-Choice Learning and the Environment* (pp. 57–76). AltaMira Press.

Ar, H., Kadim, F., & Gülçubuk, B. (2012). *İklim değişikliği ve yoksulluğun en çok etkilenenleri: Kırsal alandaki kadınlar* [Paper presentation]. Türkiye X. Tarım Ekonomisi Kongresi.

Arias, F. (2012). *El proyecto de investigacion*. Episteme.

Aristotle. (350 BCE). *Nicomachean ethics* (W. D. Ross, Trans.). Oxford University Press.

Armstrong, A. K., & Krasny, M. E. (2020). Tracing Paths from Research to Practice in Climate Change Education. *Sustainability, 12*(4779), 1–21. doi:10.3390u12114779

Arnold, R. D., & Wade, J. P. (2015). A definition of systems thinking: A systems approach. *Procedia Computer Science, 44*, 669–678. doi:10.1016/j.procs.2015.03.050

Aryal, J. P., Sapkota, T. B., Rahut, D. B., Krupnik, T. J., Shahrin, S., Jat, M. L., & Stirling, C. M. (2020). Major climate risks and adaptation strategies of smallholder farmers in coastal Bangladesh. *Environmental Management, 66*(1), 105–120. doi:10.100700267-020-01291-8 PMID:32388655

Ashrafuzzaman, M. (2023). Local Context of Climate Change Adaptation in the South-Western Coastal Region of Bangladesh. *Sustainability, 15*(8), 6664. doi:10.3390/su15086664

Association for the Development of Education in Africa. (2001). What works and what is plans in Education: Africa Speaks. ADEA.

Aubrecht, K. B. (2018). Teaching relevant climate change topics in undergraduate chemistry courses: Motivations, student misconceptions, and resources. *Current Opinion in Green and Sustainable Chemistry, 13*, 44–49. doi:10.1016/j.cogsc.2018.03.008

Ayiro, L. P., & Sang, J. K. (2017). Provision of Education to the 'Hard to Reach' Amidst Discontinuity in Nomadic Communities in Kenya. *FIRE: Forum for International Research in Education, 3*(3). Advance online publication. doi:10.18275/fire201603031070

Aykut, S. C., Comby, J.-B., & Guillemot, H. (2012). Climate Change Controversies in French Mass Media 1990-2010. *Journalism Studies, 13*(2), 154–174. doi:10.1080/1461670X.2011.646395

Balzacq, T. (2005). The Three Faces of Securitization: Political Agency, Audience and Context. *European Journal of International Relations, 11*(2), 171-201. https://journals.sagepub.com/doi/10.1177/1354066105052960

Banerjee, A. V., & Bell, C. R. (2007). Why Does Poverty Persist? *The American Economic Review, 97*(5), 1742–1759.

Bangay, C., & Blum, N. (2010). Education responses to climate change and quality: Two parts of the same agenda? *International Journal of Educational Development, 30*(4), 359–368. doi:10.1016/j.ijedudev.2009.11.011

Bansal, P., & Song, H. C. (2017). Similar but not the same: Differentiating corporate sustainability from corporate responsibility. *The Academy of Management Annals, 11*(1), 105–149. doi:10.5465/annals.2015.0095

Barrett, H. R. (1997). Women, occupation and health in rural Africa: Adaptation to a changing socioeconomic climate. *Journal of Occupational Science, 4*(3), 93–105. doi:10.1080/14427591.1997.9686425

BarrickR. K. (2011). *MEAS Project*. creativecommons.org/licenses/by/3.0/

Barry, C. (2012). *The Ethics of Educational Leadership*. Routledge.

Bartelmus, P. (1999). *Environment, Growth and Development: The Concepts and Strategies of Sustainability*. Routledge.

Barwell, R. (2013). The mathematical formatting of climate change: Critical mathematics education and post-normal science. *Research in Mathematics Education, 15*(1), 1–16. doi:10.1080/14794802.2012.756633

Bashir, F. M., Ahmad, M. H., & Hamid, M. (2013). Design Studio as Problem Based Learning in Architectural Education in Universiti Teknologi Malaysia. *4th International Research Symposium on Problem-Based Learning (IRSPBL) 2013*, 373–379.

Bebbington, A., Bury, J., & Warnaars, X. (2019). Hidden voices: From adaptation as a neoliberal burden to the collective potential of social practice theories. *Environmental Science & Policy, 100*, 29–36.

Beiswenger, H. (2020). *Proyecto de saneamiento de las fuentes hídricas del río Portoviejo. Informe final. Municipio de Portoviejo*. Portoviejo - Ecuador.

Bell, S., Tyrväinen, L., Sievänen, T., & Pröbstl-Haider, U. (2018). Connecting the social and the ecological in European peri-urban spaces. *Landscape and Urban Planning, 170*, 300–308.

Berger, P., Gerum, N., & Moon, M. (2015). "Role up your sleeves and get at it": Climate Change Education in Teacher Education. *Canadian Journal of Environmental Education, 20*, 152–174.

Berry, H. L., Waite, T. D., Dear, K. B., Capon, A. G., & Murray, V. (2018). The case for systems thinking about climate change and mental health. *Nature Climate Change, 8*(4), 282–290. doi:10.103841558-018-0102-4

Berry, P., Clarke, K.-L., Fleury, M. D., & Parker, S. (2014). Human health. In F. J. Warren & D. S. Lemmen (Eds.), *Canada in changing climate: Sector perspectives on impacts and adaptation* (pp. 191–232). Government of Canada.

Besnili Memiş, O. (2019). Kadınların iklim değişikliği ile ilgili algılarının belirlenmesi [Determination of women's perceptions related to climate change]. *Journal of Academic Value Studies, 5*(4), 700–718. doi:10.23929/javs.948

Bhadwal, S., Sharma, G., Gorti, G., & Sen, S. M. (2019). Livelihoods, gender and climate change in the Eastern Himalayas. *Environmental Development, 31*, 68–77. doi:10.1016/j.envdev.2019.04.008

Bhattarai, S. (2021, January 29). *Underreporting the climate emergency*. Retrieved from Nepali Times: https://www.nepalitimes.com/latest/underreporting-the-climate-emergency/

Biermann, F., & Boas, I. (2010). Preparing for a warmer world: Towards a global governance system to protect climate refugees. *Global Environmental Politics, 10*(1), 60–88. doi:10.1162/glep.2010.10.1.60

Biermann, F., Pattberg, P., van Asselt, H., & Zelli, F. (2009). The Fragmentation of Global Governance Architectures: A Framework for Analysis. *Global Environmental Politics, 9*(4), 14–40. doi:10.1162/glep.2009.9.4.14

Biesbroek, R., Ford, L., Ford, J., Tanabe, A., Austin, S., & Lesnikowski, A. (2018, November). Data, concepts and methods for large-n comparative climate change adaptation policy research: A systematic literature review. *Wiley Interdisciplinary Reviews: Climate Change, John Wiley & Sons, 9*(6). Advance online publication. doi:10.1002/wcc.548

Biesta, G. J. J., & Tedder, M. (2007). Agency and learning in the lifecourse: Towards an ecological perspective. *Studies in the Education of Adults, 39*(2), 132–149. doi:.2007.11661545 doi:10.1080/02660830

Biggs, J., & Tang, C. (2009). *Teaching for Quality Learning at University* (3rd ed.). Society for Research into Higher Education/Open University Press.

Biseko, J. M., & Jae, W. L. (2022). Relationships between transmission of malaria in Africa and climate factors. *Scientific Reports, 12*(1), 14392. Advance online publication. doi:10.103841598-022-18782-9 PMID:35999450

Blumler, J. G., & Katz, E. (1974). The Uses of Mass Communications: Current Perspectives on Gratifications Research. *American Journal of Sociology*, 3.

Blum, N., Nazir, J., Breiting, S., Chuan Goh, K., & Pedretti, E. (2013). Balancing the Tensions and Meeting the Conceptual Challenges of Education for Sustainable Development and Climate Change. *Environmental Education Research, 19*(2), 206–217. doi:10.1080/13504622.2013.780588

Boakye, C. (2015). Climate Change Education: The Role of Pre-Tertiary Science Curricula in Ghana. *SAGE Open, 5*(4), 1–10. doi:10.1177/2158244015614611

Boon, H. J. (2016). Pre-Service teachers and climate change: A Stalemate? *The Australian Journal of Teacher Education, 41*(4), 39–63. Advance online publication. doi:10.14221/ajte.2016v41n4.3

Bordeleau, G., Savard, M. M., Martel, R., Ampleman, G., & Thiboutot, S. (2008). Determination of the origin of groundwater nitrate at an air weapons range using the dual isotope approach. *Journal of Contaminant Hydrology, 98*(3), 97–105. doi:10.1016/j.jconhyd.2008.03.004 PMID:18499297

Bord, R. J., O'Connor, R. E., & Fisher, A. (2000). In what sense does the public need to understand global climate change? *Public Understanding of Science (Bristol, England), 9*(3), 205–218. doi:10.1088/0963-6625/9/3/301

Borg, C., Gericke, N., Höglund, H. O., & Bergman, E. (2013). Subject- and experience-bound differences in teachers' conceptual understanding of sustainable development. *Environmental Education Research, 20*(4), 526–551. doi:10.1080/13504622.2013.833584

Borucka, J., & Macikowski, B. (2017). Teaching Architecture – Contemporary Challenges and Threats in the Complexity of Built Environment. *IOP Conference Series. Materials Science and Engineering, 245*(8), 082058. doi:10.1088/1757-899X/245/8/082058

Bostrom, A., Morgan, M. G., Fischhoff, B., & Read, D.Mental Models. (1994). What do people know about global climate change? *Mental models. Risk Analysis, 14*(6), 959–970. doi:10.1111/j.1539-6924.1994.tb00065.x

Bourque, F., & Cunsolo Willox, A. (2014). Climate change: The next challenge for public mental health? *International Review of Psychiatry (Abingdon, England), 26*(4), 415–422. doi:10.3109/09540261.2014.925851 PMID:25137107

Boykoff, M. T., & Boykoff, J. M. (2004). Balance as bias: Global warming and the US prestige press. *Global Environmental Change, 14*(2), 125–136. doi:10.1016/j.gloenvcha.2003.10.001

Boykoff, M. T., & Roberts, J. T. (2007). *Media Coverage of Climate Change: Current Trends, Strengths, Weaknesses*. UNDP.

Boykoff, M., & Smith, J. (2010). Media representations of climate change. In *Handbook of Climate Change and Society* (pp. 210–218). Routledge.

Boyko, M., Desak, T., Fleming, C., Diplock, K., & Pons, W. (2020). Climate change education: The need for comprehensive climate change education in environmental public health curriculum. *Environmental Health Review, 63*(4), 114–120. doi:10.5864/d2020-029

Braithwaite, J. (2017). *Economic Analysis and the Design of Policy: Lessons from the New Institutional Economics*. Routledge.

Brechin, S. R., Wilshusen, P. R., Fortwangler, C. L., & West, P. C. (2003). Beyond the square wheel: Toward a more comprehensive understanding of biodiversity conservation as social and political process. *Society & Natural Resources, 16*(10), 905–920.

Brody, A., Demetriades, J., & Esplen, E. (2008). Gender and climate change: Mapping the linkages- A scoping study on knowledge and gaps. Institute of Development Studies, 1-27.

Brooks, C. (2006). Geographical knowledge and teaching geography. *International Research in Geographical and Environmental Education, 15*(4), 353369. doi:10.2167/irg200.0

Brossard, D., Shanahan, J., & McComas, K. (2004). Are Issue-Cycles Culturally Constructed? A Comparison of Global Climate Change. *Mass Communication & Society, 7*(3), 359–377. doi:10.120715327825mcs0703_6

Brown, A. (2018). *Sustainability and Business Ethics*. Routledge.

Brown, K., & Kasser, T. (2005). Are Psychological and Ecological Well-being Compatible? The Role of Values, Mindfulness, and Lifestyle. *Social Indicators Research, 74*(2), 349–368. doi:10.100711205-004-8207-8

Brownlee, M. T. J., Robert, B. P., & Jeffery, C. H. (2013). A review of the foundational processes that influence beliefs in climate change: Opportunities for environmental education research. *Environmental Education Research, 19*(1), 1–20. doi:10.1080/13504622.2012.683389

Brown, S. (2005). Assessment for learning. *Learning and Teaching in Higher Education, 1*, 81–89.

Brüggemann, M. (2017). Post-normal Journalism: Climate Journalism and Its Changing Contribution to an Unsustainable Debate. In P. Berglez, U. Olausson, & M. Ots (Eds.), *What is Sustainable Journalism? Integrating the Environmental, Social, and Economic Challenges of Journalism* (pp. 57–73). Peter Lang.

Buckingham, S. (2004). Ecofeminism in the twenty-first century. *The Geographical Journal, 170*(2), 146–154. doi:10.1111/j.0016-7398.2004.00116.x

Burbano, N., Becerra, S., & Pasquel, E. (2011). *Caracterización hidrogeolóica de las Cuencas Portoviejo - Chone.* Retrieved from INAMHI: https://issuu.com/inamhi/docs/caract._hidrogeologica-manabi/7

Burbano, S., Becerra, P., Efrén, S., & Pérez, L. (2015). Introducción a la Hidrología del Ecuador. Academic Press.

Burns, H., Diamond-Vaught, H., & Bauman, C. (2015). Leadership for sustainability: Theoretical foundations and pedagogical practices that foster change. *International Journal of Leadership Studies*, *9*(1), 88–100.

Büscher, B., & Davidov, V. (2013). States of exception: Conservation, militarization, and 'democracy' in the forested borderlands of Poland and Belarus. *Conservation & Society*, *11*(1), 31–43.

Busch, K. C., Henderson, J. A., & Stevenson, K. T. (2019). Broadening epistemologies and methodologies in climate change education research. *Environmental Education Research*, *25*(6), 955–971. doi:10.1080/13504622.2018.1514588

Callicott, J. B. (1989). Intrinsic Value, Quantum Theory, and Environmental Ethics. *Environmental Ethics*, *11*(1), 53–82.

Camacho, L., & Astudillo, F. (2021). Evaluación del aprovechamiento empírico del agua subterránea somera del cantón Morona-Ecuador. *Rev. Dominio las Ciencias*, *6*(2), 460–473. https://dominiodelasciencias.com/ojs/index.php/es/article/view/1179/html

Camagni, R., & Capello, R. (2013). Regional competitiveness and territorial capital: A conceptual approach and empirical evidence from the European Union. *Regional Studies*, *37*(9), 891–904. doi:10.1080/00343404.2012.681640

Cameron, C., & Norrington-Davies, G. (2010). *Children and disaster risk reduction in Asia and the Pacific: A way forward.* Discussion Paper presented at the High-Level Meeting on Cooperation for Child Rights in the Asia-Pacific Region. http:// www.unicef.org/eapro/Children_and_Disaster_Risk_Reduction.pdf

Campbell-Lendrum, D., Wheeler, N., Maiero, M., Villalobos Prats, E., & Nevelle, T. (2018). *World Health Organization COP24 special report on health and climate change.* WHO. https://www.who.int/publications/i/item/cop24-special-report-health-climate-change

Cankardaş, S., & Sofuoğlu, Z. (2021). İklim değişikliği ve birey üzerindeki etkilerinin gözden geçirilmesi. *Nesne Psikoloji Dergisi*, *9*(19), 139–146. doi:10.7816/nesne-09-19-11

Cannon, T. (2002). Gender and climate hazards in Bangladesh. Climate Change and Gender Justice. *Gender and Development*, *10*(2), 45–50. doi:10.1080/13552070215906

Cantell, H., Tolppanen, S., Aarnio-Linnanvuori, E., & Lehtonen, A. (2019). Bicycle model on climate change education: Presenting and evaluating a model. *Environmental Education Research*, *25*(5), 717–731. doi:10.1080/13504622.2019.1570487

Cantor, A., Delauer, V., Martina, D., & Rogan, J. (2015). Training interdisciplinary "wickedproblem" solvers: Applying lessons from HERO in community-based research experiences from undergraduates. *Journal of Geography in Higher Education, 39*(3), 407–419. doi:10.1080/03098265.2015.1048508

Carey, R. O., Migliaccio, K. W., & Brown, M. T. (2011). Nutrient discharges to Biscayne Bay, Florida: Trends, loads, and a pollutant index. *The Science of the Total Environment, 409*(3), 530–539. doi:10.1016/j.scitotenv.2010.10.029 PMID:21111457

Carleton, T. A., & Hsiang, S. M. (2016). Social and economic impacts of climate. *Science, 353*(6304), aad9837. doi:10.1126cience.aad9837 PMID:27609899

Carlson, A. (2000). *Aesthetics and the Environment: The Appreciation of Nature, Art, and Architecture*. Routledge.

Carter, C. R., & Rogers, D. S. (2008). A Framework of Sustainable Supply Chain Management: Moving Toward New Theory. *International Journal of Physical Distribution & Logistics Management, 38*(5), 360–387. doi:10.1108/09600030810882816

Carvalho, A., & Burgess, J. (2005). Cultural Circuits of Climate Change in U.K. Broadsheet Newspapers, 1985–2003. *Risk Analysis, 25*(6), 1457–1469. doi:10.1111/j.1539-6924.2005.00692.x PMID:16506975

Celadyn, W. (2018). Postgraduate studies and sustainable architecture. *Global Journal of Engineering Education, 20*(1), 54–58. https://www.scopus.com/inward/record.uri?eid=2-s2.0-85041920632&partnerID=40&md5=173be32b0608a8b3a41039d80aef3daf

Celadyn, W. (2019). Controversy over the visions of the architectural profession. *World Transactions on Engineering and Technology Education, 17*, 71–75.

Centre for Science and Environment. (1985). The state of India's environment 1984-1985: The second citizens' report. Author.

Chambers, R. (1983). *Rural Development: Putting the Last First*. Longman.

Chang, C., & Pascua, L. (2017). The state of climate change education – reflections from a selection of studies around the world. *International Research in Geographical and Environmental Education*.

Chang, C. H. (2015). Teaching Climate Change – A Fad or a Necessity? *International Research in Geographical and Environmental Education, 24*(3), 181–183. doi:10.1080/10382046.2015.1043763

Chang, C. H., & Pascua, L. (2016). Singapore students' misconceptions of climate change. *International Research in Geographical and Environmental Education, 25*(1), 84–96. doi:10.1080/10382046.2015.1106206

Chang, C., & Liberty, P. L. (2017). The state of climate change education – reflections from a selection of studies around the world. *International Research in Geographical and Environmental Education, 26*(3), 177–179. doi:10.1080/10382046.2017.1331569

Chemhuru, M. (2019). Interpreting ecofeminist environmentalism in African communitarian philosophy and Ubuntu: An alternative to anthropocentrism. *Philosophical Papers, 48*(2), 241-264. . doi:10.1080/05568641.2018.1450643

Chetty, K., Devadas, V., & Fleming, J. (2015). The framing of climate change in New Zealand newspapers from June 2009 to June 2010. *Journal of the Royal Society of New Zealand, 45*(1), 1-20. https://www.tandfonline.com/doi/full/10.1080/03036758.2014.996234

Cheung, Y. T.-Y., Chow, C.-F., & So, W. W.-M. (2018). A train-the-trainer design for green ambassadors in an environmental education programme on plastic waste recycling. *International Research in Geographical and Environmental Education, 27*(1), 24–42. doi:10.1080/10382046.2017.1285138

Chibvongodze, D. T. (2016). Ubuntu is not only about the human! An analysis of the role of African philosophy and ethics in environment management. *Journal of Human Ecology (Delhi, India), 53*(2), 157–166. doi:10.1080/09709274.2016.11906968

Chigangaidze, R. K. (2021). An exposition of humanistic- existential social work in light of ubuntu philosophy: Towards theorizing ubuntu in social work practice. *Journal of Religion & Spirituality in Social Work, 40*(2), 146–165. doi:10.1080/15426432.2020.1859431

Chigangaidze, R. K. (2022). The environment has rights: Eco-spiritual social work through ubuntu philosophy and Pachamama: A commentary. *International Social Work.* Advance online publication. doi:10.1177/00208728211073382

Chigwanda, E. (2016). A framework for building resilience to climate change through girls' education programming. Brookings.

Christian Aid. (2007). *Human tide: The real migration crisis.* Christian Aid Report.

Christians, C. G., Glasser, T. L., McQuail, D., Nordenstreng, K., & White, R. A. (2009). *Normative Theories of the Media.* University of Illinois Press.

Chua, Y. T. (2019, June 25). *Communication.* Retrieved from Oxford Research Encyclopedias: https://oxfordre.com/communication/view/10.1093/acrefore/9780190228613.001.0001/acrefore-9780190228613-e-799

Ciano, T., Ferrara, M., Gangemi, M., Merenda, D. S., & Pansera, B. A. (2021). Climate Change Sustainability: From Bargaining to Cooperative Balanced Approach. *Games, 12*(2), 45. doi:10.3390/g12020045

Climate Change 2014 Synthesis Report Commission for University Education. (2020, June 2). *Incorporation of Climate Change Education in University Curricula* [Press release]. Retrieved from https://www.cue.or.ke/

Coêlho, A. E., Adair, J. G., & Mocellin, J. S. (2004). Psychological responses to drought in northeastern Brazil. *Revista Interamericana de Psicologia. Interamerican Journal of Psychology, 38*(1).

Colucci-Gray, L. (2013). The inclusion of environmental education in science teacher education. *Educational Review, 65*(2), 246-247. https://doi.org.ezproxy.library.und.edu/10.1080/0013191 1.2011.634173

Connell, J. P., & Klem, A. K. (2000). You can get there from here: Using a theory of change approach to plan urban education reform. *Journal of Educational & Psychological Consultation, 11*(0), 93–120. doi:10.12071532768Xjepc1101_06

Connolly, M., & Seymour, E. (2009). *Why theories of change matter.* http://mobilizingstem. wceruw.org/documents/Why%20Theories%20of%20Change%20Matter.pdf

Constitución de la República del Ecuador. (2008). *Constitucion de la Republica del Ecuador.* Montecristi.

Cooke, B., & Kothari, U. (2001). *Participation: The New Tyranny?* Zed Books.

Correa, L., López, L., & Vergara, M. (2020). Investigaciones de representaciones sociales sobre fuentes hídricas y condiciones para la preservación hídrica (2013-2019). *Revista Espacios, 41*(29), 329–344.

Cortés, I., Cortés, M., & Cisneros, H. (2015). Recursos hídricos. *Rev. Cultura científica y tecnológica, 31*(6). Retrieved from http://erevistas.uacj.mx/ojs/index.php/culcyt/article/view/344

Costa, A., Teixeira, C., Do, N., Oliveira, M., Oliveira, Q., & Silva, J. (2012). Recursos hídricos. Cuadernos de Ciencias exactas y tecnología. *UNIT-SERGIPE, 1*(1), 67-73. Retrieved from https:// periodicos.set.edu.br/cadernoexatas/article/view/201/126

Costello, A., Abbas, M., Allen, A., Ball, S., Bellamy, R., Friel, S., Groce, N., Johnson, A., Kett, M., Lee, M., Levy, C., Maslin, M., McCoy, D., McGuire, B., Montgomery, H., Napier, D., Pagel, C., Patel, J., ... Patterson, C. (2009). Managing the health effects of climate change. *Lancet, 373*(9676), 1693–1733. doi:10.1016/S0140-6736(09)60935-1 PMID:19447250

Council Conclusion on Policy Coherence for Development. (2009). In *2974th External relations council meeting.* Council of European Union.

Cowan, S., & Van Der Ryn, S. (1994). *Ecological Design.* Island Pres.

Crim, M., Moseley, C., & Desjean-Perrotta, B. (2017). Strategies toward the inclusion of environmental education in educator reparation programs: Results from a national survey: Strategies toward the inclusion of EE. *School Science and Mathematics, 117*(3-4), 104–114. doi:10.1111sm.12211

Crimmins, A., Balbus, J., Gamble, J. L., Beard, C. B., Bell, J. E., Dodgen, D., Eisen, R. J., Fann, N., Hawkins, M. D., Herring, S. C., Jantarasami, L., Mills, D. M., Saha, S., Sarofim, M. C., Trtanj, J., & Ziska, L. (2016). *The Impacts of Climate Change on Human Health in the United States: A Scientific Assessment*. Executive Summary. U.S. Global Change Research Program. doi:10.7930/J00P0WXS

Cruz, P. M., & Bodnar, Z. (2011). El nuevo paradigma del derecho en la pos-modernidad. Revista de Estudios Constitucionales, Hermenéutica y teoría del derecho, 3, 75-83. doi:10.4013/rechtd.2011.31.08

Cruz, J., & Marques, P. (2011). Sustainable Transportation and Corporate Social Responsibility. *Procedia: Social and Behavioral Sciences*, *20*, 100–110.

Cutter-Mackenzie, A., & Rousell, D. (2019). Education for what? Shaping the field of climate change education with children and young people as co-researchers. *Children's Geographies*, *17*(1), 90–104. Advance online publication. doi:10.1080/14733285.2018.1467556

Dahan, S., & Kashiwase, H. (2016). El *ODS 6 sobre agua y saneamiento es esencial para el desarrollo sostenible.* Obtenido de Banco Mundial BLOGS. Retrieved from https://blogs.worldbank.org/es/opendata/el-ods-6-sobre-agua-y-saneamiento-es-esencial-para-el-desarrollo-sostenible

Daly, H. E. (2015). *The Unraveling of the Growth Economy and the Transition to a Sustainable Era*. University of Chicago Press.

Dankelman, I. (2010). Climate Change, Human Security and Gender. In Gender and climate change: An introduction. Routledge.

Dankelman, I. (2002). Climate change: Learning from gender analysis and women's experiences of organising for sustainable development. *Gender and Development*, *10*(2), 21–29. doi:10.1080/13552070215899

Dankelman, I. (2010). *Gender and climate change: Women as agents of change*. Zed Books.

Dankelman, I., & Jansen, W. (2013). Gender, environment and climate change: understanding the linkages. In *Gender and climate change: An introduction* (pp. 49–82). Routledge.

Dannetta, V. (2002). What factors influence a teacher's commitment to student learning? *Leadership and Policy in Schools*, *1*(2), 144–171. doi:10.1076/lpos.1.2.144.5398

Danni, S. O., Bouchaou, L., Elmouden, A., Brahim, Y. A., & N'da, B. (2019). Assessment of water quality and nitrate source in the Massa catchment (Morocco) using δ15N and δ18O tracers. *Applied Radiation and Isotopes*, *154*, 108859. doi:10.1016/j.apradiso.2019.108859 PMID:31539708

Davies, M., & Devlin, M. (2007). *Interdisciplinary higher education: Implications for teaching and learning*. Centre for the Study of Higher Education. University of Melbourne.

Davis, D. (2010). Integrating sustainability into studio design curriculum. *ASEE Annual Conference and Exposition, Conference Proceedings*.

De Kadt, E. (1979). Poverty: The Ethical Dilemma. *World Development, 7*(1-2), 81–87.

Denton, F., Wilbanks, T. J., Abeysinghe, A. C., Burton, I., Gao, Q., Lemos, M. C., . . . Rahman, A. (2019). Chapter 11: Strengthening and implementing the global response. In V. Masson-Delmotte, H.-O. Pörtner, D. Roberts, J. Skea, P.R. Shukla, A. Pirani, W. Moufouma-Okia, C. Péan, R. Pidcock, S. Connors, J.B.R. Matthews, Y. Chen, X. Zhou, M.I. Gomis, E. Lonnoy, T. Maycock, M. Tignor, & T. Waterfield (Eds.), Global warming of 1.5°C. An IPCC Special Report on the impacts of global warming of 1.5°C above pre-industrial levels and related global greenhouse gas emission pathways, in the context of strengthening the global response to the threat of climate change, sustainable development, and efforts to eradicate poverty (pp. 641-775). World Meteorological Organization.

Deschenes, O. (2014). Temperature, human health, and adaptation: A review of the empirical literature. *Energy Economics, 46*, 606–619. doi:10.1016/j.eneco.2013.10.013

DesJardins, J. R. (2013). *Environmental Ethics: An Introduction to Environmental Philosophy* (5th ed.). Wadsworth Publishing.

Desjean-Perrotta, B., Moseley, C., & Cantu, L. E. (2008). Preservice Teachers' Perceptions of the Environment: Does Ethnicity or Dominant Residential Experience Matter? *The Journal of Environmental Education, 39*(2), 21–32. doi:10.3200/JOEE.39.2.21-32

Dianne, F. (2009). Sustainable development: The response from education - Australian country report. Melbourne Graduate School of Education.

Dillon, J. (2003). On Learners and Learning in Environmental Education: Missing Theories, Ignored Communities. *Environmental Education Research, 9*(2), 215–226. doi:10.1080/13504620303480

Dillon, J., Stevenson, R., & Wals, A. (2016). Introduction to special section: Moving from citizen to civic science to address wicked conservation problems. *Conservation Biology, 30*(3), 450–455. doi:10.1111/cobi.12689 PMID:27153527

Directive 2005/36/EC of the European Parliament and of the Council of 7 September 2005 on the recognition of professional qualifications (Text with EEA relevance), CONSIL, EP, 255 OJ L (2005). https://data.europa.eu/eli/dir/2005/36/oj/eng

Djoudi, H., Locatelli, B., Vaast, C., Asher, K., Brockhaus, M., & Basnett Sijapati, B. (2016). Beyond dichotomies: Gender and intersecting inequalities in climate change studies. *Ambio, 45*(S3, Suppl 3), 248–262. doi:10.100713280-016-0825-2 PMID:27878531

Dominelli, L. (2013). Environmental justice at the heart of social work practice: Greening the profession. *International Journal of Social Welfare, 22*(4), 431–439. doi:10.1111/ijsw.12024

Dominelli, L. (Ed.). (2018). *The Routledge handbook of green social work*. Routledge. doi:10.4324/9781315183213

Dossou-Yovo, E. R., Zinsou, C. B., & Dovonou, F. (2019). Women's Roles in Climate Change Adaptation in Sub-Saharan Africa: A Systematic Review. *Gender and Development, 27*(2), 217–235.

Dragoni, L., Oh, I. S., Vankatwyk, P., & Tesluk, P. E. (2011). Developing executive leader: The relative contribution of cognitive ability, personality, and the accumulation of work experience in predicting strategic thinking competency. *Personnel Psychology, 64*(4), 829–864. doi:10.1111/j.1744-6570.2011.01229.x

Dresner, S. (2002). *The principles of sustainability*. Earthscan.

Druckman, J. N. (2001). Framing: Toward Clarification of a Fractured Paradigm. *Political Behavior, 23*(3), 225–256. doi:10.1023/A:1015006907312

Dumitru, D. (2017). Reorienting higher education pedagogical and professional development curricula toward sustainability– a Romanian perspective. *International Journal of Sustainability in Higher Education, 18*(6), 894–907. doi:10.1108/IJSHE-03-2016-0046

Dündar, T., & Özsoy, S. (2020). İklim değişikliğinin kadın üreme sağlığına etkileri [Climate change effects on female reproductive health]. *Arşiv Kaynak Tarama Dergisi, 29*(3), 190–198.

Dunn, K. (2022, March 30). *The impact of climate change in the Himalayas matters. It's a global and regional issue*. Retrieved from Reuters Institute: https://reutersinstitute.politics.ox.ac.uk/news/impact-climate-change-himalayas-matters-its-global-and-regional-issue

Ebrahim, A., & Rangan, V. K. (2010). The Limits of Nonprofit Impact: A Contingency Framework for Measuring Social Performance. *Harvard Business School Working Paper, 10*(99), 1–36.

Eccles, R. G., & Krzus, M. P. (2010). *One Report: Integrated Reporting for a Sustainable Strategy*. John Wiley & Sons.

Eckstein, D., Kunzel, V., & Schafer, L. (2021). *Global Climate Risk Index 2021*. Berlin: Germanwatch. Retrieved from https://www.germanwatch.org/sites/default/files/Global%20Climate%20Risk%20Index%202021_2.pdf

Edame, G. E., Ekpenyong, A. B., Fonta, W. M., & Duru, E. J. C. (2011). Climate change, food security and agricultural productivity in Africa: Issues and policy directions. *International Journal of Humanities and Social Science, 1*(21), 205–223.

Eilam, E. (2022). Climate change education: The problem with walking away from disciplines. *Studies in Science Education, 58*(2), 231–264. doi:10.1080/03057267.2021.2011589

Ekinci, E., & Ökde, F. (2021). İklim değişikliğinin kırsal alanlara etkisinin muhtarlar üzerinden değerlendirilmesi: Yozgat köyleri örneğinde. *Afet ve Risk Dergisi, 4*(2), 300–318. doi:10.35341/afet.993555

Elkington, J. (1997). *Cannibals with Forks: The Triple Bottom Line of 21st Century Business*. Capstone.

Elliott, B., & Crosswell, L. (2002). *Teacher commitment and engagement: The dimensions of ideology and practice associated with teacher commitment and engagement within an Australian perspective*. Available: https://www.aare.edu.au/02pap/cro02522.htm

Enarson, E. (2012). *Women confronting natural disaster: From vulnerability to resilience*. Lynne Rienner Publishers. doi:10.1515/9781588269560

Entman, R. M. (1993). Framing: Toward Clarification of a Fractured Paradigm. *Journal of Communication*, *43*(4), 51–58. doi:10.1111/j.1460-2466.1993.tb01304.x

EPA. (2021). *Información sobre la protección de las fuentes de agua.* Retrieved from Agencia de Protección Ambiental de Estados Unidos. Retrieved from https://bit.ly/3O9d9Xm

Eriksson, E., & Hilty, L. M. (2016). ICT for Sustainability: An Emerging Research Field. *Environmental Impact Assessment Review*, *59*, 1–16.

Escobar, A. (1996). Construction Nature: Elements for a Post-structuralist Political Ecology. *Futures*, *28*(4), 325–343. doi:10.1016/0016-3287(96)00011-0

Espinoza, T., & Rodríguez, C. (2016). Notas técnicas: Índice de calidad de agua ICA, de los ríos Aroa Yaracuy de los estados de Falcón y Yaracuy. *Rev. Ingeniería UC.*, *23*(3), 381–386.

Eugenio-Gozalbo, M., Aragón, L., & Ortega-Cubero, I. (2020). Gardens as science learning contexts across educational stages: Learning assessment based on students' graphic representations. *Frontiers in Psychology*, *11*, 2226–2226. doi:10.3389/fpsyg.2020.02226 PMID:33013580

European Commission. (2022). *Consequences of climate change*. https://climate.ec.europa.eu/climate-change/consequences-climate-change_en

Evans, N. (2020). What ought to be done to promote "Education for Sustainability" in teacher education? *Journal of Philosophy of Education*, *54*(4), 817–824. doi:10.1111/1467-9752.12482

Fadhullah, W., Yaccob, N. S., Syakir, M. I., Muhammad, S. A., Yue, F.-J., & Li, S.-L. (2020). Nitrate sources and processes in the surface water of a tropical reservoir by stable isotopes and mixing model. *The Science of the Total Environment*, *700*(134517), 134517. doi:10.1016/j.scitotenv.2019.134517 PMID:31629263

Fahey, S. J. (2012). Curriculum change and climate change: Inside outside pressures in higher education. *Journal of Curriculum Studies*, *44*(5), 703–722. doi:10.1080/00220272.2012.679011

Fahey, S., Verstraten, L., & Berry, A. J. (2016). Education for Sustainable Development: Enhancing Climate Change Adaptation Expertise in Developing Countries. *Journal of Education for Sustainable*, *10*(1), 54–67. doi:10.1177/0973408215625534

Faldez, R. L. (2020). *Conservación del agua: importancia y técnicas*. Retrieved from Ecología verde: https://bit.ly/3I5WJuU

FAO. (2017). *The Future of Food and Agriculture – Trends and Challenges*. https://www.fao.org/3/a-i6583e.pdf

Faria, F., Kelly, K., Daniel, P., & Azevedo, I. M. L. (2015). A New Approach of Science, Technology, Engineering, and Mathematics Outreach in Climate Change, Energy, and Environmental Decision Making." Sustainability. *The Journal of Record*, *8*(5), 261–271.

Farmer, G., & Guy, S. (2010). Making morality: Sustainable architecture and the pragmatic imagination. *Building Research and Information, 38*, 368–378. doi:10.1080/09613218.2010.482236

Farrell, K., Devlin, M., & James, R. (2017). *Nine Principles Guiding Teaching and Learning*. University of Melbourne Centre for the Study of Higher Education.

Favier, T., Van Gorp, B., Cyvin, J. B., & Cyvin, J. (2021). Learning to teach climate change: Students in teacher training and their progression in pedagogical content knowledge. *Journal of Geography in Higher Education, 45*(4), 594–620. doi:10.1080/03098265.2021.1900080

Feinstein, N. W., & Mach, K. J. (2020). Three roles for education in climate change adaptation. *Climate Policy, 20*(3), 317–322. doi:10.1080/14693062.2019.1701975

Feola, G. (2015). Societal transformation in response to global environmental change: A review of emerging concepts. *Ambio, 44*(5), 376–390. doi:10.100713280-014-0582-z PMID:25431335

Ferguson, T., Roofe, C., & Cook, L. D. (2021). Teachers' perspectives on sustainable development: The implications for education for sustainable development. *Environmental Education Research, 27*(9), 1343–1359. doi:10.1080/13504622.2021.1921113

Field, E., Schwartzberg, P., & Berger, P. (2019). *Formal Report for Learning for a Sustainable Future*. http://www.LSF-LST.ca/cc-survey

Fielding, K. S., & Hornsey, M. J. (2016). A Social Identity Analysis of Climate Change and Environmental Attitudes and Behaviors: Insights and Opportunities. *Frontiers in Psychology, 7*. Advance online publication. doi:10.3389/fpsyg.2016.00121 PMID:26903924

Figueiró, P. S., & Raufflet, E. (2015). Sustainability in higher education: A systematic review with focus on management education. *Bridges for a More Sustainable Future: Joining Environmental Management for Sustainable Universities (EMSU) and the European Roundtable for Sustainable Consumption and Production (ERSCP) Conferences, 106*, 22–33. 10.1016/j.jclepro.2015.04.118

Filho, W. L., Yayeh Ayal, D., Wall, T., Shiel, C., Paco, A., Pace, P., Mifsud, M., Lange Salvia, A., Skouloudis, A., Moggi, S., LeVasseur, T., Vinuesa Antonio, G., Azeiteiro, U. M., Ioannis, N., & Kovaleva, M. (2023). An assessment of attitudes and perceptions of international university students on climate change. *Climate Risk Management, 39*(100486), 1–16. doi:10.1016/j.crm.2023.100486

Flora, J., Saphir, M., Lappé, M., Roser-Renouf, C., Maibach, E. W., & Leiserowitz, A. A. (2014). Evaluation of a National High School Entertainment Education Program: The Alliance for Climate Education. *Climatic Change, 127*(3–4), 419–434. doi:10.100710584-014-1274-1

Florini, A. (2016). *The Right to Know: Transparency for an Age of Environmental Uncertainty*. Columbia University Press.

Food And Agriculture Organization. (2008). *Climate change, bioenergy and food security: Options for decision makers identified by expert meeting*. Prepared for the high-level conference World Food Security: The Challenges of Climate Change and Bioenergy, Rome, Italy.

Food And Agriculture Organization. (2011). *The global bioenergy partnership sustainability indicators for bioenergy (GBEP). Executive Summary.* https://www.fao.org/publications/card/en/c/939f849c-043f-5674-bf0c-fe4a488065ca/

Forsyth, T. (2003). *Critical Political Ecology: The politics of environmental science.* Routledge.

Foss, A. W., & Ko, Y. (2019). Barriers and opportunities for climate change education: The case of Dallas-Fort Worth in Texas. *The Journal of Environmental Education, 50*(3), 145–159. doi:10.1080/00958964.2019.1604479

Foucault, M. (1975). *The birth of the clinic: An archeology of medical perception.* Vintage/Random House.

Franzen, R. (2018). Environmental education in teacher education programs: Incorporation and use of professional guidelines. *The Journal of Sustainability Education.* http://www.susted.com/wordpress/content/environmental-education-in-teacher-education-programs-incorporation-and-use-of-professional-guidelines_2018_01/

Frederick, C. P. (2012). *Curriculum Improvement in Education for Sustainable Development: Measuring Learning Outcomes in an Introductory Urban Planning Course* [Master's Thesis]. Arizona State University. doi:10.1080/10464883.1988.10758475

Friedrich, H. (2009). *International Education Statistics.* http://www.childinfo.org.statistics

Friedrich-Alexander-Universitat (FAU). (2022). *What are the goals of climate change adaptation?* https://www.fau.eu/2022/03/15/news/

Fuertes, M., Andrés, S., Corrochano, D., Delgado, L., Herrero-Teijón, P., Ballegeer, A. M., Ferrari-Lagos, E., Fernández, R., & Ruiz, C. (2020). A proposal of a Category-Based Tool for Curriculum Analysis to Achieve the Climate Competence. *Educ. Knowl. Soc, 21*, 13. doi:10.14201/eks.22823

Fulbright-Anderson, A., Kubisch, A. C., & Connell, J. P. (1998). *New approaches to evaluating community initiatives: Theory, measurement and analysis.* Aspen Institute.

Fullan, M. (2006). Change theory: A force for school improvement. Seminar Series No. 157. Jolimont, Australia: Centre for Strategic Education.

Fullan, M. G. (2001). *The New Meaning of Educational Change* (3rd ed.). Cassell. doi:10.4324/9780203986561

Gal, A., & Gan, D. (2020). Transformative sustainability education in higher education: Activating environmental understanding and active citizenship among professional studies learners. *Journal of Transformative Education, 18*(4), 271–292. doi:10.1177/1541344620932310

Galárraga, S. (2020). *Estado de los recursos hídricos en Ecuador.* Retrieved from HidroRed. Retrieved from http://tierra.rediris.es/hidrored/basededatos/docu1.html

Gan, D. (2021). Environmental education leadership: The perceptions of elementary school principals as expressed in their drawings and explanations. *Environmental Education Research, 27*(10), 1440–1466. doi:10.1080/13504622.2021.1959899

García, G. M., Carvajal, E. Y., & Jiménez, H. (2016). La gestión integrada de los recursos hídricos como estrategia de adaptación al cambio climático. *Rev. Ingeniería competitive, 9*(1), 19-29. Retrieved from https://www.ecologiaverde.com/conservacion-del-agua-importancia-y-tecnicas-2195.html

Gardiner, S. M. (2006). A Perfect Moral Storm: Climate Change, Intergenerational Ethics, and the Problem of Moral Corruption. *Environmental Values, 15*(3), 397–413. doi:10.3197/096327106778226293

Gardiner, S. M. (2011). Ethics and Global Climate Change. In D. G. Arnold (Ed.), *The Ethics of Global Climate Change* (pp. 143–169). Cambridge University Press.

Gardner, G. T., & Stern, P. C. (2008). *The Short List: The Most Effective Actions U.S. Households Can Take to Curb Climate Change*. Taylor and Francis Group. Available http://www.environmentmagazine.org/Archives/Back%20Issues/SeptemberOctober%202008/gardner-stern-full.html

Garrison, J., & Liston, D. (2004). *Teaching, learning, and loving*. Teachers College Press.

Gasper, D. (2014). Understanding the diversity of conceptions of well-being and quality of life. *Journal of Socio-Economics, 40*(5), 559–570.

Gaventa, J., & Barrett, G. (2010). *So What Difference Does it Make? Mapping the Outcomes of Citizen Engagement*. IDS Working Paper, 347.

Genovese, A., Acquaye, A. A., Figueroa, A., & Koh, S. C. L. (2017). Sustainable supply chain management and the transition towards a circular economy: Evidence and some applications. *Omega, 66*(Part B), 344-357.

Georgeta, S. M. (2020). The Social Communication of the Street Children. *Scientia Moralitas Conference Proceedings*. 10.5281/zenodo.4385036

Gerrard, E. (2008). Climate Change and human rights: Issues and opportunities for indigenous peoples. *The University of New South Wales Law Journal, 31*(3), 941–952.

Gibb, N. (2016). *Getting climate-ready: A guide for schools on climate action*. UNESCO. doi:10.54675/MOYX1103

Giler, A., Donoso, P., & Zaldumbide, D. (2020). Manejo sostenible de inundaciones, cuencas hidrográficas y riberas en la provincia de Manabí. *Rev. La Técnica., 1*(5), 55–72.

Glasser, H. (2007). Minding the gap: The role of social learning in linking our stated desire for a more sustainable world to our everyday actions and policies. In A. E. J. Wals (Ed.), *Social learning: Toward a more sustainable world* (pp. 35–61). Wageningen Academic Publishers.

Glyphis, J. (2001). How can the architect contribute to. A sustainable world. *Proceedings of the Wingspread. Conference*, 24–26.

Gobierno autónomo Descentralizado. (2015). *Propuesta de Desarrollo estratégico del cantón Portoviejo*. Retrieved from Gobierno Autónomo Descentralizado del cantón Portoviejo. Retrieved from https://bit.ly/41veXNr

Goh, A. H. X. (2012). *A literature review of the gender-differentiated impacts of climate change on women's and men's assets and well-being in developing countries*. International Food Policy Research Institute. doi:10.2499/CAPRiWP106

Gökalp Alıca, S.S (2017). Kadın ve Çevre. *Tbb Dergisi, Özel Sayı*, 151-172.

Goldar, B., & Banerjee, N. (2004). Impact of informal regulation of pollution on water quality in rivers in India. *Journal of Environmental Management, 73*(2), 117–130. doi:10.1016/j.jenvman.2004.06.008 PMID:15380317

Gordon, M. (2009). Toward a pragmatic discourse of constructivism: Reflections on lessons from practice. *Educational Studies: A Journal of the American Education Studies Association, 45*(1), 39-58.

Gosseries, A. (2008). Intergenerational Justice. *Stanford Encyclopedia of Philosophy*. Retrieved from https://plato.stanford.edu/archives/win2008/entries/justice-intergenerational/

Government of Kenya. (2007). *Kenya Vision 2030: The popular version*. https://vision2030.go.ke/wp-content/uploads/2018/05/Vision-2030-Popular-Version.pdf

Government of Kenya. (2010). *National climate change response strategy*. https://www.greengrowthknowledge.org/sites/default/files/downloads/policydatabase//KENYA%29%20National%20Climate%20Change%20Response%20Strategy.pdf

Government of Kenya. (2015). *Addressing Climate Change: Success Stories from Kenya*. Author.

Government of Kenya. (2018). *National Climate Change Action Plan*. Author.

Government of Kenya. (2018). National Education Policy 2018 – 2022. *Vision (Basel)*.

Government of Kenya. (2018). *National Education Sector Strategy Plan 2018 – 2022*. Author.

Government Printer. (2016). *The Climate Change Act, 2016*. Nairobi: Government Printer. http://kenyalaw.org/kl/fileadmin/pdfdownloads/Acts/ClimateChangeActNo11of2016.pdf https://www.un.org/en/climatechange/what-is-climate-change http://www.greenbeltmovement.org/who-we-are/our-history https://www.kgun.org/about-us/

Govindan, K., Soleimani, H., & Kannan, D. (2015). Reverse logistics and closed-loop supply chain: A comprehensive review to explore the future. *European Journal of Operational Research, 240*(3), 603–626. doi:10.1016/j.ejor.2014.07.012

Gray, M., Coates, J., & Hetherington, T. (Eds.). (2012). *Environmental social work*. Routledge. doi:10.4324/9780203095300

Greene, J. A., & Yu, S. B. (2016). Educating critical thinkers: The role of epistemic cognition. *Policy Insights from the Behavioral and Brain Sciences*, *3*(1), 45–53. doi:10.1177/2372732215622223

Grover, R., Emmitt, S., & Copping, A. (2020). Critical learning for sustainable architecture: Opportunities for design studio pedagogy. *Sustainable Cities and Society*, *53*, 101876. doi:10.1016/j.scs.2019.101876

Guambo, G., Torres, J., & Quiroz, S. (2022). Análisis histórico de la contaminación hídrica en el río Portoviejo. *Minerva Journal, 3*(8), 54-60. Retrieved from https://minerva.autanabooks.com/index.php/Minerva/article/view/64/211

Gucyeter, B. (2016). The Place of Sustainability in Architectural Education: Discussion and Suggestions. *American Journal of Archaeology*, *2*, 237–256.

Guenther, L., Jörges, S., Mahl, D., & Brüggemann, M. (2023). Framing as a Bridging Concept for Climate Change Communication: A Systematic Review Based on 25 Years of Literature. *Communication Research*, *0*(0). Advance online publication. doi:10.1177/00936502221137165

Gupta, J., & Bleischwitz, R. (2018). Public Choices and Ethical Dilemmas in Managing a Limited Natural Resource: The Case of Water. *Environmental Values*, *27*(4), 367–389.

Gustafsson, P., Engström, S., & Svenson, A. (2015). Teachers' View of Sustainable Development in Swedish Upper Secondary School. *Procedia: Social and Behavioral Sciences*, *167*, 7–14. doi:10.1016/j.sbspro.2014.12.635

Guy, S., & Farmer, G. (2001). Reinterpreting Sustainable Architecture: The Place of Technology. *Journal of Architectural Education*, *54*(3), 140–148. doi:10.1162/10464880152632451

Guy, S., Yoshihisa, K., Iain, W., & O'Neill, S. (2014). Investigating the effects of knowledge and ideology on climate change beliefs. *European Journal of Social Psychology*, *44*(5), 421–429. doi:10.1002/ejsp.2039

Hackett, R. A. (1984). Decline of a paradigm? Bias and objectivity in news media studies. *Critical Studies in Mass Communication, 1*(3), 229-259. https://www.tandfonline.com/doi/abs/10.1080/15295038409360036

Hailey, J. (2008). *Ubuntu: A literature review*. Paper Document. Tutu Foundation.

Hajer, M. A. (1995). *The Politics of Environmental Discourse: Ecological Modernization and the Policy Process*. Oxford University Press.

Halder, J. N. & Islam, M. N. (2015). Water pollution and its impact on the human health. *Journal of Environment and Human, 2*(1), 36-46. doi:10.15764/EH.2015.01005

Hallar, A. G., McCubbin, I. B., & Wright, J. M. (2011). CHANGE: A Place-Based Curriculum for Understanding Climate Change at Storm Peak Laboratory, Colorado. *Bulletin of the American Meteorological Society*, *92*(7), 909–918. doi:10.1175/2011BAMS3026.1

Hall, S. (1980). Encoding/decoding. In *Culture, Media, Language*. Hutchinson.

Halstead, F., Parsons, L. R., Dunhill, A., & Parsons, K. (2021). A Journey of Emotions from a Young Environmental Activist. *Area, 53*(4), 708–717. Advance online publication. doi:10.1111/area.12745

Halstead, J. M., & Taylor, M. J. (1996). *Values in education and education in values*. London: Falmer Press. Figueiredo, P., & Perkins, P. E. (2013). Women and Water Management in Times of Climate Change: Participatory and Inclusive Processes. *Journal of Cleaner Production, 60*, 188–194.

Hamilton, L. (2009). Education, politics and opinions about climate change: Evidence for interaction effects. *Climatic Change, 104*(2), 231–242. doi:10.100710584-010-9957-8

Hannula, K. A. (2022). Education about climate change. *Journal of Geoscience Education, 70*(4), 423. doi:10.1080/10899995.2022.2115763

Haque, A. T. M. S., Kumar, L., & Bhullar, N. (2023, February 20). Kumar. L & Bhullar, N. (2023) Gendered perceptions of climate change and agricultural adaptation practices: A systematic review. *Climate and Development*, 1–18. Advance online publication. doi:10.1080/17565529.2023.2176185

Harding, S. (2004). *The feminist standpoint theory reader: Intellectual and political controversies*. Routledge.

Hargittai, E. (2002). Second-Level Digital Divide: Differences in People's Online Skills. *First Monday, 7*(4). Advance online publication. doi:10.5210/fm.v7i4.942

Harris, F. (2018). Outdoor learning spaces: The case of forest school. *Area, 50*(2), 222–231. doi:10.1111/area.12360

Hartman, H. (2012). Is Sustainability Just Another "Ism"? *Architectural Design*, 136–140. https://doi.acm.org/10.1002/ad.1444

Hart, S. L. (2005). *Capitalism at the Crossroads: Aligning Business, Earth, and Humanity*. Wharton School Publishing.

Harvey, D. (2005). *A Brief History of Neoliberalism*. Oxford University Press. doi:10.1093/oso/9780199283262.001.0001

Hase, V., Mahl, D., Schäfer, M. S., & Keller, T. R. (2021). Climate change in news media across the globe: An automated analysis of issue attention and themes in climate change coverage in 10 countries (2006–2018). *Global Environmental Change, 70*, 102353. Advance online publication. doi:10.1016/j.gloenvcha.2021.102353

Havelaar, A. H. & Melse, J. M. (2003). *Quantifying public health risk in the WHO guidelines for drinking-water quality: a burden of disease approach*. Academic Press.

Healthy Green Schools & Colleges. (n.d.). https://www.healthygreenschools.org/

Hegarty, K., Thomas, I., Kriewaldt, C., Holdsworth, S., & Bekessy, S. (2011). Insights into the value of a 'stand-alone' course for sustainability education. *Environmental Education Research, 17*(4), 451–469. doi:10.1080/13504622.2010.547931

Hengrasmee, S., & Chansomsak, S. (2016). A novel approach to architectural education for sustainability: A quest for reformation and transformation. *Global Journal of Engineering Education, 18*(3), 160–166. https://www.scopus.com/inward/record.uri?eid=2-s2.0-84994635754&partnerID=40&md5=64352ee45c35192231936be993344a4f

Herr, C. (2014). *Radical constructivist structural design education for large cohorts of Chinese learners*. Academic Press.

Hoagland, B., Schmidt, C., Russo, T. A., Adams, R., & Kaye, J. (2019). Controls on nitrogen transformation rates on restored floodplains along the Cosumnes River. California. *The Science of the Total Environment, 649*, 979–994. doi:10.1016/j.scitotenv.2018.08.379 PMID:30179826

Hofer, B. K. (2000). Dimensionality and disciplinary differences in personal epistemology. *Contemporary Educational Psychology, 25*(4), 378–405. doi:10.1006/ceps.1999.1026 PMID:11001783

Hoffmann, R., Dimitrova, A., Muttarak, R., Crespo Cuaresma, J., & Peisker, J. (2020). A meta-analysis of country-level studies on environmental change and migration. *Nature Climate Change, 10*(10), 904–912. doi:10.103841558-020-0898-6

Hofste, W., Reig, R., & Schleifer, L. (2019). *17 countries, home to one-quarter of the world's population, face extremely high-water stress*. World Resources Institute. https://www.wri.org/blog/2019/08/17-countries-home-one-quarter-world-population-face-extremely-high-water-stress

Holling, C. S. (2018). Ethics in sustainable development. In M. Redclift & G. Woodgate (Eds.), *The International Handbook of Environmental Sociology* (pp. 127–138). Edward Elgar Publishing.

Hord, N. G. (2011). Dietary nitrates, nitrites, and cardiovascular disease. *Current Atherosclerosis Reports, 13*(6), 484–492. doi:10.100711883-011-0209-9 PMID:21968645

Hornig, S. (1992). Framing Risk: Audience and Reader Factors. *The Journalism Quarterly, 69*(3), 679–690. doi:10.1177/107769909206900316

Hountondji, Y.-C., Xu, J., & Yang, C. (2020). Exploring the Relationship between Social Capital and Climate Change Awareness in Rural Communities. *International Journal of Environmental Research and Public Health, 17*(21), 7851. PMID:33120876

Huho, J. M. (2015). Climate change knowledge gap in education system in Kenya. *International Journal of Innovation and Research in Educational Science, 2*(3), 227–232.

Huho, J. M. (2015). Climate change knowledge gap in education system in Kenya. *International Journal of Innovation and Research in Educational Sciences, 2*(3), 227–232. https://www.researchgate.net/publication/279179079_Climate_Change_Knowledge_Gapin_Education_System_in_Kenya

Hulme, M. (2009). *Why We Disagree About Climate Change: Understanding Controversy, Inaction and Opportunity*. Cambridge University Press. doi:10.1017/CBO9780511841200

Hulme, M. (2010). *Why We Disagree About Climate Change: Understanding Controversy, Inaction, and Opportunity*. Cambridge University Press.

Hung, C. C. (2012). *Climate Change Education: knowing, doing and being*. Routledge.

Hyland, C., & Grierson, D. (2012). Learning Lessons from the Scottish School Building Programme. *The International Journal of Environmental Sustainability*, *8*(1), 63–76. doi:10.18848/2325-1077/CGP/v08i01/55025

ICRW. (2022). *SRHR is a Climate Issue*. ICRW. Retrieved from https://reliefweb.int/sites/reliefweb.int/files/resources/ICRW_SRHR-is-a-Climate-Issue_04.22.pdf

IDMC. (2020). Global report on internal displacement 2020. IDMC.

IEA. (2020). *Buildings – Topics*. IEA. https://www.iea.org/topics/buildings

IFRC. (2022). *Kenya hunger crisis: Voices from the drought*. https://www.ifrc.org/article/kenya-hunger-crisis-voices-drought

IICA. (2008). *Buenas prácticas agrícolas. Guía para pequeños y medianos agroempresarios*. Instituto Interamericano de Cooperación para la Agricultura. Retrieved from http://repiica.iica.int/docs/B0737e/B0737e.PDF

Ikeda, K. (2009). How women's concerns are shaped in community-based disaster risk management in Bangladesh. *Contemporary South Asia*, *17*(1), 65–78. doi:10.1080/09584930802624679

Ikonomidis, S., Papanastasious, D., Mela, D., & Avgoloupis, S. (2012). Teacher preparation and climate change curriculum at university level in Nigeria. *International Journal of Multidisciplinary Academic Research*, *21*, 768–779.

Ilahi, N., & Grimard, F. (2000). Public infrastructure and private costs: Water supply and time allocation of women in rural Pakistan. *Economic Development and Cultural Change*, *48*(3), 793–813. doi:10.1086/452490

INEC. (2016). *Indicadores ODS de agua, saneamiento e higiene en Ecuador*. Retrieved from https://www.ecuadorencifras.gob.ec/documentos/web-inec/Bibliotecas/Libros/Diagnostico_ASH_pobreza_INEC_BM.pdf

INEFAN. (2017). *Plan de manejo de área de boques y vegetación protectores de las colinas que circundan la ciuda de Portoviejo*. Instituto Ecuatoriano Forestal de Áreas Naturales y Vida Silvestre.

Intergovernmental Panel on Climate Change (IPCC). (2014). *AR5 Synthesis Report: Climate Change 2014*. Contribution of working groups I, II and III to the Fifth Assessment Report of the Intergovernmental Panel on Climate Change. Geneva: IPCC. https://www.ipcc.ch/report/ar5/syr/

Intergovernmental Panel on Climate Change. (2001). *Climate change: The scientific basis. Contribution of working group I to the third assessment report of the Intergovernmental Panel on Climate Change*. Cambridge University Press.

Intergovernmental Panel on Climate Change. (2012). Glossary of Terms. In Managing the Risks of Extreme Events and Disasters to Advance Climate Change. A Special Report of Working Groups I and II of the Intergovernmental Panel on Climate Change. Cambridge University Press.

Intergovernmental Panel on Climate Change. (2014). *Climate Change: Impact, Adaptation and Vulnerability-Part A: Global and Sectoral Aspects*. Fifth Assessment Report. https://www.ipcc.ch/site/assets/uploads/2018/02/WGIIAR5-PartA_FINAL.pdf

Internal Displacement Monitoring Centre. (2021). *Internal displacement in a changing climate*. Norwegian Refugee Council. Retrieved March 13, 2023 from https://www.internal-displacement.org/sites/default/files/publications/documents/grid2021_idmc.pdf

International Federation of Social Workers. (2015). *IFSW's Policies 2015*. IFSW.

International Fund for Agricultural Development. (2014b). *Annual Report 2014*. https://www.ifad.org/documents/38714170/40252580/AR_print_2014.pdf/868aaf81-0ef8-492c-80b0-f76b6034068c

International Fund for Agricultural Development. (2016). *Rural development report 2016: Fostering inclusive rural transformation*. https://www.ifad.org/documents/38714170/39155702/Rural+development+report+2016.pdf/347402dd-a37f-41b7-9990-aa745dc113b9

Intriago, F. J., & Quiroz, F. L. (2021). Calidad del agua de la cuenca media del río Portoviejo. Estrategias para mitigar la contaminación. *Rev. Polo del conocimiento, 6*(6), 1144-1171. PMID:33414495

Inwood, H. (2020). Emerging praxis of environmental and sustainability education in teacher education in Canada. *Journal of Philosophy of Education, 54*(4), 825–831. doi:10.1111/1467-9752.12466

IOM. (2012). *International dialogue on migration 18: climate change, environmental degradation and migration*. IOM.

IPCC. (2007). *Climate Change 2007: Mitigation. Contribution of Working Group III to the Fourth Assessment Report of the Intergovernmental Panel on Climate Change*. Cambridge University Press.

IPCC. (2014). *Climate change 2014: Impacts, adaptation, and vulnerability*. Cambridge University Press.

IPCC. (2014). *Climate Change 2014: Synthesis Report. Contribution of Working Groups I, II, and III to the Fifth Assessment Report of the Intergovernmental Panel on Climate Change. Core Writing Team* (R. K. Pachauri & L. A. Meyer, Eds.). IPCC.

IPCC. (2014). Climate change 2014: synthesis report. In R. K. Pachauri & L. A. Meyer (Eds.), Contribution of working groups I, II and III to the fifth assessment report of the intergovernmental panel on climate change (pp. 1–168). Academic Press.

IPCC. (2017). *Climate Change 2007: Synthesis Report*. Intergovernmental Panel on Climate Change, Climate Change. https://www.nrc.gov/docs/ML0932/ML093220680.pdf

Irwin, R. (2008). *Heidegger, politics and climate change*. Continuum.

Irwin, R. (2020). Climate change and education. *Educational Philosophy and Theory*, *52*(5), 492–507. doi:10.1080/00131857.2019.1642196

Iyer-Raniga, U., & Dalton, T. (2017). Challenges in Aligning the Architecture Profession in Indonesia for Climate Change and Sustainability. *International High-Performance Built Environment Conference – A Sustainable Built Environment Conference 2016 Series (SBE16), IHBE 2016, 180*, 1733–1743. 10.1016/j.proeng.2017.04.336

Jabeen, S., Haq, S., Jameel, A., Hussain, A., Asif, M., Hwang, J., & Jabeen, A. (2020). Impacts of Rural Women's Traditional Economic Activities on Household Economy: Changing Economic Contributions through Empowered Women in Rural Pakistan. *Sustainability, 12*(7), 2731. doi:10.3390/su12072731

Jamali, D., Zanhour, M., & Keshishian, T. (2018). Peculiar strengths and relational attributes of SMEs in the context of CSR. *Journal of Business Ethics*, *82*(4), 355–363.

Jamieson, D. (2014). *Reason in a Dark Time: Why the Struggle Against Climate Change Failed—and What It Means for Our Future*. Oxford University Press. doi:10.1093/acprof:oso/9780199337668.001.0001

Jasanoff, S., & Kim, S. H. (2015). *Dreamscapes of Modernity: Sociotechnical Imaginaries and the Fabrication of Power*. University of Chicago Press. doi:10.7208/chicago/9780226276663.001.0001

Jia, Z.M., Li, S.Y., & Wang, L., (2018). Assessment of soil heavy metals for eco-environment and human health in a rapidly urbanization area of the upper Yangtze Basin. *Scientific Reports*, *8*(1), 1–14. https://www.nature.com/articles/s41598-018-21569-6 PMID:33414495

Johnson, M. (2017). *Environmental Philosophy: From Theory to Practice*. Bloomsbury Academic.

Jones, P. (2019). *Sustainable Supply Chain Management: Principles, Strategies, and Practices*. Routledge.

Jones-Smith, E. (2012). Theories of counseling and psychotherapy: An integrative approach. *Sage (Atlanta, Ga.)*.

Kadim, F., Ar, H., & Gülçubuk, B. (2014). *Gıda güvenilirliği açısından kırsal alandaki kadınlar ve aile çiftçliğinin önemi* [Paper presentation]. Aile Çiftçiliği Sempozyumu.

Kagawa, F., & Selby, D. (2010). Introduction. In F. Kagawa & D. Selby (Eds.), *Education and climate change: Living and learning in interesting times*. Taylor & Francis. doi:10.4324/9780203866399

Kagawa, F., & Selby, D. (2012). Ready for the storm: Education for disaster risk reduction and climate change adaptation and mitigation. *Journal of Education for Sustainable Development, 6*(2), 207–217. doi:10.1177/0973408212475200

Kagombe, J., & Gitonga, J. (2005). *Plantation establishment in Kenya: The Shamba system case study*. Kenya Forest.

Kaijer, A., & Kronsell, A. (2014). Climate change through the lens of intersectionality. *Environmental Politics, 23*(3), 417–433. doi:10.1080/09644016.2013.835203

Kallis, G. (2011). In Defence of Degrowth. *Ecological Economics, 70*(5), 873–880. doi:10.1016/j.ecolecon.2010.12.007

Kansas Department of Health and Environment Bureau of Water. (2011). *Water Quality Standards White Paper Chlorophyll-a Criteria for Public Water Supply Lakes or Reservoirs*. Retrieved April 14, 2023 from https://ehs.ku.edu/

Kant, I. (1785). *Groundwork of the Metaphysics of Morals* (M. Gregor, Trans.). Cambridge University Press.

Kaplan, S. (2019). Climate Change and Social Equity. *Annual Review of Environment and Resources, 44*, 53–80.

Kaplan, S., & Kaplan, R. (1982). *Cognition and Environment: Functioning in an Uncertain World*. Praeger.

Kaplinsky, R. (2000). Globalisation and Unequalisation: What Can Be Learned from Value Chain Analysis? *The Journal of Development Studies, 37*(2), 117–146. doi:10.1080/713600071

Kara, Y. (2021). Kesişimsellik iklim krizinin neresinde? [Where is the intersectionality of climate crises?]. *International Journal of Cultural and Social Studies, 7*(1), 1–16. doi:10.46442/intjcss.729485

Karim, N., Othman, H., Zaini, Z., Rosli, Y., Wahab, M. I. A., Al Kanta, A. M., Omar, S., & Sahani, M. (2022). Climate Change and Environmental Education: Stance from Science Teachers. *Sustainability (Switzerland), 14*(24), 1–18. doi:10.3390u142416618

Karl, H., Curtin, C., Scarlett, L., & Hopkins, W. (2011). Adapting to climate change, a wicked problem: A road map for action. In I. Linkov & T. S. Bridges (Eds.), Climate: Global change and local adaptation (pp. 237–254). Springer Science + Business Media. doi:10.1007/978-94-007-1770-1_13

Kar, N. (2009). Psychological impact of disasters on children: Review of assessment and interventions. *World Journal of Pediatrics, 5*(1), 5–11. doi:10.100712519-009-0001-x PMID:19172325

Karsgaard, C., & Davidson, D. (2023). Must we wait for youth to speak out before we listen? International youth perspectives and climate change education. *Educational Review, 75*(1), 74–92. doi:10.1080/00131911.2021.1905611

Kasim, K., & Behcet, O. (2015). Teachers' Perceptions on Culturally Responsiveness in Education. *Journal of Ethnic and Cultural Studies, 2*(2), 47-57.

Keck, M. E., & Sikkink, K. (1998). *Activists Beyond Borders: Advocacy Networks in International Politics*. Cornell University Press.

Keitsch, M. (2012). Sustainable Design: A Brief Appraisal of its Main Concepts. *Sustainable Development, 20*(3), 180–188. Advance online publication. doi:10.1002d.1534

Keller, L., Stötter, J., Oberrauch, A., Kuthe, A., Körfgen, A., & Hüfner, K. (2019). Article. *Changing Climate Change Education, 28*(1), 35–43.

Kenya Institute of Curriculum Development (KICD). (2017). *Basic Education Curriculum Framework.* Nairobi, Kenya: Republic of Kenya.

Kerns, B., Elhouar, S., Sterling, M. J., Grant, J. M., McGowan, M., Rubash, A., Neelly, K., & Wolffe, R. (2005). Ten Principles of Effective Teaching and Practical Examples for the Classroom and Blackboard. Office of Teaching Excellence and Faculty Development, Bradley University.

Khan, N. A., Gao, Q., & Abid, M. (2020). Public institutions' capacities regarding climate change adaptation and risk management support in agriculture: The case of Punjab Province, Pakistan. *Scientific Reports, 10*(1), 14111. doi:10.103841598-020-71011-z PMID:32839499

Khatri, D. B., Bhusal, R. P., Paudel, N. S., & Gurung, N. (2012). *REDD+ politics in the media: A case study from Nepal.* CIFOR.

Kılıç, C. (2009). Küresel iklim değişikliği çerçevesinde sürdürülebilir kalkınma çabaları ve Türkiye [Sustainable development efforts in the frame of global climate change and Turkey]. *C.Ü. İktisadi ve İdari Bilimler Dergisi, 10*(2), 19-41.

Kim, K., Suh, E., & Kim, S. (2018). *Sustainability in logistics and supply chain management: Selected papers from the 5th International Conference on Logistics and Maritime Systems.* Springer.

Kim, J. W., & Kim, C. (2009). *National Report for the project 'Climate Change and Sustainable Development – The Response from Education.* College of Education, Seoul National University.

Kirchherr, J., Reike, D., & Hekkert, M. (2017). Conceptualizing the Circular Economy: An Analysis of 114 Definitions. *Resources, Conservation and Recycling, 127*, 221–232. doi:10.1016/j.resconrec.2017.09.005

Kızmaz, Z. (2021). İklim değişikliğinin kırsal alandaki etkisi ve alternatif arayışlar: Sosyolojik bir yaklaşım [The impact of climate change in rural area and alternative searches: A sociological approach]. *Fırat Üniversitesi Sosyal Bilimler Dergisi, 31*(1), 431–453. doi:10.18069/firatsbed.827702

Koirala, M., & Bhatta, R. (2010). *Communities Challenging Climate Change.* CSRC and NEFEJ.

Koissaba, B. R. (2017). *Education For All: Prospects and Challenges of Mobile Schools, Mobile Education, and E-Learning for the Nomadic Pastoralists in Kenya* (2nd Ed. 2017). Research Gate. https://www.researchgate.net/publication/319630142

Koivusilta, L., Lintonen, T., & Rimpelä, A. (2021). Digital Divide and Digital Literacy in the Context of Finnish Schoolchildren. *Nordic Journal of Digital Literacy*, *16*(4), 212–230.

Kolb, D. A. (1984). *Experiential Learning: Experience as the Source of Learning and Development* (Vol. 1). Prentice-Hall.

Kolenatý, M., Kroufek, R., & Činčera, J. (2022). What Triggers Climate ˇ Action: The Impact of a Climate Change Education Program on Students' Climate Literacy and Their Willingness to Act. *Sustainability*, *14*(16), 10365. doi:10.3390u141610365

Kolk, A. (2016). The social responsibility of international business: From ethics and the environment to CSR and sustainable development. *Journal of World Business*, *51*(1), 23–34. doi:10.1016/j.jwb.2015.08.010

Kovancı, E. (2017). Kronik: Kadınların sürdürülebilir gelişme için örgütlenme deneyimlerinden ve toplumsal cinsiyet analizinden iklim değişikliğini anlamak [Climate change: Learning from gender analysis and women's experiences of organising for sustainable development]. *Ankara Üniversitesi Sbf Dergisi*, *72*(2), 493-503. (Original work published 2002) doi:10.1501/SBFder_0000002456

Krause, F., & Strang, V. (2016). Thinking relationships through water. *Society & Natural Resources*, *29*(6), 633–638. doi:10.1080/08941920.2016.1151714

Kruse, E., Read, D., Negin, J., Hunter, E., & Rosewell, A. (2020). Responding to the Needs of Rural and Remote Populations: Insights from a Rapid Global Response to Climate Change Education in Vanuatu. *Journal of Geography in Higher Education*, *44*(1), 1–19.

Kuehnast, K., Naujoks, J., Smith, D., & Wimmer, A. (2017). *Building Local Capacity for Peace and Development*. United States Institute of Peace.

Kumar, A., & Sharma, P. (2013). Impact of climate variation on agricultural productivity and food security in rural India. In *Economics Discussion Papers*. Kiel Institute for the World Economy. http://www.economics-ejournal.org/dataset/PDFs/discussionpapers_2013-43.pdf

Kumaraguru, P., & Cranor, L. F. (2010). Privacy and Security in Online Social Networking Sites. *IEEE Security and Privacy*, *8*(3), 13–15.

Kunkle, K. A., & Monroe, M. C. (2019). Cultural cognition and climate change education in the U.S.: Why consensus is not enough. *Environmental Education Research*, *25*(5), 633–655. doi:10.1080/13504622.2018.1465893

Kuthe, A., Keller, L., Körfgen, A., Stötter, H., Oberrauch, A., & Höferli, K.-M. (2019). How many young generations are there? – A typology of teenagers' climate change awareness in Germany and Austria. *The Journal of Environmental Education*, *50*(3), 172–182. doi:10.1080/00958964.2019.1598927

Kwauku, C., & Braga, A. (2017). *Three platforms for girls' education in climate strategies. Brooke Shearer Series, number 6*. Brookings Institution.

Læssøe, J., Schnac, K., Breiting, S., & Rolls, S. (2009). *Climate Change and Sustainable Development*. The Response from Education International Alliance of Leading Educational Institutes.

Lala, B. & Hagishima, H. (2023). Impact of Escalating Heat Waves on Students' Well-Being and Overall Health: A Survey of Primary School Teachers. *Open Access*, *11*(6), 126. doi:10.3390/cli11060126

Lambert, D., & Hopkin, J. (2014). A possibilist analysis of the geography national curriculum in England. *International Research in Geographical and Environmental Education*, *23*(1), 64–78. doi:10.1080/10382046.2013.858446

Landorf, H., Doscher, S., & Rocco, T. (2008). Education for sustainable human development: Towards a definition. *Theory and Research in Education*, *6*(0), 221–236. doi:10.1177/1477878508091114

Langer, A., Meleis, A., Knaul, F. M., Atun, R., Aran, M., Arreola-Ornelas, H., Bhutta, A. Z., Binagwaho, A., Bonita, R., Caglia, J. M., Claeson, M., Davies, J., Donnay, F. A., Gausman, J. M., Glickman, C., Kearns, A. D., Kendall, T., Lozano, R., Seboni, N., ... Frenk, J. (2015). Women and health: The key for sustainable development. *Lancet*, *386*(9999), 1165–1210. doi:10.1016/S0140-6736(15)60497-4 PMID:26051370

Largo-Wight, E., Guardino, C., Wludyka, P. S., Hall, K. W., Wight, J. T., & Merten, J. W. (2018). Nature contact at school: The impact of an outdoor classroom on children's well-being. *International Journal of Environmental Health Research*, *28*(6), 653–666. doi:10.1080/09603123.2018.1502415 PMID:30047798

Leach, M., Raworth, K., & Rockström, J. (2019). Between Social and Planetary Boundaries: Navigating Pathways in the Safe and Just Space for Humanity. *World Development*, 283–295.

Leal Filho, W., Azeiteiro, U. M., Alves, F., Molthan-Hill, P., Pace, P., Mifsud, M., & Caeiro, S. (2018). Reinvigorating the Sustainability Agenda: Education for Sustainable Development (ESD) for Responsible and Transformational Leadership. *Sustainability*, *10*(10), 3415.

Leal, F. W., Sima, M., Sharif, A., Luetz, J. M., Salvia, A. L., Mifsud, M., Olooto, O. F., Djekic, I., Anholon, R., Rampasso, I., Donkor, F. K., & Dinis, M. (2021). Handling Climate change education at university; an overview. *Environmental Sciences Europe*, *33*(1), 109–128. doi:10.118612302-021-00552-5 PMID:34603904

Lehnert, M., Fiedor, D., Frajer, J., Hercik, J., & Jurek, M. (2020). Czech Students and Mitigation of Global Warming: Beliefs and Willingness to Take Action. *Environmental Education Research*, *2020*(26), 864–889. doi:10.1080/13504622.2019.1694140

Lehtonen, A., Salonen, A. O., & Cantell, H. (2019). Climate change education: A new approach for a world of wicked problems. In J. W. Cook (Ed.), *Sustainability, human well-being, and the future of education* (pp. 339–374). Springer International Publishing. doi:10.1007/978-3-319-78580-6_11

Leichenko, R. M., & O'Brien, K. L. (2008). *Environmental Change and Globalization: Double Exposures*. Oxford University Press. doi:10.1093/acprof:oso/9780195177329.001.0001

Leichenko, R., & O'Brien, K. (2020). Teaching climate change in the Anthropocene: An integrative approach. *Anthropocene, 30*, 100241. Advance online publication. doi:10.1016/j.ancene.2020.100241

Leigh, K. (2009). Energy Busters: Norfolk Schools Fight Climate Change. *Environmental Education, 91*, 13–14.

Leiserowitz, A., & Smith, N. (2010). *Knowledge of Climate Change across Global Warming's Six Americas*. Yale Project on Climate Change Communication. Available http://www.nationalacademies.org/bose/Climate_Change_Education_Leiserowitz_ Workshop_paper.pdf

Ley Orgánica de recursos hídricos. (2014). *Usos y aprovechamiento del agua.* Retrieved from https://bit.ly/3W4QFZB

Liu, S.-Y., Yeh, S.-C., Liang, S.-W., Fang, W.-T., & Tsai, H.-M. (2015). A national investigation of teachers' environmental literacy as a reference for promoting environmental education in Taiwan. *The Journal of Environmental Education, 46*(2), 114–132. doi:10.1080/00958964.2014.999742

Lopatto, D. (2003). The essential features of undergraduate research. *Council on Undergraduate Research Quarterly, 24*, 139–142.

Lorentzen, L. A., & Eaton, H. (2002). Ecofeminism: An overview. In The Forum on Religion and Ecology at Yale (pp. 1-2). Academic Press.

Lotz-Sisitka, H. (2010). Climate injustice: How should education respond? In F. Kagawa & D. Selby (Eds.), *Education and climate change: Learning and living in interesting times* (pp. 71–88). Routledge.

Louv, R. (2006). *Last child in the woods: Saving our children from nature-deficit disorder.* Algonquin Books of Chapel Hill.

Lubawy, W. C. (2003). Evaluating teaching using the best practices model. *American Journal of Pharmaceutical Education, 67*(3), 87. doi:10.5688/aj670387

Lück, J., Wessler, H., Wozniak, A., & Lycarião, D. (2016). Counterbalancing global media frames with nationally colored narratives: A comparative study of news narratives and news framing in the climate change coverage of five countries. *Journalism, 19*(12). https://journals.sagepub.com/doi/10.1177/1464884916680372

Luque, A. & Jiménez, A. (2019). Procesos de sostenibilidad textil: éxito, o nueva forma de contaminar bajo un paradigma amable. *Revista Inclusiones, 6*, 259-288.

Maeseele, P., & Pepermans, Y. (2017). Climate Change Communication, Media, and Ideology. *Climate Change Communication*, 1-22.

Maestre, T. (2023). *La mayor causa de desertificación es el mal uso del suelo y del agua, no la sequía.* Retrieved from https://www.eldiario.es/ballenablanca/economia/fernando-t-maestre-ecologo-mayor-causa-desertificacion-mal-suelo-agua-no-sequia_128_10157041.html

Mahmoodi, A. S. (2001). *The design process in architecture: A pedagogic approach using interactive thinking* [PhD thesis]. University of Leeds.

Maksud, A. K., & Imtiaj, R. (2006). The Nomadic Bede Community and their Mobile School Program. What works for the Poorest: knowledge policies and Practices. Academic Press.

Manabí, P. (2021). *Predectura de Manabi, Desarrollo y Equidad.* Retrieved from https://bit.ly/3W9pBIE

Mangheni, M. N., Tufan, H. A., Boonabana, B., Musiimenta, P., Miiro, R., & Njuki, J. (2019). Building gender research capacity for non-specialists: lessons and best practices from gender short courses for agricultural researchers in subSaharan Africa. In T. S. Marcia, K. Kristy, & D. Vasilikie (Eds.), *Gender and practice: knowledge, policy, organizations* (pp. 99–118). Emerald Publishing. doi:10.1108/S1529-212620190000028006

Mart, Ç. T. (2013). A Passionate Teacher: Teacher Commitment and Dedication to Student Learning. *International Journal of Academic Research in Progressive Education and Development, 2*(1), 437-444.

Martín de Santa Olalla, F., López Fuster, P. & Belmonte, C. (2005). *Agua y agronomía.* Madrid: Universidad de Castilla.

Matéjček, T., Bartoš, J., & Kučerová, S. R. (2020). Which subjects contribute to the teaching of cross curricular topic environmental education at elementary schools in Czechia? *Acta Universitatis Carolinae. Geographica. Universita Karlova, 55*(2), 200–209. doi:10.14712/23361980.2020.14

Mavromatidis, L. (2018). Coupling architectural synthesis to applied thermal engineering, constructal thermodynamics and fractal analysis: An original pedagogic method to incorporate "sustainability" into architectural education during the initial conceptual stages. *Sustainable Cities and Society, 39*, 689–707. doi:10.1016/j.scs.2018.01.015

Maxwell, J., & Blashki, G. (2016). Teaching about climate change in medical education: An opportunity. *Journal of Public Health Research, 5*(673), 14–20. doi:10.4081/jphr.2016.673 PMID:27190980

Mayaka, B., & Truell, R. (2021). Ubuntu and its potential impact on the international social work profession. *International Social Work, 64*(5), 649–662. doi:10.1177/00208728211022787

Mbah, M. F., Shingruf, A., & Molthan-Hill, P. (2022). Policies and practices of climate change education in South Asia: Towards a support framework for an impactful climate change adaptation. *Climate Action, 1*(1), 28. doi:10.100744168-022-00028-z

McCarty, J. (2013). REAL school gardens program: Learning gardens and teacher training to improve student engagement and academic performance in low-performing elementary schools. *The Journal of Applied Research on Children*, *4*(2), 1–8. doi:10.58464/2155-5834.1169

McCright, A. M. (2010). The effects of gender on climate change knowledge and concern in the American public. *Population and Environment*, *32*(1), 66–87. doi:10.100711111-010-0113-1

McGregor, H. E., Pind, J., & Karn, S. (2021). A 'wicked problem': Rethinking history education in the Anthropocene. *Rethinking History*, *25*(4), 483–507. doi:10.1080/13642529.2021.1992159

McIntosh, M., Cacciola, K., Clermont, S., & Keniry, J. (2001). *State of the Campus Environment—EcoLeaders*. National Wildlife Federation. https://www.nwf.org/Home/EcoLeaders/Campus-Ecology-Resource-Center/Reports/State-of-the-Campus-Environment

McKenzie, M. (2021). Climate change education and communication in global review: Tracking progress through national submissions to the UNFCCC Secretariat. *Environmental Education Research*, *27*(5), 631–651. doi:10.1080/13504622.2021.1903838

McKeown-Ice, R. (2000). Environmental education in the United States: A survey of preservice teacher education programs. *The Journal of Environmental Education*, *32*(1), 4–11. doi:10.1080/00958960009598666

McKeown, R., & Hopkins, C. (2010). Rethinking climate change education. *Green Teacher (Toronto)*, *89*, 17.

Memon, F. S & Amjad, S. (2022). Understanding Women's Perceptions of Promoting Education and Policy Initiatives about Climate Change in Rural Areas of Sindh, Pakistan. *Journal of Education and Educational Development, 7*(1), 140-156. doi:10.22555/joeed.v7i1.3223

Memon, F., Husain, I., Abdullah, F., Iqbal, R., Ahmad, S., & Abdullah, M. (2022). Addressing women's climate change awareness in Sindh, Pakistan: An empirical study of rural and urban women. *Climate and Development*, *14*, 1–13. doi:10.1080/17565529.2022.2125784

Methmann, C., & Rothe, D. (2012). Politics for the day after tomorrow: The logic of apocalypse in global climate politics. *Security Dialogue*, 323-344. doi:https://journals.sagepub.com/doi/10.1177/0967010612450746

Meyiwa, T., Maseti, T., Ngubane, S., Letsekha, T., & Rozani, C. (2014). Women in selected rural municipalities: Resilience and agency against vulnerabilities to climate change. *Agenda (Durban, South Africa)*, *28*(3), 102–114. doi:10.1080/10130950.2014.955686

Meymaris, K., Henderson, S., Alaback, P., & Havens, K. (2008). *Project BudBurst: Citizen Science for All Seasons*. Paper presented at the AGU Fall Meeting.

Mies, M., & Shiva, V. (1993). *Ecofeminism*. Zed Books. doi:10.5040/9781350219786

Milberg, S. J., Smith, J. H., & Burke, S. J. (2000). Information Privacy: Corporate Management and National Regulation. *Organization Science*, *11*(1), 35–57. doi:10.1287/orsc.11.1.35.12567

Miller, C., Jackson, T., & Thompson, A. (2021). Sustainable Logistics: Past, Present, and Future. *Journal of Sustainable Supply Chain Management, 10*(2), 65–78.

Miller, T. (2004). The Well-Tempered Self. *The Philosophical Forum, 35*(3), 195–210.

Mill, J. S. (1863). *Utilitarianism* (5th ed.). Longmans, Green, and Co.

Milne, W. (2005). Changing climate, uncertain future: Considering rural women in climate changes policies and strategies. *Canadian Women's Studies, 24*(4), 49–54.

Ministerio del Ambiente. (2012). *Contaminación de fuentes hídricas.* Retrieved from https://www.youtube.com/watch?v=w60Z2z5PF9w

Ministry of Defence. (2016). *National Security Policy, 2016.* Law Books Management Board.

Ministry of Education. (2017). *Education for Sustainable Development Policy for the Education Sector*. Government Printer.

Ministry of Education. (2018). *National Education Sector Strategic Plan 2018-2022.* Government Printer.

Ministry of Environment and Forestry. (2018). *Thematic plan for climate change.* Ministry of Environment and Forestry.

Ministry of Environment and Natural Resources (MENR). (2017). *Climate Change Policy, Planning and Budgeting at National and County Level: Facilitators Manual.* Government of Kenya.

Ministry of Home Affairs. (2018). *Nepal Disaster Report, 2017: The Road to Sendai.* Kathmandu: Ministry of Home Affairs. Retrieved from http://drrportal.gov.np/uploads/document/1321.pdf

Mirandiña, A., & Endocara, M. (2019). *Evaluación de necesidades tecnológicas para el manejo de la oferta hírica en cantidad y calidad.* Manabí - Ecuador.

Mitchell, T., Tanner, T., & Lussier, K. (2007). Action Aid: 'We know what we need': South Asian women speak out on climate change adaptation. Institute of Development Studies, 1-24.

Miteva, D. A., Pattanayak, S. K., & Ferraro, P. J. (2015). Evaluation of biodiversity policy instruments: What works and what doesn't? *Oxford Review of Economic Policy, 31*(1-2), 1–34.

Mittal, R. (2012). Climate Change Coverage in Indian Print Media: A Discourse Analysis. *The International Journal of Climate Change: Impacts and Responses, 3*(2), 219–232. doi:10.18848/1835-7156/CGP/v03i02/37105

Mochizuki, Y., & Bryan, A. (2015). Climate Change Education in the Context of Education for Sustainable Development: Rationale and Principles. *Journal of Education for Sustainable Development, 9*(1), 4–26. doi:10.1177/0973408215569109

MoE. (2007). *Economic survey. Kenya National Bureau of Statistics.* Government Printer.

MoFE. (2019). *Climate Change Scenarios for Nepal: National Adaptation Plan (NAP).* Kathmandu: Ministry of Forests and Environment. Retrieved from https://lib.icimod.org/record/34554

MoFE. (2021). *Vulnerability and Risk Assessment and Identifying Adaptation Options*. Ministry of Forests and Environment.

Molthan-Hill, P., Worsfold, N., Nagy, G. J., Leal-Filho, W., & Mifsud, M. (2019). Climate change education for universities: A conceptual framework. *Journal of Cleaner Production, 226*, 1092-1101.

Molthan-Hill, P., Jackson, D., Mifsud, A., & Odell, V. (2020). *NTU carbon literacy Toolkit for all staff and students at universities- a systems thinking approach*. NTU Green Academy.

Monroe, M. C., Oxarart, A., & Richard, R. P. (2013). A role for environmental education in climate change for secondary science educators. *Applied Environmental Education and Communication, 12*(1), 4–18. doi:10.1080/1533015X.2013.795827

Monroe, M. C., Plate, R. R., Oxarart, A., Bowers, A., & Chaves, W. A. (2019). Identifyingeffective climate change education strategies: A systematic review of the research. *Environmental Education Research, 25*(6), 791–812. doi:10.1080/13504622.2017.1360842

Montilla, A., & Pacheco, H. (2017). Comportamiento temporal y espacial del bosque ribereño en el curso bajo del rio Portoviejo y la quebrada Chilán, provincia de Manabí, Ecuador. *Rev. Internacional de Contaminación Ambiental., 33*(1), 21–35. doi:10.20937/RICA.2017.33.01.02

Morton, J. F (2007). The impact of climate change on smallholder and subsistence agriculture. *Scientific Research*. . doi:10.1073/pnas.0701855104

Mugumbate, J., & Chereni, A. (2019). Using African ubuntu theory in social work with children in Zimbabwe. *African Journal of Social Work, 9*(1).

Mugumbate, J., & Nyanguru, A. (2013). Exploring African philosophy: The value of ubuntu in social work. *African Journal of Social Work, 3*(1), 82–100.

Muhammad, S., Mussawar, S., & Irfan, U. (2017). Addressing the Issue of Nomadic Communities Children Educational Exclusion through Mobile Tent Schools in Malakand, Pakistan. *Imperial Journal of Interdisciplinary Research, 3*(3). http://www.onlinejournal.in

Muir, T. (2008). Principles of Practice and Teacher Actions: Influences on Effective Teaching of Numeracy. *Mathematics Education Research Journal, 20*(3), 78–101. doi:10.1007/BF03217531

Mulvik, I., Pribuišis, K., Siarova, H., Vežikauskaite, J., Sabaliauskas, E., Tasiopoulou, E., Gras-Velazquez, A., Bajorinaite, M., Billon, N., Fronza, V., Disterhelf, A., & Finlayson, A. (2022). *Education for environmental sustainability: Policies and approaches in European union member states*. Final Report. European Commission.

Murphy, C., Mallon, B., Smith, G., Kelly, O., Pitsia, V., & Martinez Sainz, G. (2021). The influence of a teachers' professional development programme on primary school pupils' understanding of and attitudes towards sustainability. *Environmental Education Research, 27*(7), 1011–1036. doi:10.1080/13504622.2021.1889470

Murphy, R. (1994). *Rationality and Nature: A Sociological Inquiry into a Changing Relationship*. Westview Press.

Murray, P., & Cotgrave, A. (2007). Sustainability literacy: The future paradigm for construction education? *Structural Survey*, *25*(1), 7–23. doi:10.1108/02630800710740949

Nascimento, J. D. D., & Fearnside, P. M. (2020). Mega-Dams in Brazilian Amazonia: Cost-Benefit Analysis and the Persistence of Policy Stereotypes. *Sustainability*, *12*(14), 5790.

National Climate Change Policy. (2012). *Government of Pakistan*. Retrieved from http://www.climatechange.gov.pk/en/policy

National Climate Change Policy. (2021). *Government of Pakistan, Ministry of Climate Change*. Retrieved on 14 April 2023 from https://mocc.gov.pk/SiteImage/Policy/NCCP%20Report.pdf

National Council for Law Reporting. (2012). *Environmental Management and Co-ordination Act, 1999*. Nairobi: National Council for Law Reporting. http://kenyalaw.org/kl/fileadmin/pdfdownloads/Acts/EnvironmentalManagementandCo-ordinationAct_No8of1999.pdf

National Council for Science, Technology and Innovation. (2020). *Background Report on National Climate Change Priorities and Relevant Capacity Development Goals and Initiatives in Kenya 2020*. Retrieved from https://www.nccs.go.ke/

National Research Council. (2011). *Climate Change Education: Goals, Audiences, and Strategies. A Workshop Summary*. Washington, DC: The National Academies Press.

Nation, M. T., & Feldman, A. (2021). Environmental Education in the Secondary Science Classroom: How Teachers' Beliefs Influence Their Instruction of Climate Change. *Journal of Science Teacher Education*, *32*(5), 481–499. doi:10.1080/1046560X.2020.1854968

NDI. (2022). Climate Change. Kathmandu: National Democratic Institute for International Affairs (NDI).

NEFEJ. (2011). *Nepal: Strengthening Capacity for Managing Climate Change and the Environment*. NEFEJ.

Nelson, D. R., Adger, W. N., & Brown, K. (2007). Adaptation to environmental change: Contributions of a resilience framework. *Annual Review of Environment and Resources*, *32*(1), 395–419. doi:10.1146/annurev.energy.32.051807.090348

Neumayer, E., & Plümper, T. (2007). The Gendered nature of natural disasters: The Impact of catastrophic events on the gender gap in life expectancy, 1981–2002. *Annals of the Association of American Geographers*, *97*(3), 551–556. doi:10.1111/j.1467-8306.2007.00563.x

Newig, J., & Fritsch, O. (2009). Environmental governance: Participatory, multi-level – and effective? *Environmental Policy and Governance*, *19*(3), 197–214. doi:10.1002/eet.509

Ngcoya, M. (2009). *Ubuntu: Globalization, accommodation, and contestation in South Africa*. American University.

Ngugi, M. (2016). Challenges Facing Mobile Schools among Nomadic Pastoralists: A Case Study of Turkana County, Kenya. *American Journal of Educational Research*, *4*(1), 22–32. doi:10.12691/education-4-1-6

Ng, Y.-L. (2022). Uses and Gratifications of and Exposure to Nature 2.0 and Associated Interdependence With Nature and Pro-Environmental Behavior. *Social Science Computer Review*, *40*(1), 61–76. doi:10.1177/0894439320901490

Nicholls, J. (2016). *Understanding how Queensland teachers' views on climate change and climate change education shape their reported practices* [Unpublished doctoral dissertation]. James Cook University.

Nisbet, M. C. (2009). Communicating Climate Change: Why Frames Matter for Public Engagement. *Environment: Science and Policy for Sustainable Development, 51*(2), 12-23. https://www.tandfonline.com/doi/abs/10.3200/ENVT.51.2.12-23

Nisbet, M. C., & Mooney, C. (2007). Science and society. Framing Science. *Science*, *316*(5821), 56. Advance online publication. doi:10.1126cience.1142030 PMID:17412941

Nisbet, M. C., & Myers, T. (2007). The Polls - Trends: Twenty Years of Public Opinion about Global Warming. *Public Opinion Quarterly*, *71*(3), 444–470. doi:10.1093/poq/nfm031

Njino, J. (2022, June 6). Why climate change is essential for Kenya's future. *Daily Nation*. https://nation.africa/kenya/blogs-opinion/blogs/why-climate-education-is-essential-for-kenya-s-future-3839492

Noguera, C. (2020). *Desarrollo urbano sostenible. Sociología del medio ambiente*. Retrieved from www.ua.es/personal/antonio.aledo/alumnos/nogera.doc

Norgaard, K. M. (2011). *Living in Denial: Climate Change, Emotions, and Everyday Life*. The MIT Press. doi:10.7551/mitpress/9780262015448.001.0001

North American Association for Environmental Education (NAAEE). (2017). *Guidelines for excellence: Professional development of environmental educators*. North American Association for Environmental Education.

Norton, B. G. (1987). *Why Preserve Natural Variety?* Princeton University Press.

Nyahunda, L. (2021). Social work empowerment model for mainstreaming the participation of rural women in the climate change discourse. *Journal of Human Rights and Social Work*, *6*(2), 120–129. doi:10.100741134-020-00148-8

Nyatuka, B. O. (2023). Partnering to Bolster Student Achievement: A Case of the Child Friendly School Concept. In J. Keengwe (Ed.), *Handbook of Research on Race, Culture, and Student Achievement* (pp. 376–394). IGI Global. doi:10.4018/978-1-6684-5705-4.ch019

Nyerere, J. (2022, May 26). Climate change isn't a priority for Kenyan universities: It should be. *The Conversation*. https://theconversation.com/climate-change-isnt-a-priority-for-kenyan-universities-it-should-be-181530

Nyerere, J., Gatwiri, W., & Okinyi, R. (2021, July). *Kenya's climate change policy actions and the response of higher education: Transforming universities for a changing climate, working paper series, No. 4*. London: Climate-U. https://www.climate-uni.com/_files/ugd/f81108_1fe481d96de94247b5b3a5baab268f33.pdf

Nyka, L. (2019). Bridging the gap between architectural and environmental engineering education in the context of climate change. *World Transactions on Engineering and Technology Education*, *17*, 204–209.

O'Brien, K. (2012). Global environmental change II: From adaptation to deliberate transformation. *Progress in Human Geography*, *36*(5), 667–676. doi:10.1177/0309132511425767

O'Brien, K., & Hochachka, G. (2010). *The Human Relationship with Nature: Development and Culture*. MIT Press.

O'Riordan, T., & Cameron, J. (Eds.). (1994). *Interpreting the Precautionary Principle*. Earthscan Publications.

Öcal Kara, F., Eren Yalcın, G., Işgın, T., & Özel, R. (2016). İklim değişikliğinde kadının rolü ve tarımsal yayım [Paper presentation]. 7. Tarım Ekonomi Kongresi, Isparta.

OECD. (2022). *Building the future of education*. https://www.oecd.org/education/future-of-education-brochure.pdf

Ogendi, G. M. (2022). Effects of nutrients on the water quality and the algal community of River Nyakomisaro-Riana, Kisii, Kenya. *International Journal of Fisheries and Aquatic Studies*, *10*(5), 111–118. doi:10.22271/fish.2022.v10.i5b.2734

Ogendi, G. M., Getabu, A. M., Onchieku, J. M., & Babu, J. M. (2015). Assessment of the Microbial Load of Nyanchwa-Riana and Nyakomisaro-Riana Rivers, Kisii, Kenya. *International Journal of Fisheries and Aquatic Studies*, 182–192. www.fisheriesjournal.com

OIKOS. (1998). *Diagnóstico Ambiental del cantón Portoviejo. Corporación de Gestión Tecnológica y Científica sobre el Ambiente*. Proyecto Educar.

Ojala, M. (2012). Hope and climate change: The importance of hope for environmental engagement among young people. *Environmental Education Research*, *18*(5), 625–642. doi:10.1080/13504622.2011.637157

Ojala, M. (2016). Facing anxiety in climate change education: From therapeutic practice to hopeful transgressive learning. *Canadian Journal of Environmental Education*, *21*, 41–56.

Ojala, M. (2021). Safe spaces or a pedagogy of discomfort? Senior high-school teachers' meta-emotion philosophies and climate change education. *The Journal of Environmental Education*, *52*(1), 40–52. doi:10.1080/00958964.2020.1845589

Okada, A., & Gray, P. (2023). A Climate Change and Sustainability Education Movement: Networks, Open Schooling, and the 'CARE-KNOW-DO' Framework. *Sustainability (Switzerland)*, *15*(3), 2356. Advance online publication. doi:10.3390u15032356

Okoli, N. J. (2014). Teacher preparation and climate change curriculum at university in Nigeria. *International Journal of Multidisciplinary Academic Research, 2*(3).

Okutu, A. A. (2019). Nomadism and Provision of Basic Education in West Pokot County, Kenya. *Journal of African Interdisciplinary Studies*, *3*(6), 114–125.

Olguín, E., González, R., Sánchez, G., & Zamora, J. (2010). Contaminación de ríos urbanos: El caso de la subcencia del río Sordo en Xalapa, Veracrua, México. *Rev. Latinoam Biiotecnol Amb Algal*, *1*(2), 178–190.

Olmos-Gómez, M. D. C., Estrada-Vidal, L. I., Ruiz-Garzón, F., López-Cordero, R., & Mohamed-Mohand, L. (2019). Making Future Teachers More Aware of Issues Related to Sustainability: An Assessment of Best Practices. *Sustainability*, *11*(24), 7222. doi:10.3390u11247222

Omoeva, C., Hatch, R., & Moussa, W. (2018). *The Effects of Armed Conflict on Educational Attainment and Inequality.* Education Policy and Data Center Paper No. 18-03. FHI 360.

O'Neill, J. (1993). *Ecology, Policy, and Politics: Human Well-Being and the Natural World.* Routledge.

Onta, P. (2006). *Mass media in post-1990 Nepal.* Martin Chautari.

ONU-DASES. (2015). *Decenio Internacional para la acción "El agua fuente de vida".* Departamento de Asuntos Económicos y Sociales de Naciones Unidas: Retrieved from https://bit.ly/3Mpm85c

Oreskes, N. (2004). The Scientific Consensus on Climate Change. *Science*, *306*(5702), 1686. doi:10.1126cience.1103618 PMID:15576594

Organización de Naciones Unidas. (2015). *Decenio Internacional para la acción. El agua como fuente de vida. Departamento de asuntos económicos y sociales de las naciones undas.* Organización de las Naciones Unidas. Obtenido de Departamento de asuntos económicos y sociales de las Naciones Unidas. Retrieved from https://www.un.org/spanish/waterforlifedecade/

Organización de Naciones Unidas. (2016). *Agua limpia y saneamiento; Por que es importante.* Retrieved from https://bit.ly/3W6prlk

Organización de Naciones Unidas. (2023). *Desafíos globales del agua.* Retrieved from https://www.un.org/es/global-issues/water

Organización Mundial de la Salud. (2022a). *Agua para consumo humano.* Retrieved from https://bit.ly/2HxnryQ

Organización Mundial de la Salud. (2022b). *Saneamiento.* Retrieved from https://bit.ly/2UYncQK

Ortiz, A. A., Ruiz, O. M., & Rodríguez, M. J. (2017). Planificación y gestión de los recursos hídricos: Una revisión de la importancia de la viariabilidad climática. *Rev. Logos Ciencia y Tecnología*, *9*(1), 100–105.

Örüklü, C., & Çakmak, S. (2021). Sürdürülebilir kalkınma hedefleri çerçevesinde kadın sağlığı [Women's health within the framework of sustainable development goals]. *Journal of Human Sciences, 18*(3), 364–373. doi:10.14687/jhs.v18i3.6118

Ouariachi, T., & Wim, E. J. L. (2020). Escape rooms as tools for climate change education: An exploration of initiatives. *Environmental Education Research, 26*(8), 1193–1206. doi:10.1080/13504622.2020.1753659

Özçağ, M. (2004). *Sürdürülebilir kalkınma sürecinde iklim değişikliği ve Türkiye analizi* [Doctoral dissertation]. Adnan Menderes Üniversitesi.

Özcan-Elçi, D. (2022). İklim değişikliği, toplumsal cinsiyet ve kadın temalı çalışmalara dair bir literatür taraması. *Proceedings of the International (Bio) Climate Change Symposium.*

Öztürk, K. (2002). Küresel İklim Değişikliği ve Türkiye'ye Olası Etkileri [Global Climatic Changes and Their Probable Effect upon Turkey]. *Gazi Üniversitesi Gazi Eğitim Fakültesi Dergisi, 22*(1), 47–65.

Özyol, A., & Çobanoğlu, N. (2013). *Yeşil kadın* [Green women]. https://www.academia.edu/17311834/YEŞİL_KADIN

Page, E. A. (2008). *Climate Change, Justice and Future Generations.* Edward Elgar Publishing.

Painter, J. (2013). *Climate change in the media – reporting risk and uncertainty.* London: I.B. Tauris & Co. Ltd.

Pal, J. S., & Eltahir, E. A. (2015). Future temperature in southwest Asia projected to exceed a threshold for human adaptability. *Nature Climate Change, 6*(2), 197–200. doi:10.1038/nclimate2833

Pan-American Health Organization. (2002). *Gender and natural disasters.* Author.

Papadimitriou, V. (2004). Prospective primary teachers' understanding of climate change, greenhouse effect and ozone layer depletion. *Journal of Science Education and Technology, 13*(2), 299–307. doi:10.1023/B:JOST.0000031268.72848.6d

Park, I. (2005). Teacher Commitment and its Effects on Student Achievement in American High Schools. *Educational Research and Evaluation, 11*(5), 461–485. doi:10.1080/13803610500146269

Park, N.-E., Choe, S.-U., & Kim, C.-J. (2020). Analysis of Climate Change Education (CCE) Programs: Focusing on Cultivating Citizen Activists to Respond to Climate Change. *Asia-Pacific Science Education, 6*(1), 15–40. doi:10.1163/23641177-BJA00004

Parry, L., Radel, C., Adamo, S. B., Clark, N., Counterman, M., Flores-Yeffal, N., Pons, D., Romero-Lankao, P., & Vargo, J. (2019). The (in) visible health risks of climate change. *Social Science & Medicine, 241*, 112448. doi:10.1016/j.socscimed.2019.112448 PMID:31481245

Pattberg, P., & Stripple, J. (2008). Beyond the Public and Private Divide: Remapping Transnational Climate Governance in the 21st Century. *International Environmental Agreement: Politics, Law and Economics, 8*(4), 367–388. doi:10.100710784-008-9085-3

Pedagogy of Climate Change: An Introduction. (2011). Higher Education Academy.

Pepermans, Y., & Maeseele, P. (2018). Manufacturing Consent: Rereading News on Four Climate Summits (2000-2012). *Science Communication*, 621-649. https://journals.sagepub.com/doi/10.1177/1075547018798119

Pepermans, Y., & Maeseele, P. (2014). Democratic Debate and Mediated Discourses on Climate Change: From Consensus to De/politicization. *Environmental Communication*, *8*(2), 216–232. doi:10.1080/17524032.2014.906482

Pepermans, Y., & Maeselle, P. (2017). Climate Change Journalism: From Agony to Agonistic Debate. *Desenvolvimento e Meio Ambiente*, *40*, 125–140. doi:10.5380/dma.v40i0.49257

Pidgeon, N., & Fischhoff, B. (2011). The role of social and decision sciences in communicating uncertain climate risks. *Nature Climate Change*, *1*(1), 35–41. doi:10.1038/nclimate1080

Pires, I. M., Martinho, A. P., & Gonçalves, R. (2020). The Ecological Footprint of Data Centers. *Journal of Cleaner*.

Plan Integral de Desarrollo. (2020). *Plan integral de desarrollo de los recursos hídricos de la provincia de Manabí.* Secretaría ejecutiva para asuntos económicos y sociales. República del Ecuador. Retrieved from https://bit.ly/42VmXbG

Plan Nacional de Desarrollo. (2017-2021). *Toda una vida. Ecuador.* Observatorio Regional de Planificación para el desarrollo de América Latina y el Caribe. Retrieved from https://bit.ly/2t3yGFG

Pogaˇcnik, M., Žnidarˇciˇc, D., & Strgar, J. A. (2014). *School garden in biotechnical education.* Academic Press.

Pogacnik, M., Znidarcic, D., & Strgar, J. (2014). A school garden in biotechnical education. *Archives of Biological Sciences*, *66*(1), 393–400. doi:10.2298/ABS1401393P

Porter, M. E., & van der Linde, C. (1995). Toward a New Conception of the Environment-Competitiveness Relationship. *The Journal of Economic Perspectives*, *9*(4), 97–118.

Powell, K. C., & Kalina, C. J. (2009). Cognitive and Social Constructivism: Developing Tools for an Effective Classroom. *Education*, *130*(2), 241–250.

Priestley, M., & Biesta, G. (2013). Introduction: The new curriculum. In M. Priestley & G. Biesta (Eds.), *Reinventing the curriculum: New trends in curriculum policy and practice* (pp. 229–236). Bloomsbury Academic Bloomsbury Collections. doi:10.5040/9781472553195.ch-001

Priestley, M., Biesta, G. J. J., & Robinson, S. (2015). Teacher agency: What is it and why does it matter? In R. Kneyber & J. Evers (Eds.), *Flip the system: Changing education from the bottom up* (pp. 134–148). Routledge. doi:10.4324/9781315678573-15

Pruneau, D., André, D., Liette, V., Joanne, L., Eileen, O., Elizabeth, M., Gilles, M., & Gaston, B. (2006). When Teachers Adopt Environmental Behaviors in the Aim of Protecting the Climate. *The Journal of Environmental Education*, *37*(3), 3–12. doi:10.3200/JOEE.37.3.3-12

Quiroz, E. & Menéndez, C. (2016). Modelación matemática de la capacidad de autodepuración de corrientes superficiales. Caso de estudio Río Portoviejo. *Rev. Cubana de ingeniería, 6*(2), 64-70.

Rainard, M., Smith, C. J., & Pachauri, S. (2023). Gender equality and climate change mitigation: Are women a secret weapon? *Front. Clim., 5*, 946712. doi:10.3389/fclim.2023.946712

Ratinen, I. (2016). *Primary student teachers' climate change conceptualization and implementation on inquiry-based and communicative science teaching: A design research*. University of Jyvaskyla, Jyvaskyla Studies in Education, Psychology and Social Research.

Reddy, K. V., Paramesh, V., Arunachalam, V., Das, B., Ramasundaram, P., Pramanik, M., & Sridhara, S. (2022). Farmers' Perception and Efficacy of Adaptation Decisions to Climate Change. *Agronomy, 12*(5), 1023. doi:10.3390/agronomy12051023

Reed, M. S. (2008). Stakeholder Participation for Environmental Management: A Literature Review. *Biological Conservation, 141*(10), 2417–2431. doi:10.1016/j.biocon.2008.07.014

Reid, A. (2019). Climate change education and research: Possibilities and potentials versus problems and perils? *Environmental Education Research, 25*(6), 767–790. doi:10.1080/13504622.2019.1664075

Reisinger, A., Kitching, R. L., Chiew, F., Hughes, L., Newton, P. C. D., Schuster, S. S., Tait, A., & Whetton, P. (2014). Climate change vulnerability in adaptation in rural areas. In B. Fitzharris & D. Karoly (Eds.), *Climate change 2014: Impacts, adaptation, and vulnerability. Part B: Regional aspects. Contribution of working group II to the fifth assessment report of the intergovernmental panel on climate Change* (pp. 1371–1438). Cambridge University Press.

Republic of Kenya. (2010). *Child friendly School Manual*. Ministry of Education.

Republic of Kenya. (2012). *Sessional Paper No. 14 of 2012: Reforming education and training sectors in Kenya*. Ministry of Education and Ministry of Higher Education, Science and Technology.

Ribot, J. C. (2004). *Waiting for Democracy: The Politics of Choice in Natural Resource Decentralization*. World Resources Institute.

Roberts, E. (2016). Environmental Ethics and Supply Chain Management. *Business Ethics Quarterly, 26*(4), 489–512.

Robinson, Z. (2011). *Teaching climate change in higher education: Barriers and opportunities.*

Rogers, D. S., & Tibben-Lembke, R. S. (2016). Going Backwards: Reverse Logistics Trends and Practices. *Reverse Logistics Magazine, 1*(1), 20–34.

Rogerson, C. M. (2006). Ethical Aspects of Tourism: Progress and Challenges. *Tourism Management, 27*(2), 171–183.

Romanello, M., McGushin, A., Di Napoli, C., Drummond, P., Hughes, N., Jamart, L., Kennard, H., Lampard, P., Rodriguez, B. S., Arnell, R., Ayeb-Karlsson, S., Belesova, K., Cai, W., Campbell-Lendrum, D., Chambers, J., Chu, L., Ciampi, L., Dalin, C., & Hamilton, I. (2021). The 2021 report of the Lancet Countdown on health and climate change: Code red for a healthy future. *Lancet, 398*(10311), 1619–1662. doi:10.1016/S0140-6736(21)01787-6 PMID:34687662

Ross, H., Rudd, J. A., Skains, R. L., & Horry, R. (2021). How Big Is My Carbon Footprint? Understanding Young People's Engagement with Climate Change Education. *Sustainability, 13*(1961), 1–19. doi:10.3390/su13041961

Ross, A. (2000). *Curriculum: Construction and critique*. Taylor & Francis Group.

Rousell, D., & Cutter-Mackenzie-Knowles, A. (2020). A systematic review of climate change education: Giving children and young people a 'voice' and a 'hand' in redressing climate change. *Children's Geographies, 18*(2), 191–208. Advance online publication. doi:10.1080/14733285.2019.1614532

Rowe, K. (2003). *The Importance of Teacher Quality as a Key Determinant of Students' Experiences and outcomes of Schooling*. Background paper to keynote address presented at the ACER Research Conference 2003, 19-21 October 2003.

Russo, I., & Comi, A. (2019). Circular Economy and Collaborative Networks. In M. Peruzzini & R. Raffaeli (Eds.), *Collaboration in a Hyperconnected World* (pp. 297–308). Springer.

Sachs, W. (2015). Sustainable Development Goals and Human Rights. *Ethics & International Affairs, 29*(1), 103–116.

Salas, Q. N. (2018). *Siembra de árboles en la rivera del río Portoviejo*. Retrieved from https://bit.ly/3Oh9ruy

Salter, P., & Maxwell, J. (2016). The inherent vulnerability of the Australian curriculum cross-curriculum priorities. *Critical Studies in Education, 57*(3), 296–312. doi:10.1080/17508487.2015.1070363

Sanchez, D. A., Szynkiewicz, A., & Faiia, A. M. (2017). Determining sources of nitrate in the semi-arid Rio Grande using nitrogen and oxygen isotopes. *Applied Geochemistry, 86*, 59–69. doi:10.1016/j.apgeochem.2017.09.012

Sandel, M. J. (2012). *What Money Can't Buy: The Moral Limits of Markets*. Farrar, Straus and Giroux.

Savaskan, R. C., Bhattacharya, S., & Van Wassenhove, L. N. (2004). Closed-Loop Supply Chain Models with Product Remanufacturing. *Management Science, 50*(2), 239–252. doi:10.1287/mnsc.1030.0186

Saylan, C., & Blumstein, D. (2011). *The failure of environmental education (and how we can fix it)*. University of California Press. doi:10.1525/9780520948723

Schafer, M. S. (2015). Climate Change and the Media. International Encyclopedia of the Social & Behavioral Sciences, 3, 853-859.

Schmid-Petri, H., Adam, S., Schmucki, I., & Häussler, T. (2017). A changing climate of skepticism: The factors shaping climate change coverage in the US press. *Public Understanding of Science (Bristol, England)*, *26*(4), 498–513. doi:10.1177/0963662515612276 PMID:26553047

Sen, A. (1999). *Development as Freedom*. Oxford University Press.

Şener, Y., & Başer, M. (2019). Küreselleşme ve yoksulluğun kadın sağlığı üzerine etkileri [The Effects of Globalization and Poverty on Women's Health]. *ERÜ Sağlık Bilimleri Fakültesi Dergisi*, *6*(2), 6–10.

Şen, Z. (2015). Chapter One—Introduction. In Z. Şen (Ed.), *Applied Drought Modeling, Prediction, and Mitigation* (pp. 1–41). Elsevier. doi:10.1016/B978-0-12-802176-7.00001-8

Seowa, T., & Ho, L. (2016). Singapore teachers' beliefs about the purpose of climate change education and student readiness to handle controversy. *International Research in Geographical and Environmental Education*, *25*(4), 358371. doi:10.1080/10382046.2016.1207993

Sezen-Barrie, A., Miller-Rushing, A., & Hufnagel, E. (2020). "It's a gassy world": Starting with students' wondering questions to inform climate change education. *Environmental Education Research*, *26*(4), 555–576. doi:10.1080/13504622.2019.1610158

Shackleton, S., Cobban, L., & Cundill, G. (2014). A gendered perspective of vulnerability to multiple stressors, including climate change, in the rural Eastern Cape, South Africa. *Agenda (Durban, South Africa)*, *28*(3), 73–89. doi:10.1080/10130950.2014.932560

Shareef, S. S., & Farivarsadri, G. (2020). An Innovative Framework for Teaching/Learning Technical Courses in Architectural Education. *Sustainability*, *12*(22), 9514. Advance online publication. doi:10.3390u12229514

Shiva, V. (2005). *Las mujeres de Kerala contra Coca-Cola*. Retrieved from https://mondiplo.com/las-mujeres-de-kerala-contra-coca-cola

Shoemaker, P. J., & Resse, S. D. (1991). *Mediating the message: theories of influences on mass media content*. Longman.

Shore, J. H., Tatum, E. L., & Vollmer, W. M. (1986a). Evaluation of mental effects of disaster, Mount St. Helens eruption. *American Journal of Public Health*, *76*(3, Suppl), 76–83. doi:10.2105/AJPH.76.Suppl.76 PMID:3946730

Shore, J. H., Tatum, E. L., & Vollmer, W. M. (1986b). Psychiatric reactions to disaster: The Mount St. Helens experience. *The American Journal of Psychiatry*, *143*(5), 590–595. doi:10.1176/ajp.143.5.590 PMID:3963245

Shove, E. (2010). Social Theory and Climate Change. *Theory, Culture & Society*, *27*(2-3), 277–288. doi:10.1177/0263276410361498

Shriberg, M., & MacDonald, L. (2013). Sustainability leadership programs: Emerging goals, methods, & best practices. *Journal of Sustainability Education*, *5*(1), 1–21.

Siegner, A., & Stapert, N. (2020). Climate change education in humanities classroom: A case of Lowell School curriculum. *Environmental Education Research*, *26*(4), 511–531. doi:10.1080/13504622.2019.1607258

Simpson, D., Wells, P., & Mason, D. (2019). Green Logistics. In R. C. Hoyle & P. W. Daniels (Eds.), *Transport and Logistics: Past, Present and Predictions* (pp. 297–324). Elsevier.

Siriwardhana, C., & Stewart, R. (2013). Forced migration and mental health: Prolonged internal displacement, return migration and resilience. *International Health*, *5*(1), 19–23. doi:10.1093/inthealth/ihs014 PMID:24029841

Sjögren, H. (2019). More of the same: A critical analysis of the formations of teacher students through education for sustainable development. *Environmental Education Research*, *25*(11), 1620–1634. doi:10.1080/13504622.2019.1675595

Smatanova, K., & Urban, J. (2018). Outcomes of the experimental assignment in Studio Project I - Urban Design. *World Transactions on Engineering and Technology Education*, *16*, 115–120.

Smith, K. R., Woodward, A., Campbell-Lendrum, D., Chadee, D. D., Honda, Y., Liu, Q., Olwoch, J. M., Revich, B., & Sauerborn, R. (2014). Human health: impacts, adaptation, and co-benefits. In Climate Change: Impacts, Adaptation, and Vulnerability. Part A: Global and Sectoral Aspects. Contribution of Working Group II to the Fifth Assessment. Report of the Intergovernmental Panel on Climate Change (pp. 709-754). Cambridge University Press.

Smith, G. A., & Stevenson, R. B. (2017). Sustaining education for sustainability in turbulent times. *The Journal of Environmental Education*, *48*(2), 79–95. doi:10.1080/00958964.2016.1264920

Smith, R., & Johnson, L. (2020). Green Logistics: A Pathway to Sustainable Supply Chain Management. *International Journal of Logistics Management*, *31*(3), 555–574.

Sohn, E. (2014). *2014 Climate change and the rise and fall of civilizations*. Climate NASA website. https://climate.nasa.gov/news/1010/climate-change-and-the-rise-and-fall-of-civilizations/

Somech, A., & Bogler, R. (2002). Antecedents and consequences of teacher organizational and professional commitment. *Educational Administration Quarterly*, *38*(4), 555–577. doi:10.1177/001316102237672

Sorensen, C., Murray, V., Lemery, J., & Balbus, J. (2018). Climate change and women's health: Impacts and policy directions. *PLoS Medicine*, *15*(7), 1–10. doi:10.1371/journal.pmed.1002603 PMID:29990343

Sovacool, B. K., Tan-Mullins, M., & Ren, X. (2017). Sustainable, pro-poor, and ethical: Exploring moral underpinnings of low-carbon energy. *Wiley Interdisciplinary Reviews. Energy and Environment*, *6*(2), e214.

Spretnak, C. (1990). Ecofeminism: Our roots and flowering. Reweaving the world: The emergence of ecofeminism. *The Institute for Ecosophical Studies*, *3*(2), 1–10.

Srikanthan, G., & Dalrymple, J. (2003). Developing alternative perspectives for quality in higher education. *International Journal of Educational Management*, *17*(3), 126–136. doi:10.1108/09513540310467804

Stanišić, J., & Maksić, S. (2014). Environmental education in Serbian primary schools: Challenges and changes to curriculum, pedagogy, and teacher training. *The Journal of Environmental Education*, *45*(2), 118–131. doi:10.1080/00958964.2013.829019

Stapleton, S. R. (2015). Environmental Identity Development through Social Interactions, Action, and Recognition. *The Journal of Environmental Education*, *46*(2), 94–113. doi:10.1080/00958964.2014.1000813

Stecula, D. A., & Merkley, E. (2019). Framing Climate Change: Economics, Ideology, and Uncertainty in American News Media Content From 1988 to 2014. *Frontiers in Communication*, *4*, 6. Advance online publication. doi:10.3389/fcomm.2019.00006

Stein, D., & Valters, C. (2012). *Understanding theory of change in international development.* The Justice and Security Research Program. https://www.theoryofchange.org/wp-content/ uploads/toco_library/pdf/UNDERSTANDINGTHEORYOFChangeSte inValtersPN.pdf

Stenhouse, L. (1975). *An Introduction to Curriculum Research and Development.* Heinemann.

Sterling, S. (2011). Transformative Learning and Sustainability: Sketching the Conceptual Ground. *Learning and Teaching in Higher Education*, *5*(11), 17–33.

Stevenson, K. T., Nils-Peterson, M., & Bondell, H. D. (2018). Developing a Model of Climate Change Behavior among Adolescents. *Climatic Change*, *151*(3-4), 589–603. doi:10.100710584-018-2313-0

Stevenson, R. B. (2011). Environmental learning and agency in diverse educational and cultural contexts. In R. B. Stevenson & J. Dillon (Eds.), *Engaging environmental education: Learning, culture and agency* (pp. 219–238). Sense Publishers.

Stevenson, R. B., Nicholls, J., & Whitehouse, H. (2017). What Is Climate Change Education? *Curriculum Perspectives*, *37*(1), 67–71. doi:10.100741297-017-0015-9

Stupar, A., Mihajlov, V., & Simic, I. (2017). Towards the Conceptual Changes in Architectural Education: Adjusting to Climate Change. *Sustainability*, *9*(8), 1355. Advance online publication. doi:10.3390u9081355

Sureda-Negre, J., Oliver-Trobat, M., Catalan-Fernández, A., & Comas-Forgas, R. (2014). Environmental education for sustainability in the curriculum of primary teacher training in Spain. *International Research in Geographical and Environmental Education*, *23*(4), 281–293. doi:10.1080/10382046.2014.946322

Sustainable Development Policy Institute. (2019). *Gender equity in education: Fact sheet Sindh*. Retrieved from https://www.sdpi.org/publications/files/FA-85.pdf

Swain, K. A. (2021). Media Framing of Climate Change Mitigation and Adaptation. In M. Lackner & B. Sajjadi (Eds.), *W.-Y. Chen, Handbook of Climate Change Mitigation and Adaptation*. Springer. doi:10.1007/978-1-4614-6431-0_6-3

Swyngedouw, E. (2010). Apocalypse Forever? *Theory, Culture & Society*, *27*(2-3), 213–232. doi:10.1177/0263276409358728

Szczepański, J. (2019). Sustainable monument preservation in architectural education. *World Transactions on Engineering and Technology Education*, *17*, 42–47.

Talu, N. (2017). İklim değişikliği ve toplumsal cinsiyet- Politika belirleme süreçleri [Climate change and gender- Policy making processes]. *Yasama Dergisi*, *33*, 68–87.

Tatgın, E. (2019). İklim adaleti kapsamında toplumsal cinsiyet tartışması. *Kent ve Çevre Araştırmaları Dergisi*, *1*(1), 104–118.

Taylor, P. W. (1981). The Ethics of Respect for Nature. *Environmental Ethics*, *3*(3), 135–145. doi:10.5840/enviroethics19813321

Teixeira, J., Silva, C., & Neves, J. V. (2018). School mobility management case study: German School of Oporto (Deutsche Schule zu Porto*). Case Stud. Transp. Policy*, *7*(1), 13–21. doi:10.1016/j.cstp.2018.11.002

Telenko, C., Seifert, R., Amankwah-Amoah, J., & D'Amico, K. (2019). Circular Economy in Reverse Supply Chain Management: A Comprehensive Review and Framework Development. *Sustainability*, *11*(22), 6303.

Terblanché-Greeff, A. C. (2019). Ubuntu and environmental ethics: The West can learn from Africa when faced with climate change. In M. Chemhuru (Ed.), *African environmental ethics* (pp. 93–109). The International Library of Environmental, Agricultural and Food Ethics. doi:10.1007/978-3-030-18807-8_7

The Climate Change Act. (2016). Nairobi: Government Printer. http://kenyalaw.org/kl/fileadmin/pdfdownloads/Acts/ClimateChangeActNo11of2016.pdf

The Daily Nation. (2020, June 2). *Universities to teach climate change*. Retrieved from https://www.nation.co.ke/

The Nature Conservancy. (2019). Estrategia hídrica en Ecuador. *Nature*. Retrieved from https://www.nature.org/es-us/sobre-tnc/donde-trabajamos/tnc-en-latinoamerica/ecuador/estrategia-hidrica

Theobald, E. J., Alison, C., Janneke, H., Mary, P. W., & Scott, F. (2015). Women Learn More from Local than Global Examples of the Biological Impacts of Climate Change. *Frontiers in Ecology and the Environment*, *13*(3), 132–137. doi:10.1890/140261

Thompson, P. B., & Tansey, G. (2017). *The ethical implications of global environmental change. The Oxford Handbook of Food Ethics*. Oxford University Press.

Tolppanen, S., Aarnio-Linnanvuori, E., Cantell, H., & Lehtonen, A. (2017). Dealing with a Wicked Problem – A model for holistic climate change education. *Kasvatus, 48*(5), 456–468. https://erepo.uef.fi/handle/123456789/8285

Torres, P., Cruz, C., & Patiño, P. (2009). Índices de calidad de agua en fuentes superficiales utilizadas en la producción de agua para consumo humano: Una revisión crítica. *Revista Ingenierías Universidad de Medellín, 8*(15), 79–94.

Trott, C. D. (2022). Climate change education for transformation: Exploring the affective and attitudinal dimensions of children's learning and action. *Environmental Education Research, 28*(7), 1023–1042. doi:10.1080/13504622.2021.2007223

Tuana, N. (2013). Gendering climate knowledge for justice: Catalyzing a new research agenda. In M. Alston & K. Whittenbury (Eds.), *Research, action and policy: Addressing the gendered impacts of climate change* (pp. 17–31). Springer. doi:10.1007/978-94-007-5518-5_2

UK Student Climate Network. (2019). A Manifesto for Tackling the Climate Change Crisis. *The Guardian*. https://www.theguardian.com/commentisfree/2019/mar/15/uk-student-climate-network-manifestodeclare-emergency

UN Climate Change Conference. (2021). *COP 26 Goals*. Retrieved from COP 26: https://ukcop26.org/cop26-goals/

UN Human Rights Council. (2010). *Study of the Office of the United Nations High Commissioner for Human Rights on challenges and best practices in the implementation of the international framework for the protection of the rights of the child in the context of migration*. Author.

UN. (2022). *UN Agenda 2030—Directorate of Programme Co-ordination*. Directorate of Programme Co-Ordination. https://www.coe.int/en/web/programmes/un-2030-agenda

UNCC Learn. (2020, July 22). *The journey begins towards developing Kenya's climate change learning strategy*. https://www.uncclearn.org/news/the-journey-begins-towards-developing-kenyas-national-climate-change-learning-strategy/

UNDP. (2017). *Climate change, migration and displacement: The need for a risk-informed and coherent approach*. https://cdn.odi.org/media/documents/11874.pdf

UNDP-Zimbabwe Human Development Report. (2017). *Climate Change and Human Development: Towards Building a Climate Resilient Nation*. https://www.undp.org/sites/g/files/zskgke326/files/migration/zw/UNDP_ZW_2017ZHDR_Briefs---Climate-Change-and-Education.pdf

UNESCO & UNEP. (2011). Climate change starter's guidebook: An issues guide for educational planners and practitioners. Author.

UNESCO & UNFCCC. (2016). *Action for Climate Empowerment: Guidelines for Accelerating Solutions through Education, Training and Public*. https://unfccc.int/sites/default/files/action_for_climate_empowerment_guidelines.pdf

UNESCO &UNEP. (2011). *Climate change starters guide book*. UNESCO.

UNESCO. (1997). Adult Environmental Education: Awareness and Environmental Action. UNESCO.

UNESCO. (2005). *Education for All: The Leap to Equality*. EFA Global Monitoring Report'03. portal.unesco.org/education/en/ev.phpurl_id=23023&url_do=do_topic&url_section=201.html

UNESCO. (2009). *Climate Change Education Inside and Outside the Classroom*. UNESCO.

UNESCO. (2009). *Report of the International Seminar on Climate Change Education*. UNESCO.

UNESCO. (2011). *National journeys towards education for sustainable development: Reviewing experiences from Chile, Indonesia, Kenya, Netherlands, Oman*. Paris: United Nations Educational, Scientific and Cultural Organisation. https://unesdoc.unesco.org/ark:/48223/pf0000192183

UNESCO. (2012). *Climate change education for sustainable development in small islands in developing states*. UNESCO.

UNESCO. (2014). *UNESCO Roadmap for implementing the Global Action Program on Education for Sustainable Development*. United Nation Educational Scientific and Cultural Organisation.

UNESCO. (2015). *Not just hot air: Putting climate change education into practice*. Paris: UNESCO. https://unesdoc.unesco.org/ark:/48223/pf0000233083

UNESCO. (2015). *SDG4 - Education 2030: framework for action*. UNESCO.

UNESCO. (2016, February 9). *UNEP launches Kenya Green University Network*. https://www.unesco.org/en/articles/unep-launches-kenya-green-university-network

UNESCO. (2017). *Education for Sustainable Development Goal: Learning Objectives*. United Nations Educational Scientific and Cultural Organisation.

UNESCO. (2017). Education for sustainable development goals: Learning objectives. UNESCO.

UNESCO. (2017). *Gender equality in and through education: Policy guidelines*. Retrieved from https://unesdoc.unesco.org/ark:/48223/pf0000245964

UNESCO. (2018). *Ensuring the right to equitable and inclusive education: results of the ninth consultation of Member States on the implementation of the UNESCO Convention and Recommendation against Discrimination in Education*. UNESCO.

UNESCO. (2019a). *Enforcing the right to education for refugees*. UNESCO.

UNESCO. (2019b). *Indigenous peoples' right to education: overview of the measures supporting the right to education for indigenous peoples reported by Member States in the context of the Ninth Consultation on the 1960 Convention and Recommendation against Discrimination in Education*. UNESCO.

UNESCO. (2020). *Education for Sustainable Development: A Roadmap*. United Nations Educational, Scientific and Cultural Organization.

UNESCO. (2021a). *UNESCO World Conference on Education for Sustainable Education*. Berlin: UNESCO.

UNESCO. (2021b). *UNESCO declares environmental education must be a core curriculum component by 2025*. https://www.unesco.org/en/articles/unesco-declares-environmental-education-must-be-core-curriculum-component-2025

UNESCO. (2022). *Kenya: Climate change communication and education*. https://education-profiles.org/sub-saharan-africa/kenya/~climate-change-communication-and-education

UNFCCC COP 21. (2015). *The twenty-first session of the Conference of the Parties (COP) and the eleventh session of the Conference of the Parties serving as the meeting of the Parties to the Kyoto Protocol*. Author.

UNFCCC. (2002). *United Nations Framework Convention on Climate Change*. United Nations. Retrieved from https://unfccc.int/files/essential_background/background_publications_htmlpdf/application/pdf/conveng.pdf

UNICEF. (2008). *United Nations International Children Emergency Fund Curriculum report card*. Working Paper Series, Education Section, Programme Division.

UNICEF. (2011). *UNICEF'S Child friendly schools Kenya case study*. Nairobi: Management Project Office of CFS. Available at: http://cdpen. cbern.cn/cdp/tjwz /297.htm

UNICEF. (2019). *1 de cada 3 personas en el mundo no tiene acceso a agua potable*. Retrieved from https://uni.cf/2LaVcqR

UNICEF. (2019). *It is getting hot: call for education systems to respond to the climate crisis, perspectives from East Asia and the Pacific*. UNICEF East Asia and Pacific Regional Office.

UNICEF. (2021). *The climate crisis is a child rights crisis Introducing the Children's Climate Risk Index*. UNICEF Division of Communication. www.unicef.org

United Nations (UN). (1992). *United Nations Framework Convention On Climate Change (UNFCCC)*. FCCC/INFORMAL/84. GE.05-62220 (E) 200705. https://unfccc.int/resource/docs/convkp/conveng.pdf

United Nations (UN). (1993). *United Nations Conference on Environment and Development*. Rio de Janeiro: A/cnf.151/26/Rev.1.

United Nations (UN). (2023). *The United Nations World Water Development Report 2023: Partnerships and Cooperation for Water*. UNESCO. https://www.unesco.org/reports/wwdr/2023/en/download

United Nations Development Programme. (1995). *Human development report*. Oxford University Press.

United Nations Development Programme. (2006). *Global partnership for development*. Annual Report. https://www.undp.org/sites/g/files/zskgke326/files/publications/UNDP-in-action-2006-en.pdf

United Nations Development Programme. (2010). *Designing climate change adaptation initiatives: A UNDP toolkit for practitioners*. UNDP Bureau for Development Policy.

United Nations Educational, Scientific and Cultural Organization (UNESCO). (2011). *Climate Change Starters' Guidebook*. Retrieved from https://www.perlego.com/book/1669895/climate-change-starters-guidebook-an-issues-guide-for-education-planners-and-practitioners-pdf

United Nations Environment Programme (UNEP). (2008). *Nairobi City Development Strategy Top Priority for 21st Century Future of the Kenyan Capital*. https://www.unep.org/documents.multilingual/default.asp?documentid=504&articleid=5564&l=en

United Nations Framework Convention on Climate Change (UNFCC). May 9, 1992, S. Treaty Doc No. 102-38, 1771.

United Nations Framework Convention on Climate Change. (2003). *İklime özen göstermek: İklim değişikliği sözleşmesi ve Kyoto Protokolü için kılavuz*. https://unfccc.int/resource/docs/publications/caring_trk.pdf

United Nations International Children's Emergency Fund. (2015). *Level and Trends in Child Mortality Report*. United Nations Children's Fund.

United Nations Office for Disaster Risk Reduction. (2008). *Gender perspectives: Integrating disaster risk reduction into climate change adaptation*. United Nations International Strategy for Disaster Reduction. Available at: http://www.unisdr.org/preventionweb/files/3391_GenderPerspectivesIntegratingDRRCCGood2 0Practices.pdf

United Nations Women. (2016). *Rural women in climate change in Jordan. In Brief Report*. https://jordan.unwomen.org/sites/default/files/Field%20Office%20Jordan/Attachments/publications/2018/UN%20Women-RuralWOmenJordan-Brief-WEB%20EN.pdf

United Nations WomenWatch. (2009). *Fact Sheet. Women, Gender equality and climate change*. https://www.un.org/womenwatch/feature/climate_change/

United Nations. (2020). *Sustainable development goals - Guidelines for the use of the SDG logo including the colour wheel, and 17 Icons*. https://www.un.org/sustainabledevelopment/wp-content/uploads/2019/01/SDG_Guidelines_AUG_2019_Final.pdf

United Nations. (2021). *What is climate change?* https://www.un.org/en/climatechange/what-is-climate-change#

United Nations. (2022). *World Population Prospects 2022*. Department of Economic and Social Affairs Population Division. https://population.un.org/wpp/

United States Educational, Scientific and Cultural Organization (UNESCO). (2005). *Guidelines and recommendations for reorienting teacher education to address sustainability*. United Nations Educational, Scientific and Cultural Organization. https://unesdoc.unesco.org/ark:/48223/pf0000143370

United States Educational, Scientific and Cultural Organization (UNESCO). (2018). *Guidebook on education for sustainable development for educators: Effective teaching and learning in teacher education institutions in Africa*. United Nations Educational, Scientific and Cultural Organization. https://unesdoc.unesco.org/ark:/48223/pf0000367474

University of Missouri-Columbia. (2017). *Climate change misconceptions common among teachers, study finds*. ScienceDaily. www.sciencedaily.com/releases/2017/06/170607123835.htm

Urama, K. C., & Ozor, N. (2010). *Impacts of climate change on water resources in Africa: the role of adaptation*. https://www.researchgate.net/publication/267218899_Impacts_of_climate_change_on_water_resources_in_Africa_the_role_of_adaptation#fullTextFileContent

US Environmental Protection Agency (EPA). (2023). *Inflation Reduction Act: Evaluating Greenhouse Gas (GHG) Reduction Measures*. https://www.epa.gov/inflation-reduction-act/evaluating-greenhouse-gas-ghg-reduction-measures

US EPA. (2022). *Heat Island Impacts* [Overviews and Factsheets]. https://www.epa.gov/heatislands/heat-island-impacts

USAID (2013). *Alternative Education: Mobile schools for Nomadic Pastoralists in North Eastern province*. Available from http//Kenya.Usaid.gov/programs/education.and – Yo

USGCRP. (2018). *Fourth national climate assessment, Vol. 2: Impacts, risks, and adaptation in the United States*. Rep, US Glob. Change Res. Progr. https://nca2018.globalchange.gov/downloads/NCA4_2018_FullReport.pdf

Valderrama-Hernández, R., Rubio, L. A., Sánchez-Carracedo, F., Franco, D. C., González, S. S., Gil-Doménech, D., Vidal-Raméntol, S., & Miñano, R. (2019). Does the Spanish university system teach sustainability? Perception of the students of four universities. *Visión Alumn. Cuatro Univ*, *23*, 221–245.

Valkenburg, P. M., Semetko, H. A., & Vreese, C. H. (1999). The effects of news frames on readers' thoughts and recall. *Communication Research*, *26*(5), 550–569. doi:10.1177/009365099026005002

Van Breda, A. D. (2019). Developing the notion of Ubuntu as African theory for social work practice. *Social Work*, *55*(4), 439–450. doi:10.15270/52-2-762

Varela-Losada, M., Arias-Correa, A., Pérez-Rodríguez, U., & Vega-Marcote, P. (2019). How Can Teachers Be Encouraged to Commit to Sustainability? Evaluation of a Teacher-Training Experience in Spain. *Sustainability, 11*(16), 4309. doi:10.3390u11164309

Vergunst, F., & Berry, H. L. (2022). Climate Change and Children's Mental Health: A Developmental Perspective. *Clinical Psychological Science. Volume, 10*(4), 767–785. doi:10.1177/21677026211040787 PMID:35846172

Vincent, S., & Dutton, K. (2014). Integration of interdisciplinary environmental and sustainability education and research and urban sustainability. *Journal of Environmental Studies and Sciences, 7*(1), 112–120. Advance online publication. doi:10.100713412-014-0202-z

Voulvoulis, N. (2018). W*ater reuse from a circular economy perspectiva and potential risks from an unregulated approach. Retrieved from Current Opinion in Environmental Science & Healh, 2*(3), 32–45. doi:10.1016/j.coesh.2018.01.005

Vu, H. T. (2013). The online audience as gatekeeper: The influence of reader metrics on news editorial selection. *Journalism.* Advance online publication. doi:10.1177/1464884913504259

W., Tolppanen, S., Gokpinar, T., Goldman, D., & Ben-Zvi ,A. O. (2021, June). *A multi-country comparison of climate change curricula in secondary schools.* Paper presented at the Annual Australasian Science Education Research Association ASERA Conference, South Australia, University of South Australia.

Wals, A. E. (2011). Learning Our Way to Sustainability. *Journal of Education for Sustainable Development, 5*(2), 177–186. doi:10.1177/097340821100500208

Walshe, N., & Tait, V. (2019). Making connections: A conference approach to developing transformative environmental and sustainability education within initial teacher education. *Environmental Education Research, 25*(12), 1731–1750. doi:10.1080/13504622.2019.1677858

Wapner, P. (1996). Environmental Activism and World Civic Politics. *World Politics, 49*(1), 62–89.

Wapner, P. (2016). Is sustainability possible? *Environmental Sociology, 2*(1), 1–8.

Warren, K. J. (1990). The power and the promise of ecological feminism. *Environmental Ethics, 12*(2), 125–146. doi:10.5840/enviroethics199012221

Watts, N., Amann, M., Ayeb-Karlsson, S., Belesova, K., Bouley, T., Boykoff, M., Byass, P., Cai, W., Campbell-Lendrum, D., Chambers, J., Cox, P. M., Daly, M., Dasandi, N., Davies, M., Depledge, M., Depoux, A., Dominguez-Salas, P., Drummond, P., Ekins, P., ... Costello, A. (2018). The Lancet Countdown on health and climate change: From 25 years of inaction to a global transformation for public health. *Lancet, 391*(10120), 581–630. doi:10.1016/S0140-6736(17)32464-9 PMID:29096948

WCED (World Commission on Environment and Development). (1987). *Our Common Future.* Oxford University Press.

WCED. (1987). *Report of the World Commission on Environment and Development: Our Common Future*. http://www.un-documents.net/A372BC4E-993E-4C10-B368-9E75816E59DB/FinalDownload/DownloadId-1990161B7AFC5CB358F7280210D7A48F/A372BC4E-993E-4C10-B368-9E75816E59DB/our-common-future.pdf

Weber, E. P., & Khademian, A. M. (2008). Wicked problems, knowledge challenges, and collaborative capacity builders in network settings. *Public Administration Review*, *68*(2), 334–349. doi:10.1111/j.1540-6210.2007.00866.x

Wedell, M. (2009). *Planning for Educational Change: Putting people and their contexts first.* Continuum International Publishing Group.

Weimann, G. (2015). Chapter. In J. D. Wright (Ed.), International Encyclopedia of the Social & Behavioral Sciences (2nd ed., pp. 291-293). Elsevier. doi:10.1016/B978-0-08-097086-8.95051-7

Weiss-Gal, I. (2008). The person-in-environment approach: Professional ideology and practice of social workers in Israel. *Social Work*, *53*(1), 65–75. doi:10.1093w/53.1.65 PMID:18610822

Wenger, E. (2011). *Communities of practice: A brief introduction.* Retrieved April 27, 2023, from https://www.ohr.wisc.edu/cop/articles/communities_practice_intro_wenger.pdf

Wen, J., Wan, C., Ye, Q., Yan, J., & Li, W. (2023). Disaster Risk Reduction, Climate Change Adaptation and Their Linkages with Sustainable Development over the Past 30 Years: A Review. *Int J Disaster Risk Sci*, *14*(1), 1–13. doi:10.100713753-023-00472-3

West, P., Igoe, J., & Brockington, D. (2006). Parks and Peoples: The Social Impact of Protected Areas. *Annual Review of Anthropology*, *35*(1), 251–277. doi:10.1146/annurev.anthro.35.081705.123308

White, J. (2020). The climate emergency and the transformed school. *Journal of Philosophy of Education*, *54*(4), 867–873. doi:10.1111/1467-9752.12461

Wiedmann, F., & Salama, A. (2012). *The Role of Architecture in Producing Urban Qualities for Sustainability: Implications for the Future of Architectural Education*. Academic Press.

Wiek, A., Ness, B., Brand, F. S., & Schweizer-Ries, P. (2014). Fostering sustainability transitions in the energy domain through ethical leadership. *Sustainability*, *6*(12), 7865–7886.

Williams, D. R., & Brown, J. D. (2012). *Learning gardens and sustainability education: Bringing life to schools and schools to life*. Routledge.

Williams, D. R., Brule, H., Kelley, S. S., & Skinner, E. A. (2018). Science in the learning gardens (SciLG): A study of students' motivation, achievement, and science identity in low-income middle schools. *International Journal of STEM Education*, *5*(8), 1–14. doi:10.118640594-018-0104-9 PMID:30631698

Wilson, H., Leydon, J., & Wincentak, J. (2017). Fieldwork in geography education: Defining or declining? The state of fieldwork in Canadian undergraduate geography programs. *Journal of Geography in Higher Education*, *41*(1), 94–105. doi:10.1080/03098265.2016.1260098

Winston, C. N. (2015). Points of convergence and divergence between existential and humanistic psychology: A few observations. *The Humanistic Psychologist, 43*(1), 40–53. Advance online publication. doi:10.1080/08873267.2014.993067

Wise, S. B. (2010). Climate Change in the Classroom: Patterns, Motivations and Barriers to instruction among Colorado science teachers. *Journal of Geoscience Education, 58*(5), 297–309. doi:10.5408/1.3559695

Wisner, B., Blaikie, P., Cannon, T., & Davis, I. (2004). *At risk, natural hazards. People's Vulnerability and Disasters*. Routledge.

Wodon, Q., Montenegro, C., Nguyen, H., & Onagoruwa, A. (2018). *Missed Opportunities: The High Cost of Not Educating Girls. The Cost of Not Educating Girls Notes Series*. The World Bank. doi:10.1596/29956

Women, U. N. (2020). *Making gender equality central to addressing climate change*. Retrieved from https://www.unwomen.org/-/media/headquarters/attachments/sections/library/publications/2020/issue-brief-making-gender-equality-central-to-addressing-climate-change-en.pdf?la=en&vs=5651

Women, U. N. (n.d.). *Women, Gender Equality and Climate Change*. UN Women. Retrieved from https://www.un.org/womenwatch/feature/climate_change/downloads/Women_and_Climate_Change_Factsheet.pdf

World Bank. (2010). *Knowledge, Productivity and Innovation in Nigeria: Creating a New Economy*. The World Bank.

World Bank. (2018). *Groundswell: Preparing for internal climate migration*. Author.

World Declaration on Education for All. (1990). *Education for all and Program of Action*. Common Wealth.

World Health Organization. (2014a). *Gender, climate change and health*. https://apps.who.int/iris/bitstream/handle/10665/144781/9789241508186_eng.pdf

World Health Organization. (2014b). *Quantitative Risk assessment of the effects of climate change on selected causes of death, 2030s and 2050s*. Author.

World Health Organization. (2022). *Maternal and reproductive health*. https://www.who.int/data/gho/data/themes/maternal-and-reproductive-health

World Meteorological Organization. (2022). *State of the global climate 2021*. https://public.wmo.int/en/media/press-release/four-key-climate-change-indicators-break-records-2021

World Water Week (WWW). (2022). *A Rising Tide: Shared Vision for Women in Water Diplomacy*. Online & On-site session, 30 August 2022. worldwaterweek.org/event/10314-a-rising-tide-shared-vision-for-womenin-water-diplomacy.

Xue, D. M., Baets, B. D., Cleemput, O. V., Hennessy, C., Berglund, M., & Boeckx, P. (2012). Use of a Bayesian isotopemixingmodel to estimate proportional contributions of multiple nitrate sources in surface water. *J. Environ. Pollut., 161*, 43–49. doi:10.1016/j.envpol.2011.09.033 PMID:22230066

Y. K. Hanusch, F., Ramaprasad, J., Skjerdal, T., Hasim, M. S., Muchta, N., . . . Less, S. B. (2017). Journalists' Development Journalism Role Perceptions: Select countries in Southeast Asia, South Asia, and sub-Saharan Africa. *Journalism Studies*, 576-594. https://www.tandfonline.com/doi/full/10.1080/1461670X.2016.1254060

Yambo, J. M. O. (2012). *Determinants of KCSE Examination Performance in SDA Sponsored Schools: A Key to Academic Promotion to the next Level of Learning*. Lambert Academic Publishing.

Yambo, J. M. O. (2022a). Involvement of Government and Non-Governmental Organizations' Teamwork in the Advancement of Quality Education in Kisumu West Sub County, Kenya. *Journal of Advances in Education and Philosophy, 6*(4), 194–202. doi:10.36348/jaep.2022.v06i04.001

Yambo, J. M. O. (2022b). Influence of the Principals Decision Making Skills on Students' Academic outcome in Public Teachers Training Colleges in Nyanza Region, Kenya. *Journal of Research Innovation and Implications in Education, 6*(3), 136–144.

Yaro, S. A. (2018). Conflicts of Interest and Climate Change Communication in the News Media. *Science Communication, 39*(1), 8–34.

Yli-Panula, E., Jeronen, E., Koskinen, S., & Mäki, S. (2022). Finnish University Students' Views on Climate Change Education and Their Own Ability to Act as Climate Educators. *Education in Science, 12*(3), 169. doi:10.3390/educsci12030169

Young, M. (2014). Curriculum theory: What it is and why it is important. *Cadernos de Pesquisas, 44*(151), 191–201. doi:10.1590/198053142851

Zani, M. T. (2022, May 18). *Women in rural Bangladesh pay more for rising cost of climate disasters*. Thomson Reuters Foundation. https://news.trust.org/item/20220518143335-m17kx

Zhang, H., Lai, K. K., & Zhang, M. (2020). Green and Reverse Logistics. In R. R. Negenborn & Z. Lukszo (Eds.), *Smart Port Perspectives* (pp. 187–211). Springer.

Zhang, Y., Shi, P., Li, F. D., Wei, A. L., Song, J. X., & Ma, J. J. (2018). Quantification of nitrate sources and fates in rivers in an irrigated agricultural area using environmental isotopes and a Bayesian isotope mixing model. *The Journal of Chemical Thermodynamics, 208*, 493–501. doi:10.1016/j.chemosphere.2018.05.164 PMID:29886338

Zimbabwe Human Development Report. (2017). *Human Development: Towards Building a Climate Resilient Nation Climate Change and Education*. The Embassy of Sweden in Harare Zimbabwe.

About the Contributors

Jared Sagini Keengwe is a Professor at the University of North Dakota (UND) and serves as the Director of the Undergraduate Elementary Education program. He is the editor-in-chief of two IGI Global Book Series: Advances in Higher Education and Professional Development (AHEPD) and Advances in Early Childhood and K-12 Education (AECKE). Prof. Keengwe has published several books and manuscripts in refereed journals. His work was honored with the UND Foundation/McDermott Faculty Award for Excellence in Academic Advising. He was a recipient of the North Dakota Spirit Faculty Achievement Award in recognition of significant contributions in teaching, research, and service. At the national level, Prof. Keengwe was a recipient of the International Institute of Education Fellowship under the Carnegie African Diaspora Fellowship Program (CADFP) – a scholar fellowship program for educational projects at African higher education institutions. The program is funded by a grant from Carnegie Corporation of New York (CCNY).

Benard Omenge Nyatuka is an Associate Professor and Dean, School of Education and Human Resource Development, Kisii University, Kenya. He obtained Doctorate in Comparative Education (University of South Africa, 2015), M.Ed and B.Ed degrees from Egerton (2008) and Kenyatta (1997) Universities in Kenya, respectively. Previously, he was a lecturer at Masinde Muliro University of Science & Technology, and served at the Ministry of Education as a Senior Quality Assurance and Standards Officer in Kenya. He has a decade of teaching experience in Geography and Mathematics at the secondary school cycle of education. Furthermore, he has authored several articles in peer reviewed journals and similar publication platforms, as well as chapters in scholarly books in the fields of education and development. Courtesy of UNICEF, he served as the technical advisor in the development of community service learning guidelines and materials at the Kenya Institute of Curriculum Development to facilitate implementation of Competency-Based Education in the country between 2021 and 2022. His research interests include Comparative and International Education, ethics in higher education, global citizenship education, school-family-community partnerships, as well as education for sustainable development.

* * *

Bilge Abukan is an Assist. Prof. at Ondokuz Mayif University (Turkey), Faculty of Health Sciences, Department of Social Work. She received a doctorate degree at the Clinical Social Work. At the same time, she has been continuing her law education at Ondokuz Mayıs University Faculty of Law. She has training and experience in child protection in emergencies, social work in disasters, psychological first aid and psychosocial support and secondary trauma. Her main research interests include children in need of protection, clinical social work, ecological social work, social work education, women studies, early marriage and gender.

Jenny Bladow received her doctorate in Educational Practice and Leadership from the University of North Dakota in 2023. In conjunction with her doctorate, she also completed the requirements to obtain a College Teaching Certificate along with both a principal and superintendent credential. Before transitioning into higher education, Dr. Bladow taught middle school science for ten years. She has been in her role as Director of Teacher Education at the University of North Dakota since 2018. Her research interests include ways to incorporate environmental education into teacher preparation, partnerships between universities and communities, and how to support students during the teacher preparation process.

Magdalene Dimba is the Dean of the School of Humanities and Social Sciences, and a Senior lecturer in Education Management at Strathmore University, Kenya. She holds a PhD in Educational Administration and Planning from the Catholic University of Eastern Africa, a Master's in Economics of Education and a Bachelor of Education in Economics and Geography from Moi University, Kenya. Magdalene is an experienced educationist, administrator, and researcher. She has an interest in teacher education, and has researched in education quality, access and equity, together with women's education.

Ebua Jarvis Ebua is a Ph.D. candidate in the Department of Philosophy, Faculty of Arts, at the University of Bamenda, Cameroon. He is a holder of Masters of Science in Logistics and Supply Chain Management from the Higher Institute of Transport and Logistics, University of Bamenda with another Master of Arts in Philosophy from the Catholic University of Cameroon Bamenda. A passionate researcher, with a number of publications, and have facilitated a number of international conferences, a development and Humanitarian professional, and has served as Monitoring and

Evaluation Expert in the Anglophone Crisis Emergency Response Project. He is the founder of Love Union a Non-Governmental Organization, in Cameroon.

Şeyda Emekci, MSc, PhD, received her bachelor's, master and doctoral degrees from Middle East Technical University (METU). Worked as an associate at the Development Agency and the General Director of Nature Conservation and National Parks, in Turkey from 2014 to 2017. Postdoctoral researcher at East London University, in the UK in 2019. Currently teaches in the Architecture Department at Ankara Yıldırım Beyazıt University (AYBU). Has been involved in various change projects at the different architectural levels in public and private sectors and with civil associations in Turkey since 2011. Specialization areas include sustainability, energy efficiency, smart homes, architectural education, housing economics and housing policy.

Billiah Gisore (PhD) is a lecturer in the school of education at Kisii University, Kenya. She is co-author of the books: Understanding Sustainability with Pedagogical Practice, Competence Based Curriculum: A Handbook for Teachers, Handbook of Research on Innovative Digital Practices and Globalization in Higher Education and Understanding Sustainability with Pedagogical Practice: A contribution from geography education toward sustainability. Her professional background is in educational research and evaluation. Her interests are in Teacher Education, Equity, Sustainability, Climate Change Education.

Kübra Nur Görmüş is a research assistant at Maltepe University, Faculty of Humanities and Social Sciences, Department of Social Work. She received her master degrees at the İstanbul University- Cerrahpaşa. She is a PhD student in the Department of Social Work at the University of Ankara. Her research interests are child and youth welfare, school social work, social work education, domestic violence, trauma and attachment.

Esther Muthoni King'ori has a Master's degree in Communication Studies from the University of Nairobi and a PhD in Communication with a specialisation in Media Studies from Daystar University, Nairobi. Her PhD thesis investigated the ethical frameworks of news bloggers who are not mainstream journalists. Esther is a lecturer at Strathmore University, Kenya where she teaches communication and media studies. Prior to joining academia, she worked as a journalist and news editor with several media organisations in Nairobi, Kenya. Her research interests are in media ethics, film studies and media framing.

Arturo Luque is full professor at the Technical University of Manabí, Ecuador, and member of Euro-Mediterranean Observatory on Public Policies and Democratic Quality at the University King Juan Carlos (Spain). He holds a Ph.D. in Social Sciences and Law and a Master of two years about Labor Relations. He supervises several doctoral theses (Ph.D.) in URJC (Spain) and in the Basque Country (GEZKI Institute) like a director and co-director. It has 45 indexed articles and his latest publications includes topics about corporate social (Ir)responsibility (How corporate social (ir)responsibility in the textile sector is defined, and its impact on ethical sustainability: An analysis of 133 concepts), transnational companies (The transnational textile companies relationship with environment: a Delphi analysis approach) and their relations with ethics (Corruption in the transnational textile industry: an exception or the rule?), social economy and cooperative system like economic alternative (Socially responsible public management: case spinning development in Ecuador), media analysis and social conflicts (Analysis of the indigenous uprising of Ecuador, 2019).

Falak Shad Memon is Assistant professor associated with Institute of business management since last eight years. Her doctorate is in Environment and energy management. Her key research areas are Climate change, Gender studies, and Sustainable development. Her work is published in impact factor and HEC recognized journals. She has also presented her researches various international conferences.

Evelyn Njurai is currently a lecturer of Curriculum Studies and Mathematics Education courses in the Department of Curriculum, Instruction, and Media and Coordinator of Teaching Practice in the School of Education at Kisii University. She has a wide experience in qualitative research in and out of academia. Her research interests include qualitative research, Mathematics education in the context of language diversity, Curriculum, E-learning, and STEM Education. Evelyn has published internationally and locally in peer-reviewed journals and book chapters and is the author of Dictionary of Mathematics for Secondary Schools, Kenya (2019). She is a co-editor of a handbook for the Competency-Based Curriculum in Kenya and co-author of two book chapters in the book. Evelyn is a member of various professional bodies and is involved in community work through volunteer service in teaching mathematics to marginalized communities in Kenya.

George Mokua Ogendi is a Senior Lecturer and former Chairman, Department of Environment, Natural Resources Management and Aquatic Sciences in the School of Agriculture and Natural Resources Management, Kisii University, Kenya. He obtained a Doctorate in Environmental Sciences from Kisii University in 2017, MSc. Environmental Science from Arkansas State University (2009),

BSc. Agricultural Education and Extension (2006) and Diploma in Agricultural Education and Extension from Egerton (2003). Dr. Ogendi is an ISO certified auditor of Quality Management Systems and a Lead expert in Environmental Impact Assessment in Kenya. He has authored several articles in peer reviewed journals and related online platforms. He is a seasoned writer on environmental reports. Dr. Ogendi has done consultancies in community development. His research interests include Environmental Impact assessment, Water quality, Biomonitoring, Global climate change, Wetland ecology, Agroforestry, Climate smart agriculture as well as education for sustainable development.

Arkom Palamanit is Assistant Professor at Prince of Songkla University, Thailand. Dr Arkom is working on biomass conversion technologies, biomass upgrading, biomass biorefinery, Hybrid power generation, energy management and sustainable energy systems.

José Humberto Ponce Ponce is a junior researcher at Universidad Técnica de Manabí.

Cristina Raluca Gh. Popescu is Full-Professor Habil. at the University of Bucharest, Bucharest, Romania, PhD. Supervisor and Doctoral School Member at The Bucharest University of Economic Studies, Bucharest, Romania, and an award-winning Scientific Researcher. Having more than 23 years of experience, focuses on helping students learn and apply the subjects more effectively, and centers on improving scientific research excellence. Values a special environment where her in-depth knowledge of creative thinking and academic writing can help deliver accurate research results. She is focused on becoming an outstanding researcher, centered on interdisciplinary topics, in a multicultural environment, constantly driven by passion in promoting the Sustainable Development Goals (SDGs). Among the main goals, showing a positive attitude in exploring the endless possibilities of the transformative power of individuals deeply concerned with health and well-being, education, innovation, and responsible production and consumption may be found among the top priorities on her agenda.

Maureen Syallow has extensive experience in media and conflict studies. She has a Masters Degree in International Studies from the University of Nairobi and a PhD in Mass Communication (Media and Conflict) from Jomo Kenyatta Univesity of Agriculture and Technology. Among the courses she has taught are political science, international studies, media and conflict, research methods among others. Maureen is the Research Director at the School of Humanities and Social Sciences, Strathmore University, Nairobi, Kenya. She coordinates research activities in the

School. She is also in charge of identifying research collaborations between the School and other institutions. Her research interests are in media and conflict, media and diplomacy, international studies, and political science.

Kuaanan Techato received the B.Eng. (ME) degree from the Prince of Songkla University, Thailand, in 1995, the M.Eng. (IE) degree from Chulalongkorn University, in 1999, the M.Sc.(EBM) degree from Warwick University, in 1999, and the Ph.D. degree from Chulalongkorn University, in 2008. He is currently working as an Assistant Professor with the Prince of Songkla University, Hatyai campus. He is also working as the Dean of the Faculty of Environmental Management, Prince of Songkla University. He has been a keynote speaker and an invited speaker at many international conferences. He has published many technical articles to various journals and international conferences. His research interests include renewable energy, heat-pump, power systems, and control. He is also involved with many journals as an editor or an associate editor and a successful organizer for many international conferences.

Sudeep Uprety is a Kathmandu based researcher specializing in development communications research. Sudeep is currently pursuing his PhD from Prince of Songkla University, Thailand and has published several book chapters on climate change and sustainable development, international relations, mental health, among others.

Seema Yadav, Ph.D., is working as an Assistant Professor at the Bhopal School of Social Sciences in the Department of Education.

Onyango M. John Yambo, PhD, Holds B.Th (Bugema, Uganda) PGDE (Maseno) M.Ed (UEA-Baraton) PhD (Moi University). He specialises in Educational Administration and Policy Studies. He is a Senior Lecturer of Educational Administration and Policy Studies in the School of Education at Kisii University and an Associate Chaplain of Kisii University SDA Church in Kenya. He has been a postgraduate Coordinator, school of education. He has more than fifteen years of experience in the Education, research and training industry in Kenya, Tanzania, Uganda and Rwanda. He has published books, several refereed articles, book chapters, Learning Modules, conference proceedings and reports. He is also a consultant in Educational Administration and policy matters and an external examiner for various universities in East Africa.

Filiz Yildirim is a professor at Ankara University (Turkey), Faculty of Health Sciences, Department of Social Work. She was awarded a "Doctorate Research Scholarship" from TUBITAK to carry out a part of her doctorate research at the

Department of Sociology, Criminology & Law at the University of Florida, the USA in 2011. She received the "International Young Scholar Award" from the Society for Research on Adolescence in Canada in 2012. She worked as a "Visiting Research Associate" at London University, Institute of Education, Department of Childhood, Families & Health in 2014. Her main research interests include woman studies, early (child) marriage, health literacy, sustainability. She is also a post-doctoral fellow at Hacettepe University (Turkey), Department of Sociology".

Index

A

Accreditation 170, 174, 185, 218
Adaptation 19, 21, 23, 25-26, 31-32, 35-36, 39, 41-43, 51-52, 57, 59, 62, 64-65, 67-68, 73-75, 78, 80, 83, 85-86, 88-90, 94, 114, 116, 122, 129, 131, 134, 136, 141-142, 146, 165-166, 169, 175, 181, 188, 190, 193, 197, 201, 203, 208, 214-216, 219, 221-222, 224-225, 228, 230, 243-244, 248, 251, 253-254, 256, 258-259, 295, 298, 301, 304
Anthropocene Age 190, 210
Anthropogenic Climate Change 188, 210
Architectural Curricula 171, 173
Architectural Curriculum 168, 181, 185
Architectural Education 168, 170, 173-175, 178-185

B

Barriers 1, 10-11, 61, 63, 66, 68-69, 72-73, 77-78, 81-87, 90, 104, 147, 168, 177, 212, 221, 230, 234
Biodiversity Conservation 113, 115, 122
Bio-Psycho-Social Approach 237, 243, 246, 261
Bio-Psycho-Social Impacts 235, 239

C

Carbon Footprint 29, 32, 99, 101, 122, 210
Change 2-4, 7, 12-14, 16, 19-46, 48-51, 53, 55-66, 68-97, 101-105, 109-114, 116-123, 125-131, 133-148, 150-152, 154-155, 157-158, 160-166, 168-173, 175, 180-184, 187, 189-199, 201-210, 213-262, 271-273, 277, 285, 289-296, 298-306
Climate 3, 8, 12, 14, 16-46, 48-51, 53, 55-64, 66-95, 101-105, 109-114, 116-120, 122-123, 125-148, 150, 152, 155, 157-158, 160-166, 168-171, 173-175, 180-184, 187, 189-199, 201-210, 213-261, 267, 271-272, 277, 285-286, 289-296, 298-306
Climate Adaptation 122, 169, 301
Climate Change 3, 12, 14, 16, 18-46, 49-51, 53, 55-95, 101-105, 109-114, 116-120, 122-123, 125-148, 150-152, 155, 157-158, 160-166, 168-171, 173, 175, 180-184, 187, 189-199, 201-210, 213-261, 272, 277, 285, 288-296, 298-306
Climate Change Adaptation 18-19, 26, 35, 39, 59, 67-68, 73-74, 78, 83, 86, 88, 90, 116, 129, 193, 201, 203, 208, 228, 248, 251, 256, 259, 298, 304
Climate Change Education 16, 18-20, 22-35, 37-42, 44-45, 58-61, 63-66, 68-90, 92-95, 102-105, 110-113, 119, 122, 145-146, 161, 166, 187-192, 194-195, 197-210, 212, 214-216, 218, 220-221, 225-226, 228-234, 288-291, 293, 295, 299-301, 303-306
Climate Change Education Outcomes 212, 214, 216, 221
Climate Justice 26, 111, 122, 249
Community 1, 9-11, 26, 31-33, 42-44, 53, 65, 68-69, 73-74, 76-80, 83-86, 91, 95, 97, 101, 103, 114, 123, 129, 136,

151-152, 164, 167-168, 170, 191, 193, 197-198, 205-206, 213, 217, 219-221, 223, 245, 247-249, 251, 261, 266-267, 290, 294, 297-298, 302-303, 306
Community Partnerships 32
Corporate Social Responsibility (CSR) 101, 123
Curriculum 2, 4-5, 7-8, 10, 16-19, 22-24, 31-33, 39-41, 44-46, 48, 52-53, 55, 57-60, 63, 70, 72, 76, 84, 90, 92, 144, 152, 158, 167-168, 170-173, 177, 179-182, 185, 188-189, 191, 193, 195, 198, 200, 205-207, 210, 212, 216, 218-225, 227-231, 233-234, 248, 288, 291-293, 298, 305

D

Deontology 93-94, 123
Disaster 26, 73, 87-88, 90, 132, 141, 144, 167, 189, 220, 226, 228, 239, 241-242, 258-259, 295, 297
Discourse 15, 18, 20, 22, 64, 70, 125, 127, 130-131, 140-141, 143, 170, 173-174, 177, 186, 257, 290
Dynamic 45, 57, 222, 306

E

Eco-Consciousness 123
Economic Globalization 286
Eco-Philosophy 123
Eco-System Thinking 123
Education 2-50, 52-53, 55, 58-63, 65-66, 68-91, 93-95, 102-105, 110-113, 119, 122, 128, 143, 145-150, 155-158, 160-168, 170-186, 188-210, 212-218, 220-222, 224-234, 238, 240, 243, 267-268, 290-306
Education for Sustainable Development (ESD) 6, 24, 29, 41-42, 64, 119, 188, 197, 210, 288
Education for Sustainable Education 60, 62
Empowerment 54, 68, 73, 79, 83, 91, 104, 219, 232, 242, 246, 257, 262
Environment 1-2, 4, 7, 9, 12-13, 17, 20-22, 24, 27-28, 30, 35, 37, 39, 42-43, 46, 49-57, 60-62, 65, 71, 75, 83, 93-95, 98, 100-101, 103, 106, 110, 112-115, 118, 121-124, 127-129, 131-132, 134-137, 141, 144, 151, 163, 168-169, 171-172, 174-175, 177-181, 183-185, 189-190, 193, 196-197, 200-203, 205-206, 211, 213, 216, 220, 225-226, 228, 230-231, 235-236, 238-239, 243-245, 248, 250, 252-253, 256, 263, 270, 272, 275-276, 279-282, 288, 290-294, 296-299
Environmental Conservation 20, 39, 52, 95, 100, 103, 111, 169, 292
Environmental Education 1-17, 24, 31, 35-38, 42, 47, 58, 60-62, 86-89, 166, 183, 191-194, 200, 202, 207-210, 217, 225-226, 229-232, 291, 298, 301, 303
Environmental Ethics 8, 93-95, 97-100, 110, 115-116, 121, 123, 259-260
Environmental Knowledge 4, 166, 202
Environmental Literacy 6, 15, 217
Environmental Philosophy 116, 118
Environmental Stewardship 100, 110, 112, 123
Ethical Decision-Making 106, 123

G

Gender 29, 63-64, 66-69, 71-74, 76-91, 104, 116, 124, 143, 236, 238-240, 242, 245-248, 252-253, 255-257, 259-260, 289, 294, 299
Gender Discrepancies 71, 91
Gender Inclusiveness 63, 66, 71-72, 77-78, 80-84
Global 6, 18-19, 21, 23, 27, 29, 36, 41-43, 46, 50, 58-60, 64-65, 68, 87, 89, 93, 97, 101, 109, 111, 113, 115-116, 119-121, 127-129, 132-133, 136, 138-141, 145, 147, 149-150, 160-163, 165-166, 168-172, 180-181, 183, 187, 189-191, 193, 196, 199, 202-203, 209, 212-213, 215-216, 225, 227, 229, 231, 236-237, 239, 243, 248, 251, 253-255, 257-261, 265-266, 273, 279, 285-286, 288, 291-292, 298, 301, 303-306
Globalization 119, 257-258, 286
Green House Gas Emission 210

Green Logistics 121, 123
Green Supply Chain Management 98-99, 123

H

Higher Education 4, 8, 11-12, 14, 17-18, 20, 23-24, 26, 32, 35, 38-39, 43, 90, 119, 166, 170, 172-173, 181-182, 186, 192, 225, 227, 230-231, 233, 302, 304

I

Impact 7, 15, 18, 20, 23, 32, 34, 41, 43-44, 52, 54, 66, 75, 77, 82-83, 86, 91-93, 95, 100-104, 116, 122-123, 129, 134, 139, 146, 150, 156, 164, 167, 169-170, 173-174, 177, 189, 195, 198, 201, 204-205, 207, 212-213, 221-222, 235-239, 241, 254-257, 265, 271, 279-280, 282, 288, 290-292, 295, 301, 303
Incorporating 1, 41, 67, 69, 73, 75-77, 93, 95, 98, 108, 110, 112, 162, 170-171, 188, 198, 234, 243, 250, 294
Industrialization 50, 188, 210, 265
Innovative Architectural Discourse 170, 186
Interdisciplinary Collaboration 61, 172, 186
Intergenerational Ethics 93-94, 116, 123
Intersectionality 124, 239, 255

J

Journalism 125-129, 134, 137-140, 142-143

K

Kenyan Curriculum 19, 23-24, 33, 39

L

Livelihood 47, 73-74, 77, 83, 91, 107, 147
Living Beings 93-94, 98, 269, 271
Logistics 92, 95-99, 112-113, 115, 117-119, 121-124

M

Media 21, 36, 41, 90, 122, 125-143, 164, 194, 221, 260, 282
Mitigation 18-19, 21, 23, 26, 31-32, 34-35, 41-43, 52-53, 57-58, 62, 64-65, 67-68, 73-75, 78, 80, 83, 86, 89, 91, 94, 131, 133, 136, 142, 146-147, 169, 175, 181, 183-184, 207, 214-216, 219, 222-225, 228, 268, 291, 301, 303

N

Nature 1-2, 8-9, 15, 21, 42, 45, 93, 98, 100, 110-111, 115-116, 120-121, 123, 127-128, 130-133, 135-137, 141-142, 163, 170, 174-175, 177, 179-180, 192-193, 201-203, 206, 216, 218, 236, 239, 243, 251, 256-257, 263, 265, 271, 283, 285-286
Nepal 125-127, 130, 135-138, 140-141
Nomadic Pastoralist 306

P

P-12 Education 1
P-12 Schools 7, 9-10, 17
Pakistan 63, 65-71, 73, 77, 82, 84, 88-89, 292, 295, 304
Partnerships and Cooperation for Water 285-286
Passion 300, 306
Pedagogical Strategies 18, 202, 212, 223, 234
Pedagogies 8, 31, 41-42, 44, 47, 54-55, 57-58, 62, 185, 216, 219, 222
Pedagogy 4, 16-17, 39, 44, 56, 182, 185, 188, 206, 209, 212, 214, 217, 223, 233, 294, 301
Perceptions 14, 33, 40-42, 47, 55, 57-58, 66, 72, 77, 88-89, 102, 143, 199, 208, 222, 226, 251, 303
Policy 21-22, 24-29, 36-38, 44, 51, 64, 66-67, 69-70, 72, 76, 79-83, 85-86, 88-90, 101, 104, 110, 114-115, 119-

120, 134-136, 140-141, 147, 161, 188, 190, 193, 195, 198-199, 201, 203-204, 208, 221, 227, 230-231, 233, 243-244, 253-254, 258-259, 286, 288, 295-296, 298, 302, 304
Preservice Teachers 1-7, 10-13, 17, 216, 226
Pre-Service Teachers 40-41, 43-50, 52, 54-58, 62
Professional Development 2, 5-6, 10, 16, 28, 181-182, 194, 205-206, 217, 224
Public Policy 25, 286

R

Recourses 91
Renewable Energy 28-29, 74, 85, 94, 111, 124
Resilience 29, 31, 65-69, 73-77, 80, 83-85, 87, 91, 122, 161, 164, 174, 203, 230, 256, 286, 295, 297
Reverse Logistics 96-99, 117, 121-122, 124
Rural 49, 63, 66, 68, 70, 72-79, 83-85, 88-91, 115, 117, 119, 129, 135, 148, 234, 236-242, 244-248, 250-251, 255-258, 260, 262, 275, 300
Rural Women 73-74, 79, 88, 235, 237-239, 241-242, 245-248, 250-251, 256-257, 260, 262

S

School 2, 6-9, 11-12, 14-17, 19, 22, 29, 31-33, 41, 43, 45, 47-48, 51-52, 54-57, 60, 69, 76, 78-79, 116-117, 146-147, 150-151, 154-162, 164, 166-167, 183, 186-187, 189-191, 195-196, 198-201, 203, 208, 210, 215, 219-221, 227, 230-231, 233, 242, 251, 288, 290-293, 297-304
School Education 43, 167
Science Education 6, 59, 166, 194, 200, 206, 208-209, 232
Sindh 63, 66, 69, 86, 89-90
Skills 2, 4, 19, 26-33, 40-44, 46, 54-58, 61-62, 64-66, 68, 73-75, 79, 82-83, 85, 91, 110, 112, 117, 122, 167, 170, 174, 178, 185-189, 191, 193-194, 199, 201, 203, 206, 210, 214-220, 223-224, 234, 237, 248, 290, 296-297, 306
Social Work 235-237, 242-254, 256-257, 260-262
Social Work With Women 242, 262
Socio Cultural 91
Stake Holder 91
Street Children 144, 148-150, 156, 158, 163, 167
Street Children or Urchins 167
Sustainability 2-6, 9, 11-17, 22, 26-27, 33-34, 39, 42, 74-76, 81, 85-86, 88, 93-99, 101-103, 105-106, 108, 111-112, 114-116, 118-119, 121-124, 139, 166, 171-175, 177-189, 192, 197, 201, 207-210, 216, 222, 227, 229-234, 248, 254, 261, 265-266, 271, 275, 286, 301, 303
Sustainability Education 5, 14-17, 171-172, 183, 185, 188-189, 197, 209, 234
Sustainable Development 3, 5-6, 14, 16-17, 19, 22, 24, 26, 29-30, 32, 35, 37-38, 41-42, 60-61, 64, 67, 69-70, 72-76, 80, 83-84, 86-87, 89-94, 99-100, 102, 105-110, 112-113, 117-119, 121, 124, 141, 145, 162, 169, 171, 177-178, 181-183, 188, 191-192, 197, 200, 208-210, 214, 225, 228-229, 232, 234, 237, 243, 246, 253, 255-257, 259, 261, 264, 271, 286, 288, 302-303
Sustainable Development Goals (SDGs) 22, 191, 210, 214
Sustainable Future 107, 170, 172, 182, 187, 210, 227, 288
Sustainable Lifestyle 211
Sustainable Practices 74-75, 77, 93, 95, 97, 99-101, 104-106, 110, 124, 212
Symbiotic Relationship 173, 186

T

Teacher Education 4-5, 10-17, 44, 58, 62, 196, 209, 222, 224, 292, 301
Teacher Preparation Programs 1-2, 4-6, 9-13, 17
Teaching 2, 4-6, 8-12, 15, 17-18, 23, 26, 28, 30-32, 37, 39, 41, 43-44, 47-48,

53-54, 56-58, 62, 64-65, 68, 72, 111, 148, 151, 155, 159-160, 162, 171, 173, 177, 181, 184, 189-191, 194, 196-197, 200-202, 204-205, 207-209, 213-214, 216-218, 221-226, 230-231, 234, 243, 245, 262, 288-294, 296-297, 299-304, 306
Teaching and Learning Facilities 306

U

Ubuntu 235-237, 242-247, 249-250, 252, 254, 256-257, 259-260, 262
Urchins 148-149, 167
Utilitarianism 93-94, 119, 124

V

Values 10, 30, 42, 44, 46, 61-62, 91, 95, 103, 110-111, 113, 115-117, 123, 137, 180, 191-193, 204, 206, 210, 213-215, 217, 224, 228, 234, 236-237, 243-248, 250, 299
Virtue Ethics 93-94, 124

W

Water 18-19, 28, 30, 49-52, 55, 75, 88, 108, 117, 124, 126, 128, 136, 145-148, 150-151, 157-160, 163-165, 167, 169, 176, 228, 238, 242, 263-287, 296, 306
Water Sources 160, 263-265, 267-273, 275, 278-279, 286
Waterborne Diseases 50, 144-145, 157, 167
Women 20-21, 32, 63, 66-69, 71-80, 82-90, 104, 116, 143, 150, 228, 231, 236-242, 244-247, 250-251, 253-258, 260, 262, 285

www.ingramcontent.com/pod-product-compliance
Lightning Source LLC
LaVergne TN
LVHW080605300926
849162LV00005B/1054